ARTIFICIAL INTELLIGENCE-BASED ENERGY MANAGEMENT SYSTEMS FOR SMART MICROGRIDS

ARTIFICIAL INTELLIGENCE-BASED ENERGY MANAGEMENT SYSTEMS FOR SMART MICROGRIDS

Edited by
Baseem Khan, Sanjeevikumar Padmanaban,
Hassan Haes Alhelou, Om Prakash Mahela, and
S. Rajkumar

CRC Press
Taylor & Francis Group
Boca Raton London New York

CRC Press is an imprint of the
Taylor & Francis Group, an **informa** business

First edition published 2022
by CRC Press

6000 Broken Sound Parkway NW, Suite 300, Boca Raton, FL 33487-2742

and by CRC Press
4 Park Square, Milton Park, Abingdon, Oxon, OX14 4RN

CRC Press is an imprint of Taylor & Francis Group, LLC

Library of Congress Cataloguing-in-Publication Data
Names: Khan, Baseem, 1987- editor. | Padmanaban, Sanjeevikumar, editor. | Alhelou, Hassan Haes, 1988- editor. | Mahela, Om Prakash, 1977- editor. | Rajkumar, S., editor.
Title: Artificial intelligence-based energy management systems for smart micro grids / edited by Baseem Khan, Sanjeevikumar Padmanaban, Hassan Haes Alhelou, Om Prakash Mahela, and S. Rajkumar.
Description: Seventh edition. | Boca Raton, FL : CRC Press, 2022. | Includes bibliographical references and index.
Identifiers: LCCN 2021058070 | ISBN 9780367754341 (hbk) | ISBN 9781032268835 (pbk) | ISBN 9781003290346 (ebk)
Subjects: LCSH: Microgrids (Smart power grids)
Classification: LCC TK3105 .A78 2022 | DDC 621.31--dc23/eng/20220110
LC record available at https://lccn.loc.gov/2021058070

ISBN: 978-0-367-75434-1 (hbk)
ISBN: 978-1-032-26883-5 (pbk)
ISBN: 978-1-003-29034-6 (ebk)

DOI: 10.1201/b22884

Typset in Times
by MPS Limited, Dehradun

Contents

Acknowledgments

This book is an outcome of the inspiration and encouragement given by many individuals for whom these words of thanks are only a token of our gratitude and appreciation for them.

Our sincere gratitude goes to the people who contributed their time and expertise to this book. We highly appreciate their efforts in achieving this project. The editors would like to acknowledge the help of all the people involved in this project and, more specifically, the editors would like to thank each one of the authors for their contributions and the editorial board/reviewers regarding the improvement of quality, coherence, and the content presentation of this book.

Second, the editors would like to express their sincere thanks to CRC Press for their continuous support and giving us an opportunity to edit this book.

The editors are thankful to our family members for their prayers, encouragement, and care shown towards us during the completion of this book on deregulated electricity structures and smart grids. Thank you all!!!

We also express our gratitude to GOD for all the blessings!!!

Baseem Khan
Hawassa University, Hawassa, Ethiopia

Sanjeevikumar Padmanaban
Aarhus University, Herning 7400, Denmark

Om Prakash Mahela
Power System Planning Division, RRVPNL, Jaipur, India

S. Rajkumar
Hawassa University, Hawassa, Ethiopia

Hassan Haes Alhelou
Tishreen University, Lattakia, Syria

Editors

Baseem Khan (M'16) received his bachelor of engineering degree in electrical engineering from Rajiv Gandhi Technological University, Bhopal, India, in 2008. He received his master of technology and doctor of philosophy degrees in electrical engineering from the Maulana Azad National Institute of Technology, Bhopal, India, in 2010 and 2014, respectively. Currently, he is working as faculty member at Hawassa University, Ethiopia. His research interests include power system restructuring, power system planning, smart grid technologies, meta-heuristic optimization techniques, reliability analysis of renewable energy system, power quality analysis, and renewable energy integration.

Sanjeevikumar Padmanaban is SMIEEE'15, FIETE'18, FIE'18, FIET'19, and a charted engineer (CEng., India); he received his bachelor's (First Class), master's (Distinction), and Ph.D. degree in electrical engineering from the University of Madras, Pondicherry University (India, 2002, 2006), and University of Bologna (Italy, 2012). His research work is focused in the field of power electronics (multi-phase/multilevel converters). He is on the editorial board/associate editor of the *IEEE Systems Journal*, IET PEL, subject editor of IET RPG, subject editor of *IET GTD, IEEE Access, Turkish Journal of Electrical Engineering & Computer Science, Journal of Power Electronics* (JPE-Korea), and *FACETS* (Canada).

Om Prakash Mahela (M'13) was born in Sabalpura, Kuchaman City, Rajasthan, India, in 1977. He received a B.E. degree from the College of Technology and Engineering, Udaipur, India, in 2002, a M.Tech. degree from Jagannath University, Jaipur, India, in 2013, and a Ph.D. degree from the Indian Institute of Technology Jodhpur, India, in 2018, all in electrical engineering. From 2002 to 2004, he was an assistant professor with the Rajasthan Institute of Engineering and Technology, Jaipur, India. From 2004 to 2014, he was a junior engineer with the Rajasthan Rajya Vidyut Prasaran Nigam Ltd., India and has been an assistant engineer since July 20014. He has authored more than 100 research articles and book chapters. His research interests include power quality, power system planning, grid integration of renewable energy sources, FACTS devices, transmission line protection, and condition monitoring. He was a recipient of the university rank certificate in 2002, gold medal in 2013, best research paper award from IEEE ICSEDPS 2018, C.V. Raman Gold Medal in 2019, and outstanding reviewer awards from Elsevier Journals.

Dr. S. Rajkumar is working as an assistant professor in the faculty of manufacturing, institute of technology, Hawassa University, Ethiopia. He has acquired 14 years of teaching experience in various reputed engineering colleges and universities. He has a Ph.D. from Anna University, Chennai, from 2014. He has published more than 40 research papers in international journals and international conferences. He actively participates in more than ten professional societies. He was a reviewer and editorial board member in various reputed international journals.

Hassan Haes Alhelou (S'15) is a faculty member at Tisheen University, Lattakia, Syria. He is also a Ph.D. researcher at the Isfahan University of Technology (IUT), Isfahan, Iran. He is included in the 2018 Publons list of the top 1% best reviewers and researchers in the field of engineering. He was the recipient of the Outstanding Reviewer Award from the *Energy Conversion and Management Journal* in 2016, *ISA Transactions Journal* in 2018, *Applied Energy Journal* in 2019, and many other awards. He was the recipient of the best young researcher in the Arab Student Forum Creative among 61 researchers from 16 countries at Alexandria University, Egypt, 2011. He has published more than 30 research papers in high-quality peer-reviewed journals and international conferences. He has also performed more than 160 reviews for highly prestigious journals including *IEEE Transactions on Industrial Informatics, IEEE Transactions on Industrial Electronics, Energy Conversion and Management, Applied Energy,* and *International Journal of Electrical Power and Energy Systems.* He has participated in more than 15 industrial projects. His major research interests are power systems, power system dynamics, power system operation and control, dynamic state estimation, frequency control, smart grids, microgrids, demand response, load shedding, and power system protection.

1 Flexibility of Microgrids with Energy Management Systems

Hooman Firoozi, Hosna Khajeh, and Hannu Laaksonen

Flexible Energy Resource, School of Technology and Innovations, University of Vaasa, Vaasa, Finland

CONTENTS

DOI: 10.1201/b22884-1

1

1.1 INTRODUCTION

In recent years, approaches towards energy transition and sustainable development have been ever-increasing due to the need for mitigating climate change issues and the efficient utilization of existed energy resources. With this regard, state-of-the-art technologies and infrastructures along with active operation and control of different energy resources would become crucial. Amongst all energy resources, microgrids (MGs) are believed to be one of the highly potent resources to deal with the issues of electrical systems. In other words, active operation and control of MGs in which there exist different kinds of demands and energy resources (e.g., energy storages, micro-generation units, etc.) would be beneficial not only for MG stakeholders in terms of cost-benefit efficiency but also for power system operators in terms of MGs' contribution to grid's flexibility [1–5].

In order to unlock the active utilization of MGs, cutting-edge technologies along with efficient infrastructure are a necessity. These technologies together in communication with the MGs' energy resources are known as energy management systems (EMSs). EMSs are intelligent automated systems that contribute to, for instance, lowering/shifting energy consumption in critical moments along with a reduction in the MGs' costs. Although the utilization of EMS might consider other objectives such as CO_2 emission reduction or self-sufficiency, they mostly employ optimization techniques either as single-objective or multi-objective approaches. EMSs can also enable either the bidirectional energy exchange with the network in grid-connected mode, or stand-alone operation of MGs in islanded-mode [6–10].

In this chapter, the focus of the study is on the MGs equipped with an EMS. There have been introduced several approaches to the energy management of MGs. However, in most of them, economic aspects, i.e., cost reduction, are the top priority desire of the problem from the MG stakeholders' point of view. This could be done in different ways. On the one hand, reducing the total costs of the MGs by maximum utilization of self-production facilities (PV panels, wind turbines, etc.) as well as changing the energy consumption over time from peak hours to off-peak hours during the day. On the other hand, exploiting MGs' flexibility so as to help the upstream grid in critical moments for monetary profits in return. Accordingly, the authors first present an introduction to flexible energy resources (FERs) in MG along with their characteristics in Section 1.2. Afterward, the MGEM modeling approaches are widely presented in Section 1.3. In this section, first, the different kinds of management method deployed in the MGs are illustrated. Then, various objectives for energy management in MGs will be introduced. Regarding this section, we introduce a number of approaches based on well-known optimization algorithms considering different MG-related as well as grid-related constraints. Microgrids' constraints are related to the physics and limitations of the MG's resources whilst the constraints of the grid are related to the limitation of energy exchange with the upstream grid (e.g., congestion management, emission reduction, and/or energy loss reduction). Moreover, the application of the MGEM system in MGs with FERs such as energy storages, electric vehicles (EVs), and thermo-statically controllable loads (TCLs) which exchange energy and flexibility with the grid will be discussed as well which is followed by the flexibility services that MGs could provide to the different levels of power system. Finally, this chapter will be summarized and concluded in Section 1.4.

1.2   FLEXIBLE ENERGY RESOURCES IN MICROGRIDS

There could be various types of energy sources in MGs. They might be supplied by either fossil fuels or renewable sources such as wind, solar, etc. [11]. Figure 1.1 depicts the most common energy resources in MGs. In general, any energy resource that is located in the MG's demand-/generation-side, and enables the MG's reacting to the needs, could be defined as flexible energy resources. However, based on the amount of flexibility, these energy resources could be divided into two main categories, namely high-flexible energy resources and low-flexible energy resources.

The low-flexible energy resources in the MG consist of renewable generation units which their output power is not fully controllable. This originates from the fact that renewable energy sources such as wind, solar radiation, etc. depending on meteorological conditions. In certain situations, the only way to take action for low-flexible energy resources is to curtail their generation from MG's generation side. Therefore, in MGEM systems, they could be entitled to low-flexible energy resources.

On the contrary, there might be some flexible energy resources in MGs, which could help to increase the flexibility of the MG. These types of energy sources could be entitled to high-flexible energy resources. The high-flexible resources could be utilized either in demand-side or generation-side of the MG. In other words, they might be among the consumers' assets, consumers' load, or as a part of a bulk PV system.

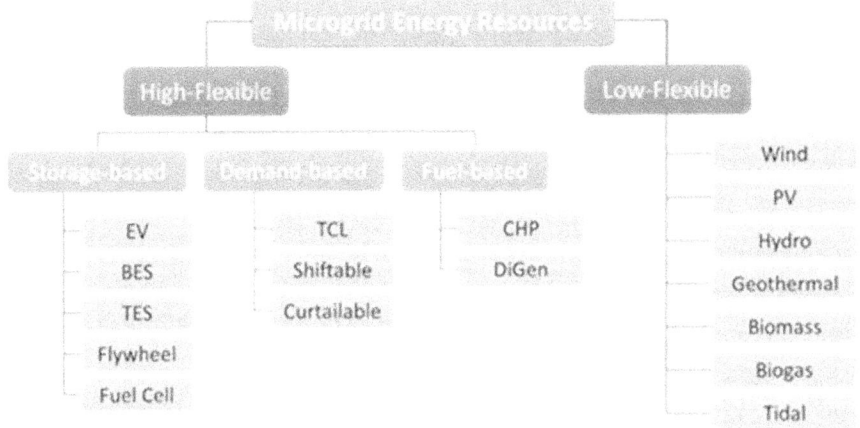

FIGURE 1.1   An overview of microgrids' energy resources.

Having employed the high-flexible energy resources in MGs, the MGEM system could take advantage of them to enhance the flexibility of MG by the active utilization of these resources. Therefore, the focus of this section is on the high-flexible energy resources in MGs. In the following subsections, the explanation about the characteristics of the high-flexible energy resources in MGs is presented.

1.2.1   STORAGE-BASED FLEXIBLE RESOURCES

The storage-based FERs refers to the devices that could store the energy in different shapes (i.e., electrical, thermal, mechanical, etc.) in order to utilize it when there is a shortage or in critical moments. Energy storages could help the MGs to enhance their flexibility by injecting the power back to the MG, especially in islanding situations. Storage-based energy resources could be mostly categorized as electric vehicles, battery energy storages, thermal energy storages, flywheels, fuel cells, etc. In the following subsections, the most common storage-based flexible resources are illustrated.

1.2.2   ELECTRIC VEHICLES (EVs)

Electric vehicles as one of the ever-increasing types of flexible energy resources in future smart grids are believed to be among the potential solutions to systems' flexibility. These FERs are adjustable, shiftable, and fast-response which could considerably enhance the MG's flexibility in flexibility services provision. Moreover, EVs could be charged when the prices are at the lowest level meaning that they could be contributing to the cost-reduction target at all system levels. Although the EVs act as the load consumption when they are in charging mode, the recent version of EVs with new charging facilities makes the EVs capable of injecting power back to the grid (i.e., vehicle-to-grid mode) when it is needed. The equations related to the EV's operation are defined as follows. Equation (1.1) presents the energy stored in the EV's battery at time t:

$$E_t = E_{t-1} + \begin{cases} \eta^{ch} P_t^{ch} \Delta t & \text{charging} \\ \frac{P_t^{dis}}{\eta^{dis}} \Delta t & \text{discharging} \\ 0 & \text{unplugged} \end{cases} \quad (1.1)$$

where η^{ch}/η^{dis} are the charging/discharging efficiency and P_t^{ch}/P_t^{dis} are the power of charging/discharging, respectively. Δt is the duration time at which the EV is being charge or discharge. According to this equation, the energy of EV's battery depends on its current level of energy as well as its current mode of charging. The EVs' battery is mostly chosen from Li-Ion technology batteries since they are highly efficient compared to the other types of batteries. However, the capital cost of these batteries is nowadays high. Therefore, it is recommended to restrict the lower and upper levels of the battery's stored energy in MGEM systems. This restriction could be considered a constraint in MGEM problems as in (1.2), which helps to reduce the number of charging or discharging cycles over a time span.

$$E^{min} \leq E_t \leq E^{max} \tag{1.2}$$

$$0 \leq P_t^{ch}, P_t^{dis} \leq P^{max} \tag{1.3}$$

Another constraint related to the EVs' battery could be found in (1.3). This one similarly helps to limit the charging/discharging power of the battery to avoid the battery from early depreciation. Note that in order to take advantage of flexibility provision by EVs, the EVs, as well as the charging facilities, must have the capability of working in the vehicle-to-grid mode.

1.2.2.1 Battery Energy Storage (BES)

Battery energy storages are one of the best solutions for future smart grids. They could be centrally controlled, they have a very rapid response, and also they could be useful in remote local energy systems, islanded MGs, or in power shortage situations. Moreover, they could effectively help the grid in terms of stability, resiliency, and flexibility. BESs could be found in different sizes, from domestic level to MG level or even grid levels. There have been several materials introduced andused in manufacturing BESs such as Li-ion, vanadium redox flow, etc. The equations related to BES are similar to those mentioned in the previous subsection. However, all the BESs support the bidirectional power flow since they are meant to be discharged when it is needed.

1.2.2.2 Thermal Energy Storage (TES)

Thermal energy storages are used to store the thermal energy and utilized it when it is required. These storages could be beneficial in MGEM systems in order to store the heat in off-peak low-price times over night for use in high-price moments. The heat might be coming from combined heat and power (CHP) units, the waste heat from biomass/biogas units, or the exhausting heat from industrial units. They can be also beneficial not only for storing heat in summers but also for preserving the cold in the winters and reverting it to the MG's facilities in summers (i.e., seasonal TES). Thermal storages could be various in size and also in response time. Table 1.1 presents the typical types of thermal storages with their characteristics [12].

1.2.2.3 Flywheel

A flywheel is mechanical energy storage that consists of a rotational part and other facilities for connecting to the system. In charging mode, a flywheel is speeding up to its nominal rotational speed and store energy as kinetic type. Afterward, the stored kinetic energy is preserved in standby mode. When energy is required, the flywheel starts to discharge the stored energy back to the grid [13].

Figure 1.2 depicts a flywheel storage utilization in a hybrid grid-connected MG. In order to calculate the energy of a flywheel storage, the following well-known formula is [13]:

TABLE 1.1

The Typical Types of Thermal Storages with Their Characteristics

Technology	Capacity (kWh)	Power (kW)	Efficiency (%)	Cost (€/kWh)	Storage Time
Sensible	10–50	1–10,000	50–90	0.1–10	days-month
Phase-change	50–150	1–1,000	75–90	10–50	hours-month
Chemical	120–250	10–1,000	75–100	8–100	hours-days

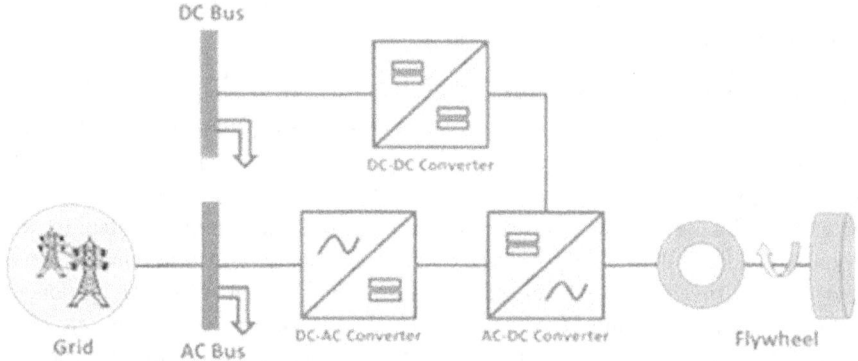

FIGURE 1.2   Flywheel storage utilization in a hybrid grid-connected MG.

$$E = \frac{1}{2}mr^2\omega^2 \tag{1.4}$$

In Equation (1.4), E is the kinetic energy stored in the flywheel and m, r and ω are the mass of the cylinder, radius, and rotational speed, respectively.

1.2.2.4   Fuel Cell (FC)

A fuel cell (FC) is another high-flexible energy resource that could also be utilized in MG applications. FCs can produce electricity by converting the chemical energy originating from hydrogen-oxygen reactions into electrical energy. The capacity of a FC could be different based on their applications from 100 kW to 100 MW [14]. In solid-oxide FC, for example, anode supplies hydrogen and catalytically split it into a number of protons and electrons. The electrons are then flowing towards the positive side (i.e., the cathode) by flowing through the external circuit. The oxygen then reacts with the protons and also the electrons flowing in the circuit, forming water formula [15]. Solid-oxide FC can operate in parallel with MG's PV panels, meaning that it can be integrated with solar power as a hybrid PV system since they can store the produced energy for hours [14]. Therefore, in the nighttime, when PV panels cannot produce electricity, the FC could be employed to supply the demand. This could help the MG to have the flexibility to reduce the power exchange with the connected grid aiming at a generation cost reduction or provision of flexibility to the grid for monetary profits in return.

1.2.3 DEMAND-BASED FLEXIBLE RESOURCES

The demand-based FERs refer to the devices that only consume energy. In residential MGs, for example, they might be found in the residential home appliances [16]. In other types of MGs, they might be in the shape of an industrial unit's demand or a commercial building's load. In general, all the devices on the demand side of their consumption power could be controlled are considered as demand-based FERs. Since a great number of demand-based FERs are widely being utilized in residential/commercial units, it would be beneficial for MGs to unlock the flexibility that could be emerged from these resources. Thereby, the demand-based FERs in the consumers' premises that are capable of controlling, changing, or shifting are at the center of attention for MGs' manager/operator. Note that the demand response programs [17] could be a key factor for incentivizing small-scale consumers in MGs for flexibility provision. In the following subsections, some of these demand-based FERs that could enhance the flexibility of MG are presented.

1.2.3.1 Thermostatically Controllable Load (TCL)

Thermostatically controllable loads refer to the loads that their power consumption could be adjusted by sending command signals to their thermostat. These loads have a great portion of the total demand. In summers, the power consumption is used for cooling, while in winters, the power consumption is utilized to heat the internal spaces of houses, offices, etc., for instance, electric water heaters (EWHs); heating, cooling, air conditioning systems (HVACs); and refrigerators could be categorized in TCLs. These appliances are closely intertwined with the thermal comfortness and other consumers' preferences. Therefore, in MGEM systems, in addition to the operational constraints of appliances, the constraints regarding TCLs must also be considered. One of these constraints is thermal comfort of the users for HVACs and EWHs, which could be found in (1.5)–(1.6):

$$\theta_i^{min} \le \theta_{i,t} \le \theta_i^{max} \qquad (1.5)$$

$$\theta_i^{w,min} \le \theta_{i,t}^w \le \theta_i^{w,max} \qquad (1.6)$$

where θ_i^{min} and θ_i^{max} are the minimum and maximum desired temperature requested by user i, respectively. $\theta_i^{w,min}$ and $\theta_i^{w,max}$ are the minimum and maximum desired temperature of hot water requested by user i, respectively. Finally, $\theta_{i,t}$ and $\theta_{i,t}^w$ are the interior and hot water temperature of user i, respectively.

In order to unlock the flexibility from TCLs in an MG, they must be aggregated. Aggregating several TCLs in an MG could help to reduce or increase the power consumption in certain moments for the provision of upward or downward flexibility services to the connected upstream network. However, it is worth mentioning that the response time of some TCLs is a bit low. Therefore, they might not be suitable for all kinds of flexibility services but still beneficial for those services that have a slow activation time.

1.2.3.2   Shiftable Load

Shiftable loads are those that their consumption power cannot be controlled; however, the operating time could be shifted from high-price to low-price hours. These loads also need to work for a constant cycle and they could not be disconnected once they started which means shiftable loads must run a cycle completely. Therefore, the MGEM system can only schedule the related start time. Dishwashers, washing machines, and clothes dryers, as the devices which might be found mostly in residential households in the MG, could be grouped in shiftable loads' category [18]. A MG operator could take advantage of shiftable loads for energy scheduling during a day considering the consumers' comfortness constraints.

1.2.3.3   Curtailable Load

The power consumption of curtailable loads could be adjusted, usually without any significant effect on consumers' comfortness. These adjustments limit the energy consumption of devices by changing the settings thorough a command signal, without any consequences. As an example, lighting devices could be curtailed by a command during the day or as an automated function of natural daylight [18].

1.2.4   Fuel-Based Flexible Resources

The fuel-based FERs are those that could be categorized in the generation side. Naturally, these FERs produce power by using fossil fuels as input energy sources. However, their output generating power could be regulated according to the system's needs. The output power generation of fuel-based FERs could be adjusted by changing the amount of intake fuel. Therefore, these FERs could contribute to increasing the MG's flexibility. Although these resources could not be categorized as a totally sustainable energy resource, their power production could be controlled in critical moments for flexibility provision targets. In the following subsections, a brief overview of the most important fuel-based FERs is provided.

1.2.4.1   Combined Heat and Power (CHP)

The most famous fuel-based flexible resource in MGs is combined heat and power (CHP) unit. A CHP unit is generally a power generation unit which combines the heat production with electricity generation. CHPs could be regarded as a decentralized distributed generator located at the MG level. This DG has the ability to produce heat and electricity simultaneously which could be beneficial in increasing the efficiency and flexibility of the MG. The exhausted heat from the power generation cycle in the CHP could be utilized to provide the required energy for the heating load as well as hot water within the MG internal network. In the MG level, the size of the CHPs depends on the size of the MG which could be found up to 20 MW. CHPs, however, could also be installed in customer-level applications with a maximum capacity of 15 kW [19].

Although the energy efficiency of the CHP unit can be assumed to be constant, the CHP unit's efficiency practically differs with dynamic operation due to the variation of output power. This could help the MGs to enhance their flexibility or providing flexibility services to the upstream grid. Note that ramping constraints need to be considered in MGEM problems since the CHP unit requires some time to reach the steady-state condition after changing its set point [20].

1.2.4.2 Diesel Generator (DiGen)

A diesel generator as one of the fuel-based FERs that could be beneficial in MGEMs. This FER may be utilized when there is a power shortage in the MG. They could also be considered as flexibility sources when upward flexibility is needed from the upstream network. However, the sizing of the diesel generators in the MG is quite crucial since the ramping rate of the generator should be adequate for fulfilling the MG's/network's need. In fact, diesel generators play a quite important role when, for example in an islanded MG, the power generation is not enough and the energy storages are almost discharged. Therefore, these FERs are also called backup units in local energy systems. They could be different in size from 5 kW to 5 MW or more. The equation regarding fuel consumption of the diesel generator can be calculated from (7) that should be considered in MGEM optimization problems [21].

$$Cost = c_1 \times P^{DG} + c_2 \times P^n \qquad (1.7)$$

In Equation (1.7), P^{DG} and P^n are the produced power and the nominal power of the generator, respectively. c_1 and c_2 are the coefficients related to fuel consumption's curve which typically considered $c_1 = 0.246$ l/kW and $c_2 = 0.08145$ l/kW, respectively [22].

1.3 MODELING THE MICROGRID ENERGY MANAGEMENT

A microgrid, as one of the potential solutions to the future smart grids, usually confronts the lack of power generation. This is due to the variability and intermittency both from generation and demand sides [23]. Energy management methods have been believed as one of the solutions to this issue. The most important target of energy management is to find the optimal operation point of different kinds of energy resources in order to supply the requested demand constantly and efficiently [24]. It should be mentioned that the main objective of these studies is reducing consumption costs while taking advantage of the MGs' flexible capacity for the provision of energy and flexibility services. However, there might be various approaches and tools towards this target. Before discussing the MGEM tools and techniques, the MG management methods are briefly illustrated in the next subsection.

1.3.1 MICROGRID ENERGY MANAGEMENT METHODS

The approaches toward the control and operation of MG resources as well as dispatchable loads are known as MG management methods. This management method could be deployed by having an agreement between the MG operator and the MG's members/stakeholders. The microgrid energy management (MGEM) could be defined in three perspectives [25]:

1. Decentralized energy management
2. Centralized energy management
3. Distributed energy management

In decentralized MGEM, the control and operation of FERs located at the MG have more degree of freedom. This means the FERs' adjustability in this management method helps more to meet the preferences of the stakeholders/consumers. In centralized MGEM, however, a central controller decides how the FERs and generation units should be operated. It has to be mentioned that in both centralized and decentralized management methods, the technical constraints of the MG must be taken into account. The most important constraint would be the balance between load and production within the MG [26].

Distributed MGEM as another management method in MGs is presented in the literature as well. This type of management method is mostly based on game-theoretic approaches. In distributed MGEM, game players, as the agents in the MG, seek the best management method for their own objectives taking into account the overall goal of the MG regarding energy management considerations [27].

Having mentioned the above approaches, the MGEM problems generally aim at scheduling the operation of generation units, storage systems, and even controllable loads in the MG [28]. These problems have been presented with various objectives. In the following section, some of these objectives with the related considerations are elaborated.

1.3.2 MICROGRID ENERGY MANAGEMENT OBJECTIVES

1.3.2.1 Cost Reduction/Profit Maximization

One of the most important objectives of energy management is reducing the total operation cost of MGs' components. This operational cost includes, for instance, the fuel cost, cost of purchasing energy from the grid, degradation cost of battery energy storages, etc. [29]. The cost reduction in an MGEM could be over different time spans from real time to daily, monthly, or even yearly periods. However, energy management sometimes might be defined for real-time operation. In this case, the real-time operational cost of MG is the objective of the problem. Accordingly, the generation and demand as well as the scheduling of the FERs should be in a way that the overall cost of the MG tends to be minimized in real time [30]. Accordingly, the MGEM system is in charge of scheduling the generation and flexible loads so that the total cost of energy purchasing from the grid as well as the operational costs of the DGs become minimized as in Equation (1.8).

$$\min \text{Cost}^{\text{MG}} = \sum_t \left(\text{Cost}_t^{EN} + \sum_i \text{Cost}_t^{DG_i} + \sum_j \text{Cost}_t^{BES_j} + \sum_j \text{Cost}_k^{EV_k} \right) \quad (1.8)$$

- Subject to: Constraint {DGs, FERs}

In Equation (1.8), $Cost_t^{EN}$ is the cost of purchasing energy from the grid at time t. Accordingly, the total temporal operational cost of DGs, BESs, and degradation cost of EVs that must be paid to the EV owners for vehicle-to-grid contribution are $\sum_i Cost_t^{DG_i}$, $\sum_j Cost_t^{BES_j}$ and $\sum_j Cost_k^{EV_k}$, respectively.

This objective could also be considered in a different shape, which says the MGEM objective is to maximize the total profit of the MG instead of operational cost. The monetary profit for a MG mostly comes from selling energy to the grid or providing flexibility services to balancing responsible parties. Accordingly, the objective function of the MG could be defined considering the following formulation:

$$\max \text{Profit}^{MG} = \sum_t (P_t^{sell} \lambda_t^{sell} - P_t^{buy} \lambda_t^{buy} - Cost_t^{MG}) \qquad (1.9a)$$

$$\max \text{Profit}^{MG} = \sum_t (F_t^{up} \lambda_t^{up} + F_t^{dn} \lambda_t^{dn} - P_t^{EN} \lambda_t^{EN} - Cost_t^{MG}) \qquad (1.9b)$$

- Subject to: Constraint {DGs, FERs, Grid Limits}

Equation (1.9a) indicates the objective function of MGEM problems for a grid-connected MG, which only exchange energy with the grid while Equation (1.9b) presents the objective for a flexibility provider MG. In Equation (1.9a)–(1.9b), P_t^{sell} and λ_t^{sell} are the exported power to the grid and the price of selling to the grid, respectively. P_t^{buy} and λ_t^{buy} are the imported power from the grid and the price of energy to the grid, respectively. F_t^{up} and F_t^{dn} are the upward and downward flexibility provided to the network, respectively. λ_t^{up} and λ_t^{dn} are the price of upward and downward flexibility, respectively. P_t^{EN} and λ_t^{EN} are the quantity and price of purchased energy from the grid, respectively. Finally, $Cost_t^{MG}$ refers to the total operational cost of the MG, which includes degradation cost of energy storages, EVs as well the operational cost of generation-side resources. Note that, in both definitions, the constraints related to the operational consideration of the assets as well as members' comfortness must be taken into account in MGEM problems.

1.3.2.2 Self-Sufficiency

One of the important targets of MGEM in MG is self-sufficiency. A MG is self-sufficient when there is a balance between the generation and consumption within the MG. In other words, the power produced by the MG's resources could fulfill its demand over a period of time. This objective becomes pivotal mostly when an islanding situation is predictable since, in that case, the MG becomes disconnected from the grid and the stability of the MG becomes critical. The MGEM with an objective of self-sufficiency could be tackled by reducing the peak demand, load shedding as well as discharging the storage-based flexible resources. In this way, based on the level of emergency, the MGEM should define a priority for the utilization of fast-response

FERs located in the MG. However, this objective could have other targets inside itself. For example, the self-sufficiency of MG in moments at which the renewable energy resources have production and the energy storages have a sufficient level of charge could be deployed to reduce energy purchasing from the grid. Therefore, fewer greenhouse gases emission as well as cost reduction could also be considered as the results of self-sufficiency objective. The objective function regarding the self-sufficiency in MGEM systems must satisfy the following constraint:

$$G_t^{MG} \geq D_t^{MG} \tag{1.10}$$

$$G_t^{MG} = \sum_i P_t^{DG_i} + \sum_j P_t^{dis_j} \tag{1.11}$$

$$D_t^{MG} = P_t^{BL} + \sum_i P_t^{FL_i} + \sum_j P_t^{ch_j} \tag{1.12}$$

In the above equation, G_t^{MG} is the MG total generation and D_t^{MG} is the MG total demand at time t. In Equations (1.10–1.12), the $P_t^{DG_i}$ is the production of DG i at time t. P_t^{BL} and $P_t^{FL_i}$ are the baseline load and power consumption of flexible load i in the MG at time t, respectively. $P_t^{dis_j}$ and $P_t^{ch_j}$ are the discharging and charging power of storage-based resources j at time t, respectively. It has to be mentioned that the other constraint regarding the simultaneous charging/discharging limitation and the operational constraints of DGs also must be taken into account.

1.3.2.3   Flexibility Provision

As the traditional power systems have been experiencing a fast and vast transition to the smart, local, and decentralized ones, the flexibility services concept has been introduced in order to cover the whole system-related issues. In this light, MGs as the local energy systems is believed to be a suitable choice in providing local and system-wide flexibility. Flexibility services could be provided by MGs through the effective utilization of FERs and also distributed energy resources in MG by using MGEM systems. However, before mentioning the flexibility services provision by the MGs, the definition of flexibility in an electrical system should be clarified. A comprehensive definition of the flexibility of electrical systems could be the ability of the system to adjust its operating point continuously and also to resist the predicted and unpredicted differentiations happening in operating conditions. Accordingly, a flexible electrical system must adapt to the possible changes both in generation and consumption in a temporal manner [31]. Therefore, another possible objective of MGEM might be providing flexibility services by MGs to the connected upstream networks. These services could appear in different shapes. An overview of the flexibility services (e.g., in Nordic countries [32]) that MGs can provide to other entities are presented in Figure 1.3.

1.3.2.4   TSO-Level Flexibility Services

The transmission system operator is the responsible party for transmission system operator (TSO)-level balancing issues that could be addressed by the contribution of

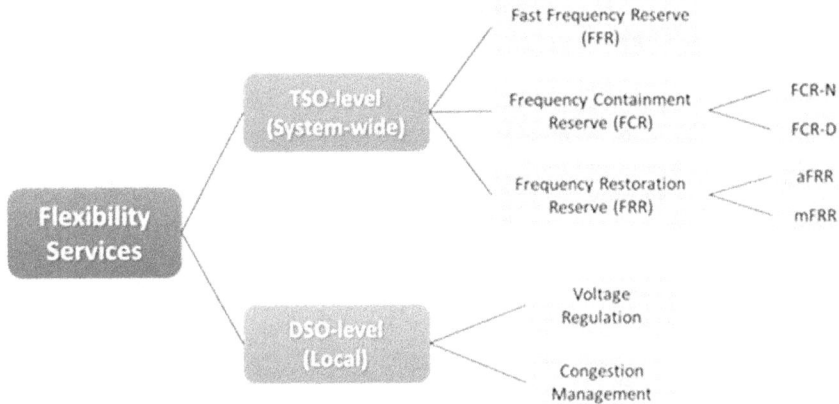

FIGURE 1.3 An overview of the flexibility services (in Nordic).

TABLE 1.2
The Characteristics of the Nordic Flexibility Services

Service	FFR$^{(new)}$	FCR-D	FCR-N	aFRR	mFRR
Application	In very low-inertia situations	In big frequency deviations	Always in use	In certain hours	Incidents/imbalance of balancing parties
Activation	1 sec.	Less than 1 min.	1–5 min.	5 min.	15 min.
Min. Bid Size	Not defined yet	1 MW	0.1 MW	5 MW	5 MW

all system-level flexible resources, e.g., MGs. There are three types of services that local energy systems can contribute to flexibility provision to transmission-level needs, which are fast frequency reserve (FFR), frequency containment reserve (FCR), and frequency restoration reserve (FRR). TSO-level flexibility services have been the conventional generation units' responsibility. However, recently and more increasingly in future power systems, MGs as the potent sources of flexibility, would be among the flexibility service responsible parties. Depending on the size of grid-connected MGs and the flexibility needs of the upstream entities, MGs could contribute to one or more specific flexibility services in singular or aggregated manners. The TSO-level flexibility services (e.g., in Finland [33]) with their characteristics are summarized as in Table 1.2. The flexibility services in this table are categorized as reserve product services [33].

- **FFR:** The FFR service as the recently introduced flexibility service in Nordic will be utilized in extremely low-inertia situations when there are ±0.5 Hz frequency fluctuations. The maximum amount of FFR services needed in Nordic is estimated at 300 MW.

- **FCR-D:** The FCR-D service is needed in huge frequency deviations with at least 50% of it needs to be activated in 5 seconds and the rest is required to be activated in 30 seconds. The system's need in this service is only for under-frequency situations (i.e., increase in generation or decrease in demand).
- **FCR-N:** The FCR-N service is for normal operation of the system and is being activated all the time. The system's need in this service is only for both under-frequency and over-frequency situations (i.e., increase/decrease in generation or demand). Note that this service is symmetrical, which means the flexibility providers like MGs must be able to provide the flexibility needs equally for upward and downward flexibility.
- **aFRR:** The aFRR service is activated when in this service, unlike FCR, the asymmetrical bids are also accepted, which means the upward and downward flexibility bids could be submitted separately. Note that the activation price to the service providers will be paid according to the price of balancing energy market.
- **mFRR:** The mFRR service is activated manually in 15 minutes. Bids are needed to be delivered 45 minutes prior to the activation hour. In this service, like aFRR, the upward/downward flexibility bids are being submitted separately. Note that in this service the prices are constantly greater than day-ahead energy prices so that it is quite beneficial for flexibility providers like MGs and energy communities [34].

1.3.2.5 DSO-Level Flexibility Services

There are two types of flexibility services that are introduced in the electrical systems namely voltage regulation and congestion management in which MGs can contribute as distribution system operator (DSO)-level flexibility providers. Voltage regulation services could be provided by MG's power electronic devices like FERs' converters and also by injected active power control through the point of common coupling (PCC) with the distribution grid. The power electronic devices are able to control the reactive power, which is effective in voltage regulation applications. Similarly, congestion management services could be provided by the mentioned FERs. In DSO-level flexibility provision by MGs, along with the MG-related constraints, the distribution network's limitations such as active and reactive power and injected current should also be taken into account.

1.3.3 MICROGRID ENERGY MANAGEMENT TOOLS AND TECHNIQUES

There have been introduced various types of MGEM modeling techniques in the previous literature. These techniques include the optimization approaches along with intelligence control tools such as model predictive control, game theory methods, etc. In the following subsections, some of the most popular tools and methods will be introduced.

1.3.3.1 Optimization Methods

The basic approach to energy management problems would be based on optimization algorithms. This originates from the nature of the energy management since

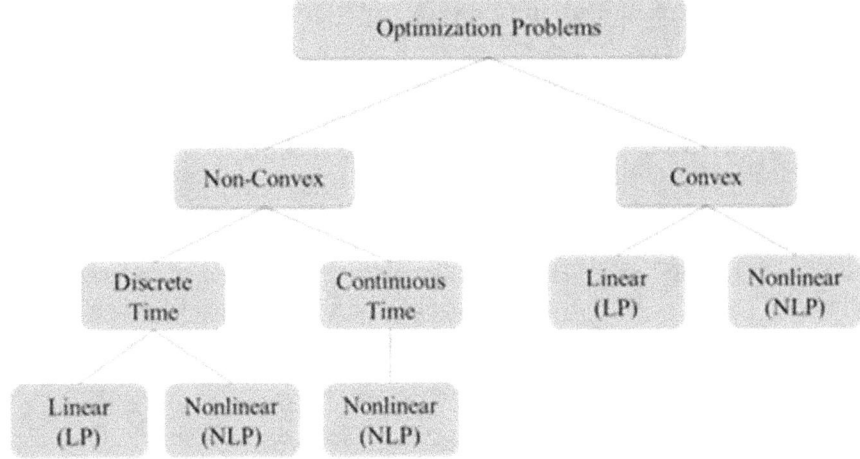

FIGURE 1.4 General categorization of optimization problems.

it is supposed to minimize or maximize an objective depending on the targets of MG's stakeholders as well as the method of asset management [35]. There have been introduced many optimization techniques which could be employed correctly depending on the type of the problem. In general, the optimization techniques could be split into two main categories, convex and non-convex. Figure 1.4, presents a general overview of the proposed types of optimization problems [36].

The type of optimization technique could be chosen correctly depending on the problem's characteristics. In MGEM, the type of problems mostly is convex or the problems are defined in a way that they could be solved with convex techniques (i.e. convex relaxation). This is due to the fact that the convex problems give better convergence compared to the non-convex ones [36]. Furthermore, there could be many uncertainties in MG operation due to the nature of MG components. For instance, the intermittent renewable components such as wind turbines, photovoltaic units, and on the top of them, the unpredictable demand could create the mentioned uncertainties [37]. These uncertainties, however, could be addressed by means of some well-known mathematical and statistical techniques during the definition of optimization problems for MGEM. In order to analyze the impact of uncertainty of data, the following three types of optimization methods have been proposed:

1. Deterministic optimization
2. Stochastic optimization
3. Robust optimization

Although the deterministic approach for defining an optimization problem could be beneficial for comparing the results of the problem with other approaches, robust and stochastic optimization solutions are believed as the most effective techniques in energy management problems which are illustrated in details in the following subsections [38].

1.3.3.2 Deterministic Optimization

Deterministic problems are those that have a unique output for any kind of input [36]. As an example, wind speed or solar radiations which are variable over a time span could directly affect the output power of wind turbines or PV systems. Therefore, for different values of wind speed and solar radiation, the power generation of these renewable units would change. However, the power generation function of these renewable units could be defined deterministic meaning that, for any wind speed and solar radiation, the output power of wind and PV units are considered as certain values. This type of problem formulation would not take place in reality, however, there are some applications for deterministic approaches. Furthermore, this method could be beneficial when the aim is to have an idea about the overall operating points of the system in a certain condition. In MGEM problems, there is some component in the system that has strong stochasticity (e.g., renewables, demand, etc.) and could not be modeled as deterministic functions. Consequently, the other types of optimization techniques (i.e., stochastic and robust) are usually recommended that will be introduced in the following subsections.

1.3.3.3 Stochastic Optimization

In stochastic optimization, the problem of energy management could be presented by a statistical objective function. In this light, the uncertain parameters of the problem such as the output power of renewable energy units could be modeled as the well-known probability distribution functions. These distribution functions might be different due to the difference in the nature of renewable sources such as solar irradiation or wind power. They also might be different due to the uncertainties stemming from the stochastic behavior of consumers such as the behavior of EV owners and the pattern of charging their vehicles. However, all these uncertainties are can be considered as the most popular distribution functions since their sources mostly follow a predictable pattern. The general formulation of a stochastic optimization problem could be found in Equation (1.13):

$$\min_{x \in \chi} \sum_{\omega \in \Omega} \pi(\omega) F(x, \omega) \qquad (1.13)$$

In Equation (1.13), $\pi(\omega)$ is the probability of scenario ω, Ω is the set of scenarios, χ is the set of decision variables. The function $F(x, \omega)$ could be different based on the objective of the MGEM problem. According to the stochastic optimization method, the main objective function of the problem could be written as the sum of the objective function of each scenario multiplied by the scenario probability. In this method, the value of uncertain variables in the system is considered by defining several possible scenarios. Note that, for each scenario, a probability of occurrence should be considered in a way that the summation of the probability of all scenarios must be equal to one as follows:

$$\sum_{\omega} \pi(\omega) = 1 \qquad (1.14)$$

There have been presented a number of computational methods for generating the above scenarios. Amongst these methods, one of the most popular scenario generation techniques is the Monte-Carlo method, which is widely employed in literature [39]. The number of the considered scenarios for the problem has a direct impact on the accuracy of the problem. In other words, by increasing the number of these scenarios the result would be more accurate. However, a large number of considered scenarios for the stochastic optimization problems could result in a huge computational cost. In order to reduce the computational cost of solving, the number of scenarios should be reduced. Therefore, one could use a mathematical method to limit the possible scenarios. For instance, the K-means clustering technique is proposed in order to tackle a large number of scenarios [40]. In the following subsections, uncertainty modeling methods for different sources of stochasticity are illustrated.

1.3.3.4 Robust Optimization

The robust optimization method was first introduced in 1973 [41]. This method has been introduced and employed in many research as one of the most powerful approaches towards energy management in order to act as an alternative for modeling the problems with uncertain parameters. The robust optimization is employed when the energy management problems confront a limited amount of data but at the same time several uncertainties. In this optimization method, unlike the stochastic optimization with many possible scenarios, we consider only one scenario which means this optimization does not need any kind of probability distribution function [42]. This scenario is assumed to be the worst case regarding the uncertain situations in the optimization procedure. In energy management problems, the worst-case scenario is the one that is believed to have the most severe outcome that happens in the real situation. In other words, in this method, it is assumed that the uncertain parameters are in their worst condition [41]. This could help to have a realistic paradigm towards the occurring scenario and if possible, it could improve the results of the optimization in comparison with stochastic methods [43].

In this method, the optimal result of the optimization has two features. First, less data is needed for uncertain parameters here which means only minimum, maximum, and mean of the uncertain parameter is required. Second, the optimal solution of the problem is feasible for all conditions which could be quite beneficial in decision making.

The robust optimization approach has many applications in MGEM, such as bidding strategy [44]. These applications could be dealing with the uncertainty of renewable energy units' generation, consumers' load, and energy/flexibility market prices. Figure 1.5, provides a summarized overview of the most important applications of the robust optimization techniques in MGs.

In general, the basic formulation of a robust optimization problem would be as follows:

$$\min_{x \in \chi} \max_{\omega \in \Omega} \mathbb{C}(x, \omega) \tag{1.15}$$

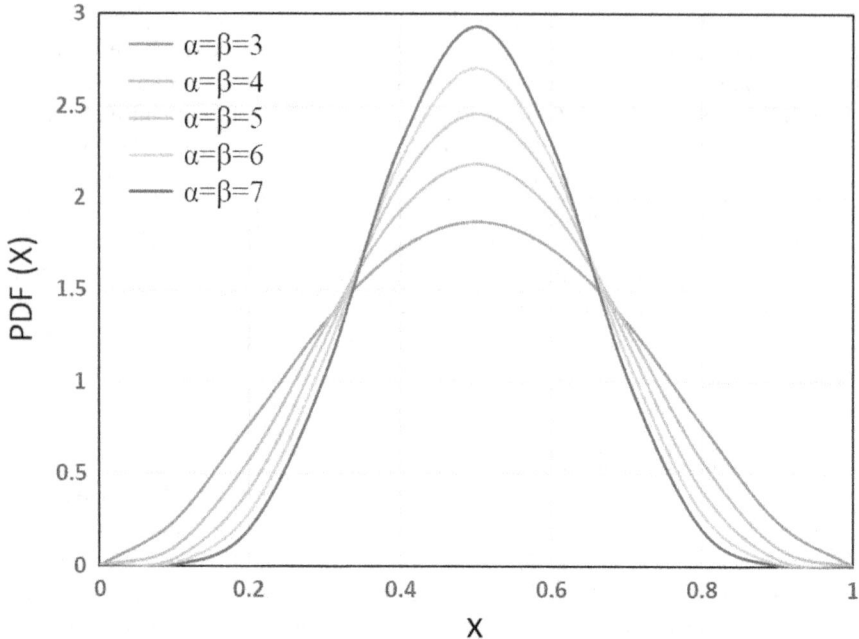

FIGURE 1.5 Truncated distribution of EV's SOC.

where χ is the set of uncertainties and Ω is the decision variables' space [45]. In MGEM problems, for instance, the robust optimization technique could be employed in order to minimize the cost function of MG (i.e., $\mathbb{C}$), while the baseline demand of the MG is at the highest possible value. In the following subsection, the uncertainty characterization methods are presented.

1.3.3.5   Uncertainty Characterization

1.3.3.5.1   Uncertainty of Wind Units

The uncertainty of the wind power units is due to the variable nature of wind speed in different weather conditions. The uncertainty related to the generation output of wind power units has been specifically studied in previous works like [46], [47], and [48]. In order to model the uncertainty of wind power production, the well-known Weibull distribution function is proposed [47]. The introduced formula of the Weibull probability distribution function is as follows:

$$f_s(s) = \left(\frac{k}{c}\right)\left(\frac{s}{c}\right)^{k-1} e^{-\left(\frac{v}{c}\right)^k} k, \, c > 0 \tag{1.16}$$

In Equation (1.16), c and k refers to the scale factor and shape factor, respectively. This distribution function could be divided into N_{sc} scenarios in which the probability of occurrence of each scenario could be defined and written as follows:

$$\pi_\omega = \int_{S_\omega}^{S_{\omega+1}} f_s(s)ds \quad \omega = 1, 2, \ldots, N_{sc} \tag{1.17}$$

In Equation (1.17), the S_w denoted the wind speed of the scenario w. Accordingly, the output power of the wind, P^{WT}, unit could be obtained by using the following equation:

$$P^{WT} = \begin{cases} 0 & 0 \leq S_\omega < S_i \\ P_r(A + S_\omega B + S_\omega^2 C) & S_i \leq S_\omega < S_r \\ P_r & S_r \leq S_\omega < S_o \\ 0 & S_\omega \geq S_o \end{cases} \tag{1.18}$$

The power generation curve of wind units is found in Figure 1.6.

The generation of the wind unit directly depends on the wind speed in a specific time-step. In Equation (1.18), the constant values A, B, and C could be achieved, for example from [49], and are related to the characteristics of the wind turbine. Note that S_i, S_o, S_r and P_r indicate the cut-in speed, cut-out speed, rated speeds, and rated power, respectively.

1.3.3.5.2 Uncertainty of PV Units

The uncertainty related to the photovoltaic unit stems from the variable amount of solar irradiation. Solar irradiation could be different from one location to another which results in the intermittent generation of PV units. The amount of solar radiation is dependent on the weather temperature, weather conditions, and the angle of photovoltaic panels. However, by studying the long-term patterns of solar radiation, for instance, in a specific location, it can be realized that they mostly follow a pattern. These patterns could be modeled as one of the most popular probability distribution functions. The most utilized distribution that is being used to model the generation of PV units would be beta distribution [50]. This function is introduced as the following equations:

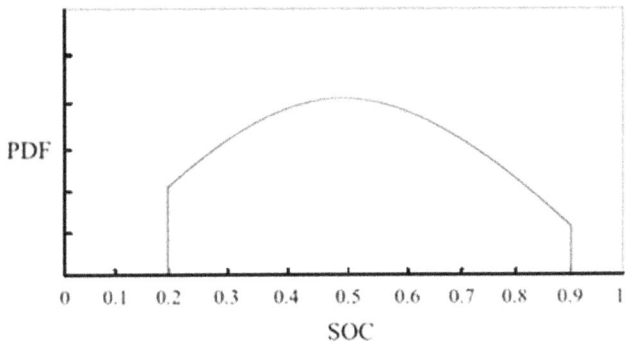

FIGURE 1.6 Applications of the robust optimization.

$$f_{\mathcal{R}} = \begin{cases} \mathcal{R}^{\alpha+1}(1 - \mathcal{R})^{\beta-1}\left(\frac{\Gamma(\alpha+\beta)}{\Gamma(\alpha)\Gamma(\beta)}\right) & 0 \leq \mathcal{R} \leq 1; \ \alpha, \ \beta \geq 0 \\ 0 & \text{otherwise} \end{cases} \tag{1.19}$$

$$\alpha = \left(\frac{\mu}{1 - \mu}\right)\beta \tag{1.20}$$

$$\beta = \frac{\mu(1 - \mu^2)}{\sigma^2} - (1 - \mu) \tag{1.21}$$

In Equations (1.19–1.21), the parameters α and β denote the features of function which can be obtained by means of Equations (1.20) and (1.21), respectively [51]. The beta distribution curve for different values of α and β is depicted in Figure 1.7. The variable $\mathcal{R}$ refers to the solar radiation in kW/m^2. Note that, the Γ refers to well-known gamma function.

In Equations (1.20) and (1.21), the mean and standard deviation of solar radiation could be denoted by μ and σ. Accordingly, the output power of the photovoltaic unit can be calculated using the following equation [50]:

$$P^{PV} = N_p\mathcal{R}(V_{oc} - k_v\theta_c)(I_{sc} + k_c\theta_c - 25k_c)\left(\frac{V_{mpp}I_{mpp}}{V_{oc}I_{sc}}\right) \tag{1.22}$$

where P^{PV} is the power generation of the PV system and the parameter N_p refers to the number of panels in the PV unit. $\mathcal{R}$ is the solar radiation at the location of the PV unit. The constants k_v and k_c are related to coefficient temperatures of voltage and current, respectively. V_{mpp} and I_{mpp} denote the respective voltage and current in the

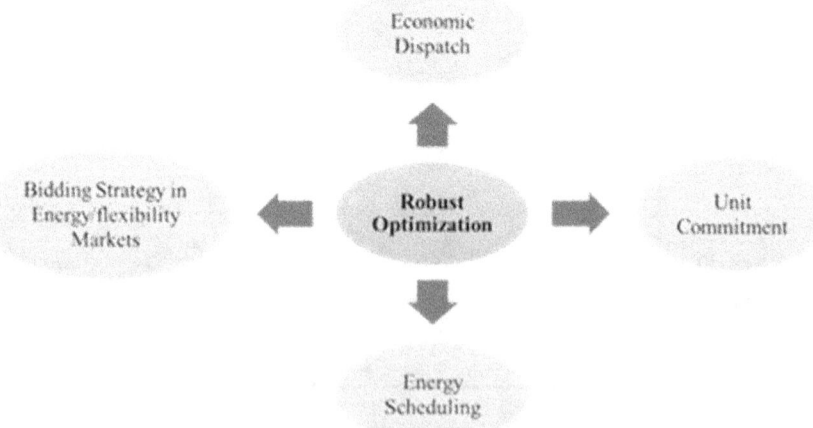

FIGURE 1.7 Output power of wind units based as a function of wind speed.

maximum power point condition. V_{oc} and I_{sc} refer to the voltage in open-circuit and current in short-circuit conditions, respectively. θ_c is the temperature of solar cells that can be approximated using the following equation:

$$\theta_c = \theta_{amb} + \frac{R(\theta_n - 20)}{0.8} \tag{1.23}$$

In Equation (1.23), the term θ_{amb} is the ambient temperature of PV panels and θ_n is the temperature in nominal operation condition.

1.3.3.5.3 Uncertainty of EV Owners' Behavior

In order to model the uncertain behavior of EV owners, for example in an MG, a distribution function is needed that could correctly represent the usage pattern of EVs. The most usual probability distribution that is used to model the uncertainty of EVs is truncated Gaussian distribution function [52], [53], and [54]. In the case of an MG, the EVs owned by the MG stakeholder, residential/commercial units, and/or a charging station could be considered in the uncertainty modeling with single or multiple probability distributions. In this light, for every single EV, the initial state-of-charge (SoC) and the availability of the EV in the understudy time horizon could be taken into account.

$$SoC_i^{ini} = f_{TG}(x, \mu^{soc}, \sigma^{soc}, SoC_i^{min}, SoC_i^{max}) \tag{1.24}$$

In Equation (1.24), the SoC_i^{ini} is the initial SoC of EV i. μ^{soc} and σ^{soc} are the mean and standard deviation of EVs' SoC, respectively. SoC_i^{min} and SoC_i^{max} are the minimum and maximum possible SoC of EVs in the MG. This equation could be utilized to generate the possible scenarios for initial SoC of the EVs.

According to the above equation, the truncated Gaussian distribution that could be used for modeling the initial SoC of an EV is depicted in Figure 1.8 [55]. In this

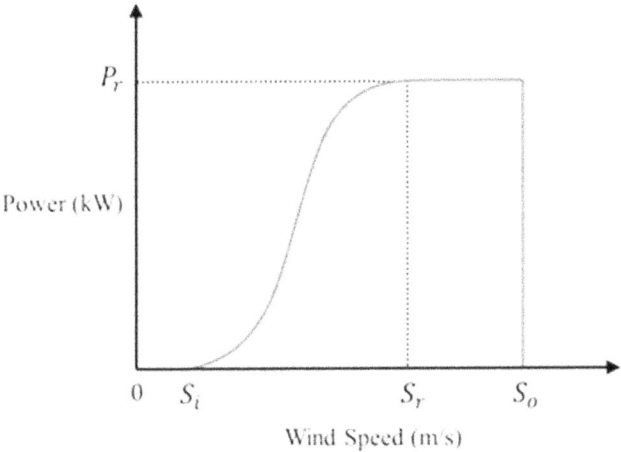

FIGURE 1.8   The beta distribution curve for different values of α and β.

exemplary figure, the minimum and maximum SoC for EVs considered 0.3 and 0.9, respectively. In Figure 1.8, the value of mean and the standard deviation are considered 0.5 and 0.25, respectively. The exact value of the parameters could be estimated by studying the historical and regular patterns of EVs' behavior.

Moreover, the most probable availability times of each EV have to be in hand for modeling the behavior of EVs. This could happen by considering the historical plug-in and plug-out times of EVs to the grid. Accordingly, the plug-in and plug-out times of EVs in the MG could be modeled with a distribution function as follows:

$$
\begin{cases}
t_i^{in} = f_{TG}\left(x, \mu^{in}, \sigma^{in}, t_i^{in,min}, t_i^{in,max}\right) & \forall\, i \\[2ex]
t_i^{out} = f_{TG}\left(x, \mu^{out}, \sigma^{out}, t_i^{out,min}, t_i^{out,max}\right) & \forall\, i
\end{cases}
\tag{1.25}
$$

In Equation (1.25), t_i^{in} and t_i^{out} are the times of plug-in and plug-out by EV i, respectively. This equation could be employed to generate several possible scenarios for plug-in and plug-out times of EVs in the MG.

1.3.3.5.4 Uncertainty of Flexibility Needs

In order to schedule and operate the MG in an efficient manner, the flexibility needs of the system should be predicted in advance. The concept of flexibility needs refers to the amount of regulation up/down needed for maintaining the system's balance in a predefined bandwidth. In fact, the flexibility need for a specific time is determined by the system's operator by the time of activation. However, the MG manager/aggregator should have the idea about the approximate values of flexibility that are supposed to be assigned from the system's operator. The flexibility need is always uncertain due to its dependency on several factors that need to be modeled by stochastic methods.

In order to model the uncertainty related to the flexibility needs, one can deploy a probability distribution function. It is obvious that the amount of flexibility need from the upstream grid that needs to be activated has a value between zero and the assigned value by the MG. In other words, the minimum activated amount of flexibility is zero when the MG is not supposed to provide any flexibility for a time step, and in contrast, the maximum value of the activated flexibility by the MG when the MG is supposed to provide the entire amount of assigned flexibility to the upstream network. However, the MG aggregator must schedule the demand and generation in a way that provides all the offered flexibility to the upstream network. With this regard, the activated amount of upward and downward flexibility from the system's operator could be modeled as uniformly distributed between zero and its maximum value [55].

$$
UF_{w,t} = f(x) = \frac{1}{F_t^{up}} \quad 0 \le x \le F_t^{up}
\tag{1.26}
$$

$$DF_{w,t} = f(x) = \frac{1}{F_t^{dn}} \quad 0 \le x \le F_t^{dn} \tag{1.27}$$

In Equations (1.26) and (1.27), $UF_{w,t}$ and $DF_{w,t}$ are the upward and downward activated flexibility. F_t^{up} and F_t^{dn} refer to the upward and downward assigned flexibility.

1.3.3.5.5 Model Predictive Control

Model predictive control (MPC) as a subfield of optimal control has several applications in electrical energy systems operation and control, especially in the energy management systems. Generally, MPC is a technique that makes a decision at a time through solving an approximate model over future horizons. There might be many engineering problems where the model is not in hand. In the MPC method, at least, an approximate model of the system is required. MPC is mostly employed to solve a problem with stochastic parameters which is modeled by means of a deterministic approximation. MPC might also utilize a stochastic model of the system in the future, however, the solution may be hard to converge [45]. In this method, the actions about the future configurations are realized by making the decisions now. Alternatively, it might utilize sampled approximations for the future, introduced as MPC in some literature, which are standard strategies in stochastic programming [56]. The overview of the MPC strategy for more clarification is depicted in Figure 1.9. This figure states that how the decision made by MPC at the current moment could predict the optimal trajectory of the system towards the future changes in the next time-steps. This procedure will iterate every time-step until the controller finds the best solution.

The MPC method is believed to be applicable in MGEM systems, especially when there is much stochasticity within the MG. Some of the most important applications of MPC method in MGs could be classified as follows [57]:

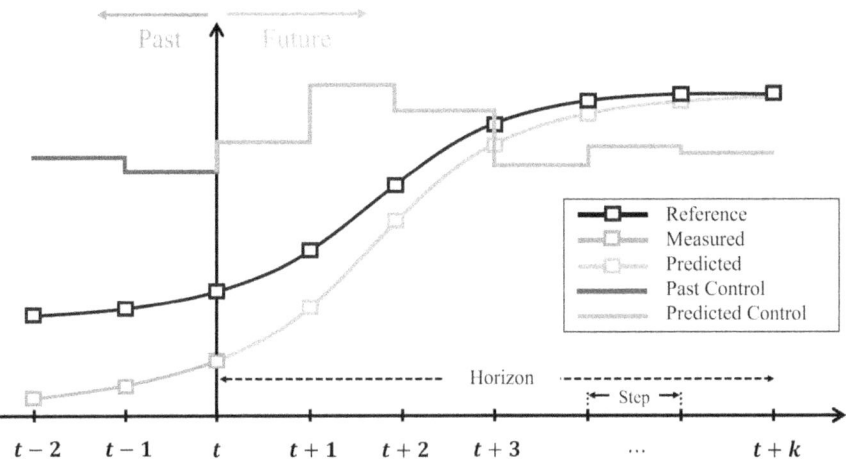

FIGURE 1.9 Model predictive control analogy.

- Providing an optimal solution to control the operation of MG's FERs for different objectives in MGEM systems.
- Providing control-based decision making to deal with the intermittency of consumers (demand, EVs, etc.) as well as renewable energy resources (wind, PV, etc.) in the MGEM optimization problems in order to tackle the stochasticity and randomness of these components.
- MPC is useful in handling some binary variables that need to be considered in MGEM problem. This could be beneficial when the situation of some components (e.g., charging mode of BESs, availability of EVs, ON-OFF mode of flexible loads) might possibly change and the new decisions should be made based on the situation.
- MPC could be beneficial in dealing with sudden changes in the MG where the new decisions should be made based on the new situation to maintain the MG in its normal operating point. This could help the MG to improve its degree of freedom in unusual conditions.
- MPC is also helpful when a distributed management method is taken within the MG. In this case, there might be several agents in the MG who make the MGEM problem complicated. Hence, MPC is believed to be a suitable choice in dealing with such problems.

Despite the above advantages and usefulness of the MPC method, it might have a high computational cost due to several optimization problems that need to be run in each time-step over a horizon. More information regarding different types of MPC formulation can be found in [57]. In the following subsection, brief overviews of the game-theory application in MGs are presented.

1.3.3.5.6   Game Theory
Another potential approach to MGEM problems is game theory. Game theory is believed to be among potential techniques in the operation of MGs due to the capability of enabling distributed management for MGs' resources [58] [59]. In order to implement a MGEM problem as a game theory problem, the resources located at the MG could be considered game participants. Despite the game theory's shortcomings in problem convergence, it is still a good choice for multi-agent-based decision-making studies.

Game theory has many applications from the energy industry to economic studies in complex systems. In general, its concept refers to mathematical techniques that model the interaction between multiple decision-makers [60]. The choice of one decision-making entity could affect the choices of the other entities. One of the applications of game theory could be in distributed management of MGs. In order to implement a MGEM problem as a game theory problem, the resources located at the MG could be considered game participants. The following four steps should be taken for a game-theoretic problem:

a. Defining players of the proposed game
b. Defining the individual and overall goal
c. Dealing with the coupled constraints
d. Finding the Nash equilibrium

Game theory methods could be split into two categories: cooperative game and non-cooperative game [61]. A cooperative game is the one with a number of entities in which the main goal of these players are in line with each other. In contrast, the non-cooperative game refers to a game in which the entities are in conflict with each other and they try to act independently regarding their goals [62]. However, the choice of the suitable game-theory approach for MGEM problems directly depends on the method of MG management and the agreements between MG operator and members.

1.4 CONCLUSION

The MG is an autonomous or semi-autonomous system that consists of DGs, RESs, and FERs as well as dispatchable loads working together which are able to operate in both grid-connected and islanded mode. MGs as the potential sources of sustainability are designed to expand the decentralization goals along with cooperation with the whole system toward flexible electrical energy systems. However, the increasing penetration of renewable sources and the decentralization in either MG or system-level networks have resulted in several stability and resiliency issues. Consequently, the EMSs are proposed to efficiently control and manage the operation of all energy resources locate in generation-side and demand-side so as to deal with the stochastic outcomes of a deregulated system. Moreover, the smart power systems in the future will confront several uncertainties due to the very low-inertia situations that must be addressed by the EMS as well as the novel control techniques in advance.

In MG concepts, the EMS also could play a pivotal role in dealing with the aforesaid issues since they could be employed to satisfy several objectives in operation and control of the MG. The EMSs could also be beneficial in providing different kinds of services to the national or regional grids. Accordingly, MGEM systems could be designed and planned carefully so that they can take into account all the individual or environmental constraints and limitations of consumers' electrification. Hence, in order to have an efficient and synergetic contribution by MGs, the MGEM problems should be defined by problem formulations in which all the uncertainty, stochasticity, restrictions, and also monetary payback for its stakeholders are considered precisely.

Different types of optimization formulations were used for the MGEM problems. Most of them focused on minimizing MG's operating costs such as fuel costs, maintenance costs, and the cost of imported energy from the grid. These optimization techniques could be categorized based on their optimization types, objective functions, constraints, and also tools that are utilized to solve MGEM problems. The most popular ones include stochastic and robust optimization techniques. Furthermore, some tools have been introduced, such as model predictive control, intelligent techniques, and game theory, in order to make the MGEM more efficient and predictable.

To sum up, the utilization of MGs equipped with the MGEM system could be a potential solution to future power systems issues. The control and optimization techniques in MGEM system enable the active and sustainable utilization of energy resources.

These energy management systems could help to enhance the flexible utilization of energy resources as well as the flexibility of the whole power system by providing different types of balancing and ancillary services. However, in order to make the most out of these MGEM plans, the smart selection of control and optimization trajectories are of the necessity.

REFERENCES

[1] B. Khan and P. Singh, "Optimal power flow techniques under characterization of conventional and renewable energy sources: A comprehensive analysis," *J. Eng.*, vol. 2017, Article ID 9539506, 16 pages, 2017.

[2] P. Singh and B. Khan, "Smart microgrid energy management using a novel artificial shark optimization," *Complexity*, vol. 2017, Article ID 2158926, 22 pages, 2017.

[3] T. Molla, B. Khan, B. Moges, H. H. Alhelou, R. Zamani and P. Siano, "Integrated optimization of smart home appliances with cost-effective energy management system," *CSEE J. Power Energy Syst.*, vol. 5, no. 2, pp. 249–258, June 2019.

[4] Z. Tang, Y. Lin, M. Vosoogh, N. Parsa, A. Baziar and B. Khan, "Securing microgrid optimal energy management using deep generative model," *IEEE Access*, vol. 9, pp. 63377–63387, 2021

[5] S. P. Bihari et al., "A comprehensive review of microgrid control mechanism and impact assessment for hybrid renewable energy integration," *IEEE Access*, vol. 9, pp. 88942–88958, 2021.

[6] B. Khan and P. Singh, "Selecting a meta-heuristic technique for smart micro-grid optimization problem: A comprehensive analysis," *IEEE Access*, vol. 5, pp. 13951–13977, 2017.

[7] O. P. Mahela et al., "Comprehensive overview of multi-agent systems for controlling smart grids," *CSEE J. Power Energy Syst.*, vol. 8, no. 1, pp. 115–131, doi: 10.17775/CSEEJPES.2020.03390

[8] Z. Wang, B. Zhang, M. Mobtahej, A. Baziar and B. Khan, "Advanced reactive power compensation of wind power plant using PMU data," *IEEE Access*, vol. 9, pp. 67006–67014, 2021.

[9] S. Padmanaban, C. Dhanamjayulu and B. Khan, "Artificial Neural Network and Newton Raphson (ANN-NR) algorithm based selective harmonic elimination in cascaded multilevel inverter for PV applications," *IEEE Access*, vol. 9, pp. 75058–75070, 2021.

[10] M. S. Fakhar et al., "Implementation of APSO and improved APSO on non-cascaded and cascaded short term hydrothermal scheduling," *IEEE Access*, vol. 9, pp. 77784–77797, 2021.

[11] H. Shayeghi, E. Shahryari, M. Moradzadeh, and P. Siano, "A survey on microgrid energy management considering flexible energy sources," *Energies*, vol. 12, no. 11, p. 2156, 2019.

[12] A. Hauer, "Storage Technology Issues and Opportunities, International Low-Carbon Energy Technology Platform," in Proceedings of the Strategic and Cross-Cutting Workshop "Energy Storage—Issues and Opportunities", Paris, France, 2011, vol. 15.

[13] F. Faraji, A. Majazi, and K. Al-Haddad, "A comprehensive review of flywheel energy storage system technology," *Renew. Sustain. Energy Rev.*, vol. 67, pp. 477–490, 2017.

[14] Greensmith, "Energy Storage enabling Flexibility in Electrical Grid Systems," EEF Dinner Debate, Strasbourg, June 13, 2017, http://www.europeanenergyforum.eu/sites/default/files/events/doc/Final%20Greensmith.pdf, Accessed on 14th June 2021.

[15] W. Bai, M. R. Abedi, and K. Y. Lee, "Distributed generation system control strategies with PV and fuel cell in microgrid operation," *Control Eng. Pract.*, vol. 53, pp. 184–193, 2016.

[16] Y. Liu, L. Xiao, G. Yao, and S. Bu, "Pricing-based demand response for a smart home with various types of household appliances considering customer satisfaction," *IEEE Access*, vol. 7, pp. 86463–86472, 2019.

[17] H. Khajeh, M. Shafie-Khah, and H. Laaksonen, "Blockchain-based demand response using prosumer scheduling," in *Blockchain-based Smart Grids*, Elsevier, 2020, pp. 131–144.

[18] J. Leitão, P. Gil, B. Ribeiro, and A. Cardoso, "A survey on home energy management," *IEEE Access*, vol. 8, pp. 5699–5722, 2020.

[19] J. Wang, S. You, Y. Zong, C. Træholt, Z. Y. Dong, and Y. Zhou, "Flexibility of combined heat and power plants: A review of technologies and operation strategies," *Appl. Energy*, vol. 252, p. 113445, 2019.

[20] D. Li, X. Xu, D. Yu, M. Dong, and H. Liu, "Rule based coordinated control of domestic combined micro-CHP and energy storage system for optimal daily cost," *Appl. Sci.*, vol. 8, no. 1, p. 8, 2018.

[21] J. M. Lujano-Rojas, C. Monteiro, R. Dufo-López, and J. L. Bernal-Agustín, "Optimum load management strategy for wind/diesel/battery hybrid power systems," *Renew. Energy*, vol. 44, pp. 288–295, 2012.

[22] T. Khatib and W. Elmenreich, "Novel simplified hourly energy flow models for photovoltaic power systems," *Energy Convers. Manag*, vol. 79, pp. 441–448, 2014.

[23] Y. Zheng, B. M. Jenkins, K. Kornbluth, A. Kendall, and C. Træholt, "Optimization of a biomass-integrated renewable energy microgrid with demand side management under uncertainty," *Appl. Energy*, vol. 230, pp. 836–844, 2018.

[24] B. Li, R. Roche, D. Paire, and A. Miraoui, "A price decision approach for multiple multi-energy-supply microgrids considering demand response," *Energy*, vol. 167, pp. 117–135, 2019.

[25] C. Dou, M. Lv, T. Zhao, Y. Ji, and H. Li, "Decentralised coordinated control of microgrid based on multi-agent system," *IET Gener. Transm. Distrib.*, vol. 9, no. 16, pp. 2474–2484, 2015.

[26] W. L. Theo, J. S. Lim, W. S. Ho, H. Hashim, and C. T. Lee, "Review of distributed generation (DG) system planning and optimisation techniques: Comparison of numerical and mathematical modelling methods," *Renew. Sustain. Energy Rev.*, vol. 67, pp. 531–573, 2017.

[27] J. Bai, Z. Zhou, S. Zhou, and M. Tariq, "Distributed energy management in smart grid with dominated electricity provider and multiple microgrids," in 2014 International Conference on Power System Technology, 2014, pp. 3249–3256.

[28] Q. Fu, A. Nasiri, V. Bhavaraju, A. Solanki, T. Abdallah, and C. Y. David, "Transition management of microgrids with high penetration of renewable energy," *IEEE Trans. Smart Grid*, vol. 5, no. 2, pp. 539–549, 2013.

[29] Y. Liu, S. Yu, Y. Zhu, D. Wang, and J. Liu, "Modeling, planning, application and management of energy systems for isolated areas: A review," *Renew. Sustain. Energy Rev.*, vol. 82, pp. 460–470, 2018.

[30] Z. Iqbal et al., "A domestic microgrid with optimized home energy management system," *Energies*, vol. 11, No. 4, p. 1002, 2018.

[31] H. Khajeh, H. Laaksonen, A. S. Gazafroudi, and M. Shafie-Khah, "Towards flexibility trading at TSO-DSO-customer levels: A review," *Energies*, vol. 13, no. 1, p. 165, 2020.

[32] "Fingrid Open Data Source." [Online]. Available: https://www.fingrid.fi/en/, Access on 15th July 2021.

[33] Fingrid Oyj, "Reserve products and marketplaces," 2020, Access on 15th July 2021.

[34] H. Firoozi, H. Khajeh, and H. Laaksonen, "Optimized operation of local energy community providing frequency restoration reserve," *IEEE Access*, vol. 8, pp. 180558–180575, Sep. 2020, doi: 10.1109/access.2020.3027710.

[35] H. Firoozi and H. Khajeh, "Optimal day-ahead scheduling of distributed generations and controllable appliances in microgrid," in 2016 Smart Grids Conference (SGC), 2016, pp. 1–6.

[36] M.-H. Lin, J.-F. Tsai, and C.-S. Yu, "A review of deterministic optimization methods in engineering and management," *Math. Probl. Eng.*, vol. 2012, pp. 1–15, 2012.

[37] Z. A. Farhath, B. Arputhamary, and L. Arockiam, "A survey on ARIMA forecasting using time series model," *Int. J. Comput. Sci. Mob. Comput*, vol. 5, pp. 104–109, 2016.

[38] W. Wei, F. Liu, S. Mei, and Y. Hou, "Robust energy and reserve dispatch under variable renewable generation," *IEEE Trans. Smart Grid*, vol. 6, no. 1, pp. 369–380, 2014.

[39] T. Baležentis and D. Streimikiene, "Multi-criteria ranking of energy generation scenarios with Monte Carlo simulation," *Appl. Energy*, vol. 185, pp. 862–871, 2017.

[40] Y. Dvorkin, Y. Wang, H. Pandzic, and D. Kirschen, "Comparison of scenario reduction techniques for the stochastic unit commitment," in 2014 IEEE PES General Meeting| Conference & Exposition, 2014, pp. 1–5.

[41] M. Nazari-Heris and B. Mohammadi-Ivatloo, "Application of robust optimization method to power system problems," in *Classical and Recent Aspects of Power System Optimization*, Elsevier, 2018, pp. 19–32.

[42] W. Hu, P. Wang, and H. B. Gooi, "Toward optimal energy management of microgrids via robust two-stage optimization," *IEEE Trans. Smart Grid*, vol. 9, no. 2, pp. 1161–1174, 2016.

[43] S. Cui, Y.-W. Wang, J.-W. Xiao, and N. Liu, "A two-stage robust energy sharing management for prosumer microgrid," *IEEE Trans. Ind. Informatics*, vol. 15, no. 5, pp. 2741–2752, 2018.

[44] H. Khajeh, A. A. Foroud, and H. Firoozi, "Robust bidding strategies and scheduling of a price-maker microgrid aggregator participating in a pool-based electricity market," *IET Gener. Transm. Distrib.*, vol. 13, no. 4, pp. 468–477, 2018.

[45] W. B. Powell, "A unified framework for stochastic optimization," *Eur. J. Oper. Res.*, vol. 275, no. 3, pp. 795–821, 2019.

[46] A. Nasri, S. J. Kazempour, A. J. Conejo, and M. Ghandhari, "Network-constrained AC unit commitment under uncertainty: A Benders' decomposition approach," *IEEE Trans. Power Syst.*, vol. 31, no. 1, pp. 412–422, 2015.

[47] R. Karki, P. Hu, and R. Billinton, "A simplified wind power generation model for reliability evaluation," *IEEE Trans. Energy Convers.*, vol. 21, no. 2, pp. 533–540, 2006.

[48] S. J. Kazempour and A. J. Conejo, "Strategic generation investment under uncertainty via Benders decomposition," *IEEE Trans. Power Syst.*, vol. 27, no. 1, pp. 424–432, 2011.

[49] P. Giorsetto and K. F. Utsurogi, "Development of a new procedure for reliability modeling of wind turbine generators," *IEEE Trans. Power Appar. Syst.*, no. 1, pp. 134–143, 1983.

[50] Y. Li and E. Zio, "Uncertainty analysis of the adequacy assessment model of a distributed generation system," *Renew. Energy*, vol. 41, pp. 235–244, 2012.

[51] Y. M. Atwa, E. F. El-Saadany, M. M. A. Salama, and R. Seethapathy, "Optimal renewable resources mix for distribution system energy loss minimization," *IEEE Trans. Power Syst.*, vol. 25, no. 1, pp. 360–370, 2009.

[52] M. Amini and A. I. Sarwat, "Optimal reliability-based placement of plug-in electric vehicles in smart distribution network," *Int. J. Energy Sci.*, vol. 4, no. 2, 2014.

[53] S. I. Vagropoulos and A. G. Bakirtzis, "Optimal bidding strategy for electric vehicle aggregators in electricity markets," *IEEE Trans. Power Syst.*, vol. 28, no. 4, pp. 4031–4041, 2013.

[54] M. H. Amini et al., "Plug-in electric vehicle owner behavior study using fuzzy systems," *Int. J. Power Energy Syst.*, vol. 35, no. 2, p. 40, 2015.

[55] M. Shafie-Khah, P. Siano, D. Z. Fitiwi, N. Mahmoudi, and J. P. S. Catalao, "An innovative two-level model for electric vehicle parking lots in distribution systems with renewable energy," *IEEE Trans. Smart Grid*, vol. 9, no. 2, pp. 1506–1520, 2017.

[56] G. Schildbach and M. Morari, "Scenario-based model predictive control for multi-echelon supply chain management," *Eur. J. Oper. Res.*, vol. 252, no. 2, pp. 540–549, 2016.

[57] C. Bordons, F. Garcia-Torres, and M. A. Ridao, *Model Predictive Control of Microgrids*, Springer, 2020.

[58] P. Li, J. Ma, and B. Zhao, "Game theory method for multi-objective optimal operation of microgrid," in *IEEE Power & Energy Society General Meeting*, 2015, pp. 1–5.

[59] E. Mojica-Nava, C. A. Macana, and N. Quijano, "Dynamic population games for optimal dispatch on hierarchical microgrid control," *IEEE Trans. Syst. Man, Cybern. Syst.*, vol. 44, no. 3, pp. 306–317, 2013.

[60] P. Aristidou, A. Dimeas, and N. Hatziargyriou, "Microgrid modelling and analysis using game theory methods," in International Conference on Energy-Efficient Computing and Networking, 2010, pp. 12–19.

[61] M. J. Osborne and A. Rubinstein, *A Course in Game Theory*. MIT Press, Cambridge, MA, 1994.

[62] W. Saad, Z. Han, H. V. Poor, and T. Basar, "Game-theoretic methods for the smart grid: An overview of microgrid systems, demand-side management, and smart grid communications," *IEEE Signal Process. Mag.*, vol. 29, no. 5, pp. 86–105, 2012.

2 Hybrid Particle Swarm Optimization – Artificial Neural Network Algorithm for Energy Management

Kranthi Kumar
Department of Electrical Engineering, KPRIT, India

Altaf Q. H. Badar
Electrical Engineering Department, National Institute of
Technology Warangal, India

Sanjeevikumar Padmanabhan
CTiF Global Capsule, Aarhus University, Department of
Business Development and Technology, Aarhus University,
Herning, Denmark

CONTENTS

DOI: 10.1201/b22884-2

2.1 INTRODUCTION

Energy is a vital entity on which the world runs. It can be transferred to an object for it to perform work or to generate heat. Energy may be preserved, based on the law of conservation of energy which states that energy can be regenerated but cannot be created or destroyed. The standard unit of energy is a Joule; alternative units are KWh, BTU, Calorie, eV, and foot-pound.

2.1.1 ENERGY MANAGEMENT (EM)

The energy can be used in alternative ways to obtain an effective output. In power systems, energy use needs to be optimized for economic operation of the system [1–10]. The terms embraced for operations related to optimal use of energy are EM, demand side management (DSM), demand response (DR), load management, fuel shift, energy efficiency and conservation, etc.. [11]. The subsections below outline energy management with additional details.

EM relates to all aspects from managing energy, including behavioral changes in energy maintenance practices, economic use of energy through instrument retrofits, fuel conservation, energy recovery, peak demand reductions (temporary/permanent) actions, and control of distributed energy resources (DERs).

DSM is another method that deals with optimal EM on the demand side [11]. It deals with utility programs or DR programs, geared towards containing rising energy use on demand side. "DSM is designing, implementation, and observing of utility activities designed to influence use of electricity on demand side, in ways which can produce desired changes in a specified pattern and time frame [11]."

2.1.2 FUEL SWITCH

A fuel switch is an important concept introduces to save/reduce requirement of certain fuel [11]. Fuel switch does not necessarily deal with reduction of load but concentrates on use of fuels that are not scarce. Thus, we try to use fuel based on its availability. This will ultimately result in lowering operating costs and long-term reliability of supply.

2.1.3 DERs

In their most general sense, DERs utilize technologies required for distributed generation (DG) (non-renewable and renewable), combined heat and power (CHP), storage systems, etc. [11]. DERs do not perform well on power quality and thus need to have equipment to deal with it specifically. The main advantage of DERs is a reduction of transmission and distribution (T&D) losses in the power systems.

2.1.4 DEMAND RESPONSE (DR)

DRs are a set of load management techniques that deals with client-side actions that quickly change the load quantity in response to cost or alternative signals from the utility. A DR is usually different from alternative load management strategies, like energy storage or energy potency enhancements which lead to a permanent reduction in energy demand [11].

2.2 ENERGY MANAGEMENT SYSTEMS (EMS)

An EMS is a combination of computer-aided tools employed by electric utility grids to monitor, manage, and optimize the performance of overall power systems. Consequently it can be implemented in smaller systems like microgrids.

EMS measures through automated systems to collect energy-related information from the sensors to present it to the users using graphics, online watching tools and energy quality analyzers that facilitate optimal control and management of energy flow. An EMS is responsible to observe in real time the variables and events happening within the electrical systems and archive them as digital storage for future analysis. The analysis of data helps in better management and economical use of available resources.

EMSs are usually employed by various business entities for live observation and management their electrical entities. EMS can centrally manage devices like HVAC and lighting systems across multiple locations, like retail, grocery, hotels, etc. EMSs may have metering, sub-metering, and other observing meters/sensors that permit the utilities and building managers to collect information and their details. This helps them take better decisions regarding energy activities across their sites.

An EMS has become a crucial issue for several utilities round the world. It is very tough to satisfy energy demands while having load shedding and other forced actions for energy control. An EMS can be used by different consumers like residential, business, and industrial sector for scale back of energy needs, thus leading to savings in the value of energy consumed that cumulatively has a large, positive impact on system operations. Energy management isn't solely necessary in distribution system only and has quite a significance on the generation side also.

Energy management is needed to make sure that a lot of necessary activities are going to be checked and optimized. Total quality management (TQM) is being adopted by a number of entities that also include management of energy in their firms/entities.

An EMS includes designing, monitoring, operation, and archiving of energy production and consumption units. EMS objectives include conservation of resources, reduced pollution, and economic operation. However, the users that have access to energy management shall try to maximize profits while reducing the cost. Some of the other objectives of EMS are [12]:

• Improve energy efficiency while reducing energy costs.
• Reduce greenhouse gas emissions (pollution)/improve air quality.

- Improve communications within the system.
- Develop and maintain effective management information systems related to energy use; thus leading to forming better management strategies.
- Optimal return on energy-related investments.
- Involve all stakeholders to take part in energy management programs.
- Reduce interruption in supply of energy in any form.

Energy tracking, targeting, and communication are a powerful techniques for energy management, used to:

- analyze past/archived energy usage
- set targets for reduction in energy
- optimize energy usage
- plan for future costs for energy usage.

This method is quite effective in achieving energy savings. The operating definitions are:

- **Energy Tracking:** In energy tracking, energy-related data is assorted and analyzed. The purpose of tracking is to compare the current consumption pattern with the existing historical pattern of energy consumption. This helps the EMS to take note of the deviation and take necessary actions.
- **Targeting:** After spotting a deviation from the predicted energy consumption pattern, the management is allocated with the objective of finding a reason for the same.
- **Communication:** Communication is the basic requirement of this technique. A closed communication loop is formed for identifying a deviation, initiated action by management, accomplishment of targets, and verification of actions taken through realization of savings.

As the world progresses, there are new opportunities for conserving energy using the latest energy storage technologies and electric vehicles (EVs) that are integrated into the power systems. The grid is also evolving into a smart grid involving latest equipments, processes, and communication techniques. A smart grid is a grid which has smart meters, EVs, communications systems, renewable energy sources, etc.

2.3 ARTIFICIAL NEURAL NETWORK (ANN)

An artificial neural network (ANN) is a computing model based on the structure and functions of biological neural networks (BNNs).

The working of neurons was introduced by Warren McCulloch, neurophysiologist and Walter Pitts, mathematician in [13]. They modeled the neurons using electrical circuits to observe the workings of neurons. In 1949, Donald Hebb [14] presented "The Organization of Behavior," which outlined a law for synaptic neuron learning called Hebbian learning. It had a very simple and straightforward method and fundamental learning rule for ANN [14]. The first functional neural network with multiple layers was published by Ivakhnenko and Lapa [15] in 1967.

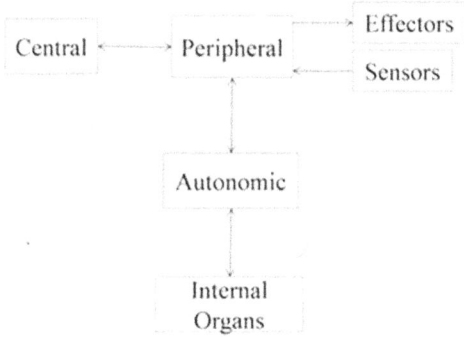

FIGURE 2.1 Basic parts of a nervous system.

2.3.1 BIOLOGICAL AND ARTIFICIAL NEURAL NETWORK

The living organism nervous system (NS) has three basic parts: (i) central, (ii) peripheral, (iii) autonomic nervous system [16], as shown in Figure 2.1.

The primary function of NS is to control the activities of a living organism. It needs to process available information to achieve this function. As seen in Figure 2.1, signals are transferred from peripheral NS, connected to effectors and sensors. The autonomic NS administrates the workings of internal organs.

In a real organism, the major part of NS is present in the brain. It has a highly composite, irregular, and side-by-side information processing system. The neurons form the basic part of neural cells in the brain. The brain has the ability to sort neurons and its activity so as to perform certain deliberations such as perception, motor control, pattern recognition, etc.

The human brain processes data in a synchronized manner. It uses a network of neurons having a very large number of densely interlinked and relatively simple decision-making elements. The human brain has 10^{11} neurons, which is nearly equal to the number of stars in the Milky Way; each neuron is connected to 10^3 to 10^4 neighboring neurons [17].

The characteristics of biological neural network (BNN) are embedded in its structure and its function. The basic unit of a BNN is called a neuron (or) nerve cell, as shown in Figure 2.2.

A nerve cell consists of soma, also called a cell body, where nucleus cells are located. Tree-like nerves called dendrites are connected with a cell body. It receives signal from other neurons. Connected to soma is an axon, a single long fiber, which connects with many other neurons through synapses.

Features of ANNs that are observed from a biological neuron:

1. Many signals are received by the processing element.
2. Modification of signal may happen through synapse weights.
3. Synapse weight strength is dynamic in nature.
4. The weighted inputs are summed by the processing element.
5. Usually the neuron will have a single output.

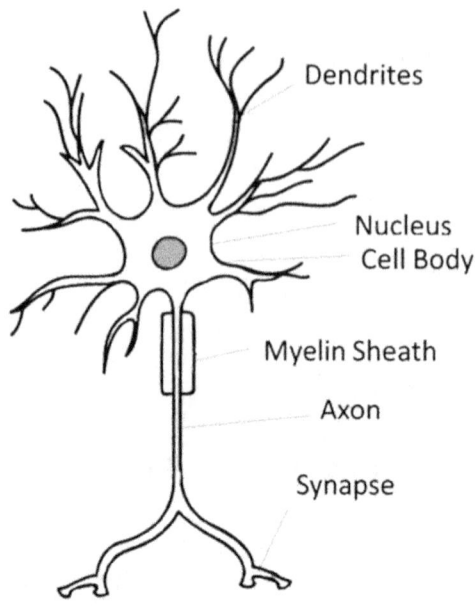

FIGURE 2.2 Nerve cell.

6. The output of a single neuron may act as an input to many other neurons.
7. Local processing of information.
8. Distributed memory.
9. Neurotransmitters for synapses may be excitatory or inhibitory.
10. Fault tolerance.

The size of a soma (cell body) for a typical neuron is in the range of 10–80 micrometers and the gap is 200 nanometers wide for a synaptic junction.

ANN represents the method modeled to mimic the method in which the human brain performs a specific task or function. It bears a resemblance to the brain in the following aspects:

• Learning process is involved to acquire knowledge.
• Synaptic weights, i.e., inter-neuron connection strengths are used to store/represent the knowledge.

The procedure of the learning process is known as a learning algorithm; it modifies the synaptic weights within the network to obtain a desired objective/output.

2.3.2 COMPARISON BETWEEN ANN AND BNN

BNN and ANN both work in a parallel mode; however, ANN is faster and BNN is slower. ANN displays an inferior performance compared to BNN. As far as the number of neurons is considered, the BNN has around 10^{11} neurons with 10^{15}

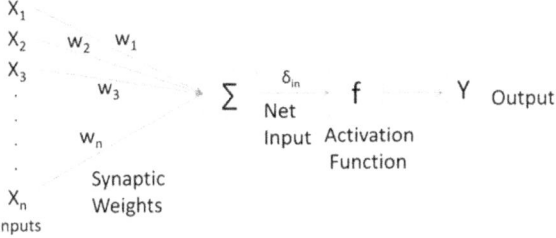

FIGURE 2.3   Generalized model of ANN.

interconnections while the ANN may have 10^2 and 10^4 nodes/neurons. In BNN, physical damage can cause performance degradation, while ANN cannot have physical damage. Even for other damage, ANN is found to be quite tolerant to faults. In BNN, ambiguity is easily handled, whereas in ANN lots of training is required to overcome the problem of ambiguity. ANN uses memory locations to store memory, whereas BNN uses the synaptic weights.

2.3.3   MODEL OF ANN

The general model of ANN is shown in Figure 2.3.

The net input shown in Figure 2.3 can be calculated as given in Eq 2.1 (a and b):

$$\delta_{in} = x_1w_1 + x_2w_2 + x_3w_3 + \ldots\ldots\ldots.. + x_nw_n \tag{2.1a}$$

$$\text{Net input } \delta_{in} = \sum_{(i=1)}^{n} (x_iw_i) \tag{2.1b}$$

The output of the model is calculated as given in Eq 2.2:

$$y = f(\delta_{in}) \tag{2.2}$$

ANN depends on three elements [18]:

- Network architecture
- Adjustments of weights
- Squashing function (or) activation function

2.3.3.1   Network Architecture

Network architecture means the sorting along with nodes and connecting lines. These are classified into three types [17]:

1. One-layer feed forward N/W
2. Many-layer feed forward N/W
3. Recurrent N/W

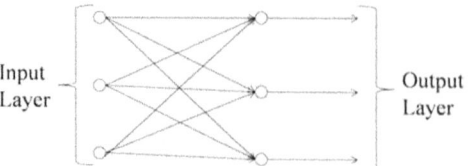

FIGURE 2.4   One-layer feed forward network.

2.3.3.2   One-Layer Feed Forward N/W

The name itself says that it's a network with one way from input to output i.e., the information flows in one direction from input to output. In this all the input layers are connected to output layer directly. The representation is shown in Figure 2.4.

2.3.3.3   Many-Layer Feed Forward N/W

In this network between input and output layers any number of hidden layers can be connected. A representation with one hidden layer is shown in Figure 2.5.

2.3.3.4   Recurrent Network

In a recurrent network, all the input layers are connected to the output layers with or without hidden layers but at least one feedback should be connected to the network. Figure 2.6 a and b shows a representation of recurrent network along with feedback.

2.3.3.5   Adjustment of Weights (or) Learning

In ANN, learning is a method that deals with the modification of connections and their weights between the neurons of a specified network. It is classified into three types [18]:

   1. Supervised learning
   2. Unsupervised learning
   3. Reinforcement learning

2.3.3.6   Supervised Learning

It represents learning under the guidance of a teacher who acts like a supervisor. When the input is applied to a neural network, an output is obtained. This output is then compared with the desired output. If any difference is present in the output, then an error signal is generated. This process is continued until both outputs i.e.,

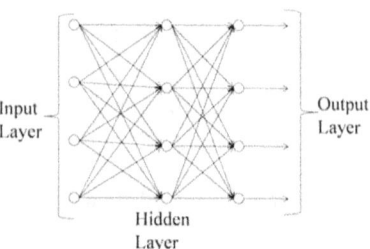

FIGURE 2.5   Many-layer feed forward network with one hidden layer.

(a)

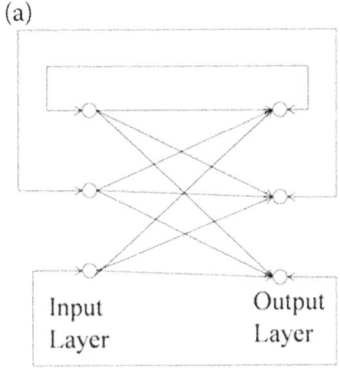

Input Layer Output Layer

(b)

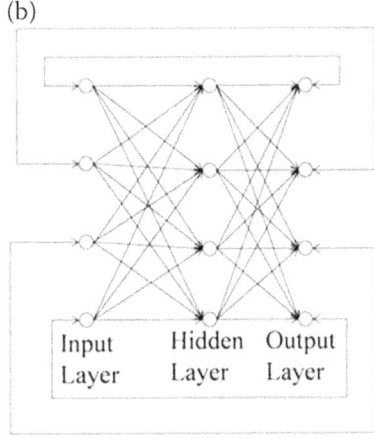

Input Layer Hidden Layer Output Layer

FIGURE 2.6 (a) Recurrent network without hidden layer, (b) recurrent network with one hidden layer.

expected and observed, are the same. The block diagram in Figure 2.7 represents supervised learning.

2.3.3.7 Unsupervised Learning

It represents learning without the guidance of a teacher; thus, unsupervised. The process is independent of other factors. In unsupervised learning, the actual output is not compared with the desired output, which means there is no feedback. The output will depend on input only and it varies as the input varies. Figure 2.8 represents unsupervised learning.

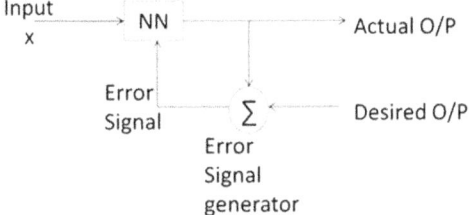

FIGURE 2.7 Block diagram for supervised learning.

FIGURE 2.8   Unsupervised learning.

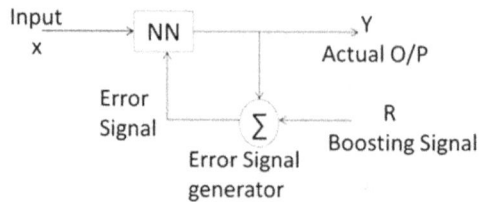

FIGURE 2.9   Reinforcement learning.

2.3.3.8   Reinforcement Learning

This process is similar to learning from one's own experience. The boost signal is used in a network to monitor the output. In supervised learning, the output data is labeled, which is not the case in reinforcement learning. The block diagram of reinforcement learning is given in Figure 2.9.

2.3.3.9   Activation Function

It is defined as the additional force applied at input to get a desired output. Some of these activation functions are linear activation function, sigmoid activation function, etc.

2.4   PARTICLE SWARM OPTIMIZATION

Particle swarm optimization (PSO) is a multi-dimensional non-linear optimization technique. It is a very robust, simple, and easily adaptable method. One can find a number of applications of PSO in various fields.

It was introduced in 1995 by Kennedy and Eberhart [19]. PSO has since been improved many number of times by various researchers. However, the initial contribution came from Shi and Eberhart [20] and M Clerc [21].

A PSO is derived from the movement of birds and fishes. The birds or fishes move together in a flock or school, respectively, to optimize their chances of finding food, escape their predators, and to find mates. PSO also imitates the thought process of a human being. A normal person thinks and decides based on two levels. The first level represents his/her own previous experiences while the second part represents the experience of his/her social contacts.

The terminology used in a PSO is:

- **Particle:** An n-dimensional vector which represents a complete solution.
- **Population:** A collection of particles.
- **Swarm:** Particles moving in random directions and still converging.
- **Particle Velocity:** The movement of particles from one position to another is given by their velocity.

- **Individual Best:** It is a position that has the best fitness function value from amongst all the positions that the particle has visited until the current iteration.
- **Global Best:** It is a position that has the best fitness function value from amongst all the positions that any of the particles in the swarm has visited until the current iteration.
- **Stopping Criteria:** It is the condition that is required to be satisfied for the search process to terminate. The different criteria to stop the process may be: a) swarm converging to a single position, or b) pre-specified number of iterations, or c) global best not improving for a pre-specified number of iterations.

A PSO can be explained in short as particles are created randomly throughout the search space. The particles move/search around in the search space, trying to find the optimal position. The movement of any particle is dependent on its previous velocity, its individual best position, and the global best position.

A PSO uses two equations: first to find the new position of a particle and the second equation to find the velocity of the particle. The velocity gives movement of the particle.

The first equation is given as Eq (2.3):

$$x_{id} = x_{id} + v_{id} \qquad (2.3)$$

The equation for velocity has been described in different ways, as listed below in Eqs (2.4) and (2.5):

$$v_{id} = \omega * v_{id} + c_1 * rand() * (p_{id} - x_{id}) + c_2 * Rand() * (p_{gd} - x_{id}) \qquad (2.4)$$

$$v_{id} = K * [v_{id} + c_1 * rand() * (p_{id} - x_{id}) + c_2 * Rand() * (p_{gd} - x_{id})] \qquad (2.5)$$

The previous two equations can be considered as the most widely implemented/derived and original velocity equations of a PSO.

In Eq (2.3):
x_{id} →particle position of ith particle having "d" dimensions
v_{id} → particle velocity of ith particle having "d" dimensions

In Eq (2.4):
ω → inertia constant
c_1, c_2 → constants representing the effect of individual and global bests
p_{id}, p_{gd} → individual and global positions, respectively
rand(), Rand() → random value between 0 and 1

In Eq (2.5):
K → constriction factor, usually taken as 0.729.

The value of "ω" is responsible for the exploitation and exploration of the search space. A high value of "ω" makes the particle overshoot its destined position and thus exploration of the search space takes place. On the other hand, a smaller value of "ω" makes the particles revolve around the individual and global best positions, which leads to exploration of the search space.

2.4.1 FLOW CHART

The flow chart for PSO is shown in Figure 2.10.

As can be seen from the flow chart, the random particles are initially generated. The particles should be within the search space. In the next step, the objective

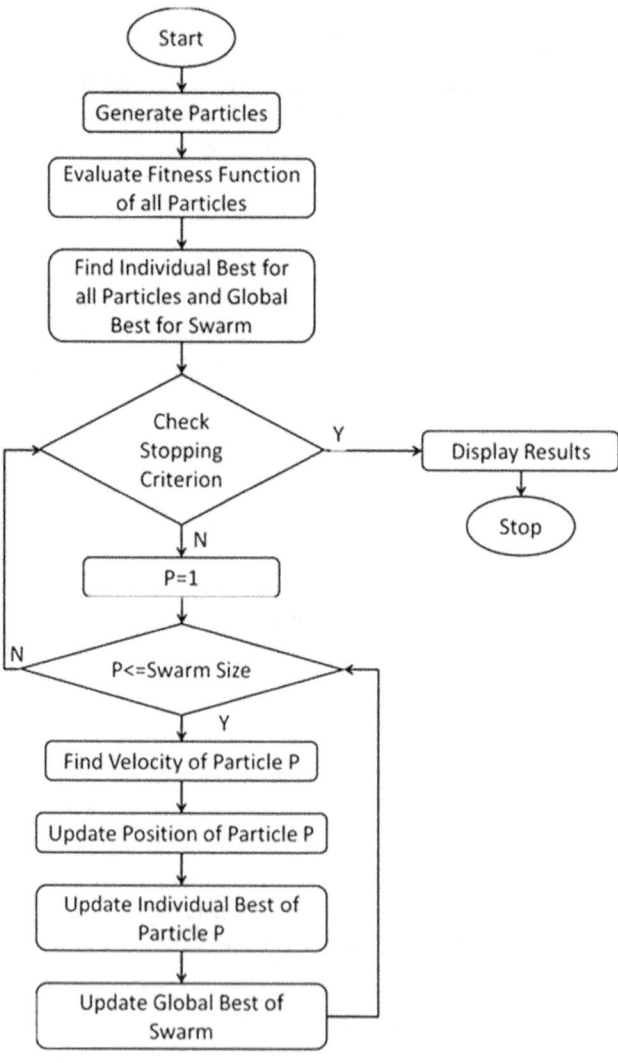

FIGURE 2.10 Flow chart of PSO.

function value for the particles are found. For the initial positions, the individual best of the particles will be same as their positions, respectively. The particle with best fitness function value in the swarm is chosen as the global best. Next, a PSO moves into the iterations and performs four tasks. Inside the loop, the velocities of the particles are evaluated. Next, the particles are moved to a new position and also checked whether they lie within the search space. The individual best positions of each particle and global best of the swarm is then updated. Lastly, the stopping criterion of the optimization technique is checked. In this case, the stopping criterion is satisfied, the iterations are stopped, and the results are printed; otherwise, the loop continues.

2.5 ANN-PSO-EMS

Nowadays, the electrical energy demand is increasing rapidly, which leads to decrease in energy efficiency. To meet the demand efficiently, monitoring and managing of electrical appliances is required. Proper energy management architecture is required to achieve this objective and this is where EMS is introduced. There are different approaches to model, simulate, and identify electrical appliances for DR to be applied in EMS. One such methodology that is considered for this purpose is the combination of ANN with PSO [22].

An ANN is a powerful computational tool to solve load forecasting problems. PSO is an evolutionary algorithm technique used because as compared to other techniques it can be computationally inexpensive, easily implemented, and does not require gradient information of an objective function. The operation time of electrical appliances is obtained using a PSO algorithm optimally while the ANN monitors the energy consumption patterns.

The ANN with PSO for home energy management system (HEMS) [23] mainly focuses on home appliances, which consume high energy and develop efficient HEMS to increase the energy efficiency on the basis of scheduled operation according to specific time. The home appliances that consume high energy are air conditioners (ACs), water heaters, refrigerators, washing machines, and electric ranges/induction stoves. A simple HEMS is presented in Figure 2.11.

DRs will be obtained from the utility or grid. These DR programs are received by smart meters transferred to HEMS or controllers that have the details of how much and duration of load is required to be controlled. The entire home appliances will get demand response signals from HEMS. Let us consider the power consumption of home appliances, which are mentioned as 2.8, 3.8, 1.2, 0.6, and 2.0 KW, respectively [24].

The priority of the home appliances is listed below in Table 2.1.

The ANN structure has six inputs with three hidden layers and five outputs. The inputs are T_r (room temperature), T_{wh} (water heater temperature), T_{tot} (total power consumption), DR signal, T_{er} (electric range temperature), and T_s (time of the system) and the outputs are AC, WH, Ref, WM, and ER [23], as shown in Figure 2.12.

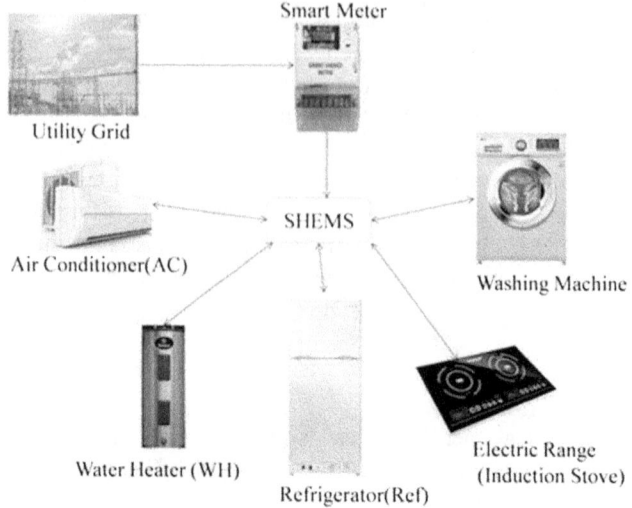

FIGURE 2.11   HEMS application diagram.

TABLE 2.1
Priority of Home Appliances

Sr. No.	Home Appliance	Priority	Load in KW
1	Water Heater	1	3.8
2	Air Conditioner	2	2.8
3	Electric Range	3	2.0
4	Refrigerator	4	1.2
5	Washing Machine	5	0.6

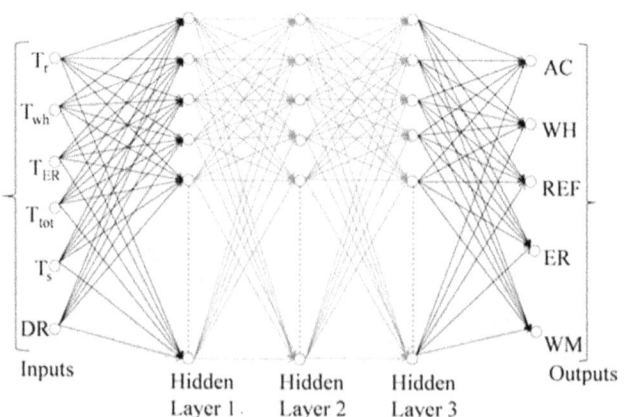

FIGURE 2.12   ANN model for SHEMS.

2.6 CONCLUSION

In the system, the sudden changes of home appliances can be forecasted or predicted by ANN hidden layers. The DRs of the system may be applied throughout the day or for certain hours in a day. In the previous case, the demand limit is assumed to be 4.5 KW [23]; if T_{tot} is higher than the demand limit, then the controller will switch off the appliances with lower priority, if not then there is no change in the working of the system. The outputs of the ANN signals are used to turn ON/OFF the home appliances according to the customer preferences and priority of appliances. The PSO is applied to obtain the optimal weights in the ANN.

REFERENCES

[1] B. Khan and P. Singh, "Optimal Power Flow Techniques under Characterization of Conventional and Renewable Energy Sources: A Comprehensive Analysis," *Journal of Engineering*, vol. 2017, Article ID 9539506, 16 pages, 2017.

[2] P. Singh and B. Khan, "Smart Microgrid Energy Management Using a Novel Artificial Shark Optimization," *Complexity*, vol. 2017, Article ID 2158926, 22 pages, 2017.

[3] T. Molla, B. Khan, B. Moges, H. H. Alhelou, R. Zamani and P. Siano, "Integrated Optimization of Smart Home Appliances with Cost-effective Energy Management System," *CSEE Journal of Power and Energy Systems*, vol. 5, no. 2, pp. 249–258, June 2019.

[4] Z. Tang, Y. Lin, M. Vosoogh, N. Parsa, A. Baziar and B. Khan, "Securing Microgrid Optimal Energy Management Using Deep Generative Model," *IEEE Access*, vol. 9, pp. 63377–63387, 2021.

[5] S. P. Bihari et al., "A Comprehensive Review of Microgrid Control Mechanism and Impact Assessment for Hybrid Renewable Energy Integration," *IEEE Access*, vol. 9, pp. 88942–88958, 2021.

[6] B. Khan and P. Singh, "Selecting a Meta-Heuristic Technique for Smart Micro-Grid Optimization Problem: A Comprehensive Analysis," *IEEE Access*, vol. 5, no., pp. 13951–13977, 2017.

[7] O. P. Mahela et al., "Comprehensive overview of multi-agent systems for controlling smart grids," *CSEE Journal of Power and Energy Systems*, vol. 8, no. 1, pp. 115–131, Jan. 2022, doi: 10.17775/CSEEJPES.2020.03390.

[8] Z. Wang, B. Zhang, M. Mobtahej, A. Baziar and B. Khan, "Advanced Reactive Power Compensation of Wind Power Plant Using PMU Data," *IEEE Access*, vol. 9, pp. 67006–67014, 2021.

[9] S. Padmanaban, C. Dhanamjayulu and B. Khan, "Artificial Neural Network and Newton Raphson (ANN-NR) Algorithm Based Selective Harmonic Elimination in Cascaded Multilevel Inverter for PV Applications," in *IEEE Access*, vol. 9, pp. 75058–75070, 2021.

[10] M. S. Fakhar et al., "Implementation of APSO and Improved APSO on Non-Cascaded and Cascaded Short Term Hydrothermal Scheduling," in *IEEE Access*, vol. 9, pp. 77784–77797, 2021.

[11] C. B. Smith, and K. E. Parmenter. *Energy, Management, Principles: Applications, Benefits, Savings*. Elsevier, 2013.

[12] B. L. Capehart, W. C. Turner, and W. J. Kennedy. *Guide to Energy Management*. The Fairmont Press, Inc, 2006.

[13] W. S. McCulloch, and W. Pitts. "A logical calculus of the ideas immanent in nervous activity." *The Bulletin of Mathematical Biophysics*, vol. 5.4, pp. 115–133, 1943. doi:10.1007/BF02478259

[14] D. Hebb. *The Organization of Behavior*. New York: Wiley, 1949. ISBN 978-1-135-63190-1.

[15] A. G. Ivakhnenko, and V. G. Lapa. *Cybernetics and Forecasting Techniques*. New York: American Elsevier Pub. Co., 1967.

[16] Z. Waszczyszyn, "Fundamentals of Artificial Neural Networks." *Neural Networks in the Analysis and Design of Structures*. Vienna: Springer, 1999, 1–51.

[17] L. V. Fausett, *Fundamentals of Neural Networks: Architectures, Algorithms and Applications*. Pearson Education India, 2006.

[18] S. Haykin. "Learning strategies." *Neural Networks: A Comprehensive Foundation*. Upper Saddle River, NJ, USA: Prentice Hall PTR, 1998, 299.

[19] J. Kennedy and R. Eberhart, "Particle Swarm Optimization," in *Neural Networks*, 1995. Proceedings., IEEE International Conference on, vol. 4, pp. 1942–1948, November 1995.

[20] Y. Shi and R. Eberhart, "A Modified Particle Swarm Optimizer," in Evolutionary Computation Proceedings, 1998. IEEE World Congress on Computational Intelligence., The 1998 IEEE International Conference on, pp. 69–73, IEEE, 1998.

[21] M. Clerc, "The Swarm and the Queen: Towards a Deterministic and Adaptive Particle Swarm Optimization," in *Evolutionary Computation*, 1999. CEC 99. Proceedings of the 1999 Congress on, vol. 3, pp. 1957, 1999.

[22] L. T. Le, H. Nguyen, J. Dou, and J. A. Zhou, Comparative Study of PSO-ANN, GA-ANN, ICA-ANN, and ABC-ANN in Estimating the Heating Load of Buildings' Energy Efficiency for Smart City Planning. *Applied Sciences*, vol. 9, no. 13, pp. 2630, 2019.

[23] M. S. Ahmed, A. Mohamed, R. Z. Homod, and H. Shareef, Hybrid LSA-ANN Based Home Energy Management Scheduling Controller for Residential Demand Response Strategy. *Energies*, vol. 9, no. 9, pp. 716, 2016.

[24] M. S. Ahmed, A. Mohamed, H. Shareef, R. Z. Homod, and J. Abd Ali, Artificial Neural Network Based Controller for Home Energy Management Considering Demand Response Events. In *2016 International Conference on Advances in Electrical, Electronic and Systems Engineering (ICAEES)* (pp. 506–509). IEEE, November 2016.

3 Community Microgrid Energy Scheduling Based on the Grey Wolf Optimization Algorithm

Md Shafiullah
Center of Research Excellence in Renewable Energy, KFUPM, Dhahran, Saudi Arabia

Md Ershadul Haque and Shorab Hossain
Mechanical Engineering Department, KFUPM, Dhahran, Saudi Arabia

Md. Sanower Hossain
University of Texas at Dallas, Texas, USA

Md Juel Rana
School of Engineering and Information Technology, University of New South Wales Canberra, Australia

CONTENTS

DOI: 10.1201/b22884-3

47

3.1 INTRODUCTION

The traditional electricity grids are the extensive networks consisting of massive fossil-fuel power plants, large transmission and distribution lines, and energy consumers. Such arrangements make the energy delivery from the sources to the consumers easy, flexible, and efficient by minimizing the complexities and power losses. The rise of carbon footprints in the atmosphere of the traditional power generating sources pushed the electric utilities to incorporate clean and environmentally friendly alternative and renewable energy resources (RER) to the existing networks throughout the world [1]. However, the complexities of the current network do not grant for a high entrance of intermittent to unconventional energy [2]. In response, the concept of microgrid (MG) came into the picture. An MG is an independent small-scale electric network that can be described as the distributed low-voltage network consisting of different distributed generators (DG), energy storage devices, and sensitive loads that can be run in both grids connected and off-grid modes [3]. It can be connected to the grid, exchanging power with the utility through the point of common coupling (PCC). If the distribution network fails, an MG converts automatically from grid-connected mode to isolated mode and proceeds to support the isolated portion [4–13]. An MG can also boost economic efficiency, reliability, minimize emissions, and encourage interaction with customers by using flexible DG and energy storage devices through efficient control strategies [14–16].

A community microgrid energy (CMG) system is a form of microgrid built to be in the service of a specific energy-consuming community. It uses DG accessible locally, such as the photovoltaic (PV) system, wind turbine (WT) system, energy storage systems, and fuel cell (FC) to fill up consumers' demand for energy insistence [17][18]. It can also take the initiative as a single administrable entity under the utility grid. The CMG consumers in the domain of energy resources prevent energy losses caused by energy transmission for long distance, which leads to efficient utilization of the produced energy. Unlike the traditional distribution grid, the CMG provides bidirectional energy/information flow capabilities. This property allows CMGs to make it easier to incorporate a significant amount of renewable energy resources (RERs). Its energy reliance on the utility grid is not only reduced but can also supply the community with electricity during the main grid blackout. This also helps end-users (consumers) to generate electricity-using renewables, such as solar rooftop PV systems. Therefore, it turns the passive energy consumers out to the prosumers (commodity manufacturer). They can sell the surplus energy to their neighbors/grid by being active in the energy market. Hence, they can either make a profit or save on their energy costs. By implementing mass-scale CMG systems, carbon emissions can be reduced significantly for the flexible and effective future electric grid. Despite being significance benefit of CMG, it emerges some financial and operational challenges. The optimization of energy management can perform to acquire the optimal operation for the CMG. These optimization problems usually consider minimization of the operational costs and contamination of the atmosphere, or maximization of the benefits, as the objective function.

To deal with energy management challenges, numerous optimization approaches have been explored, including traditional and meta-heuristic optimization techniques. Amongst the explored techniques, energy management of the CMG are solved using mixed-integer linear programming (MILP) [19], convex optimization [20], non-linear programming (NLP) [21], dynamic programming (DP) [22], differential evolution (DE) [23], and particle swarm optimization (PSO) [24]. Li et al. [25] established a model of a three-layer multi-agent system that considers the power-heat demand response and energy storage system based on the actual situation to unfold the CMG energy management problems. Results showed that the proposed optimization of particle swarms could effectively minimize costs and the number of iterations. Hossain et al. [26] suggested that the allocation of battery energy storage for the day. Results showed that the proposed method could reduce operating costs by approximately 40% compared with the standard method. Rana et al. [27] developed a heuristic enhanced differential evolution (DE-H) technique while dealing with CMG energy scheduling and compared their results with non-heuristic differential evolution (DE-NH). The results showed that it would produce a high-quality solution with appreciably less numerical effort and substantiate the effectiveness of the proposed method. Soham and Kamal [23] suggested a hybrid optimization strategy using DE and the theory of chaos explored for efficient microgrid activity consisting of both conventional and renewable energy resources. They reported that their employed methodology could yield superior cost and emission results. Ref. [28] illustrated a genetic algorithm (GA)–based solution methodology to reduce the cost of a microgrid over a 24-hour schedule. Stefano et al. [29] explored various approaches for the synthesis of the fuzzy inference system–based energy management system using a genetic algorithm to optimize the benefits from the grid electricity exchange assuming the electricity price policy for time of use. Results showed that the output was just 10% lower than the optimal solution, while the rule base structure was lowered to 30 rules [30]. However, energy storage system (ESS) degradation costs are rarely taken into consideration while forming the optimization problem for CMG energy management. By taking the mentioned notes, this chapter formulates the optimization problem for a CMG incorporating the ESS degradation costs. The key contributions of this chapter are:

- Formulation of CMG energy scheduling problem considering ESS degradation costs for its practical impact on longstanding capital costs.
- Employment of an effective meta-heuristic approach, namely the grey wolf optimization technique for the well-being of the impartiality constraint while solving the formulated CMG energy-scheduling problem.
- Comparison of the obtained results of the employed grey wolf optimization (GWO) approach with genetic algorithm and literature reported techniques.

This chapter deals with a grid-connected CMG system incorporated with various distributed generation (DG) units, consumer loads, and energy storage systems with discharging and charging characteristics. The principal focus of this chapter is the performance analysis of the proposed GWO in CMG energy scheduling, considering the volatile electricity rates, loads, and renewable generations over 24 hours.

This chapter is organized in the following sections. The CMG model and mathematical problem formulation are described in Section 3.2. Section 3.3 briefly illustrates the employed meta-heuristic approach, namely the GWO. Section 3.4 demonstrates the numerical results and discussions along with the comparative study. Finally, the conclusions and future research directions are presented in Section 3.5.

3.2   CMG ENERGY SCHEDULING PROBLEM FORMULATION

This section comprises the community microgrid model and formulation of its energy-scheduling problem as described below.

3.2.1   COMMUNITY MICROGRID SYSTEM MODEL

In general, a CMG consists of different residential loads, DG, and ESS. Common DG units used in a CMG are the wind turbines, photovoltaic systems, hydraulic powerplants (HP), microturbine powerplants (MT), fuel cells (FC), and the energy storage systems. Often it is connected to the main grid, and it operates in bi-directional ways for power exchange. It takes electricity when required from the utility grid; however, sell electricity when the production exceeds the demand. Generally, it buys electricity at the time of low electricity prices and sells electricity at a time of high price to gain profit. It is known that renewable energy resources provide intermittent power; therefore, these sources are non-dispatchable – cannot be operated according to demands. Thus, power from renewable resources is utilized fully whenever available. A comprehensive CMG model is shown in Figure 3.1. The figure displays different DG units, consumer loads, energy storage systems, and connected utility grid by indicating their energy exchange relationships.

3.2.2   PROBLEM FORMULATION

A multi-objective optimization model for the CMG energy scheduling is discussed in this section, combining the energy costs of different sources along with the degradation cost of the ESS. Among the DG units shown in Figure 3.1, the PV and WT systems are non-dispatchable and have to be assumed of zero operating costs. Yet, they have some non-negligible maintenance costs. For this reason, the decision variable "x" in the optimization model depends on only exchanged energy from the dispatchable sources and the utility grid, which can be written as in equation (3.1):

$$x = [P_{mt_t} P_{FC_t} P_{fc_t} P_{ess_t} P_{grid_t}]$$
(3.1)

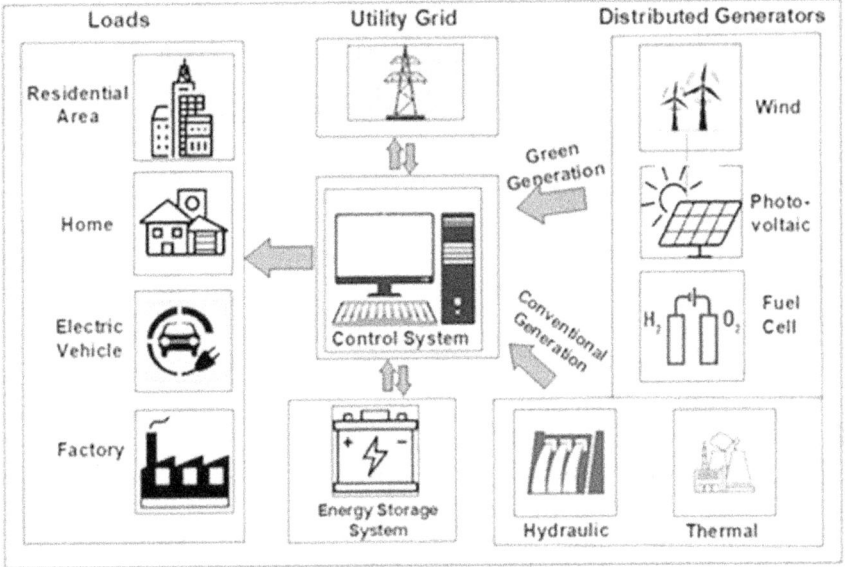

FIGURE 3.1   Community microgrid model.

Here, P_{mt_t}, P_{fc_t}, P_{ess_t}, and P_{grid_t} represent the power output of MT, FC, ESS, and utility grid power, respectively.

The combined optimization model discussed here has the main purpose of making CMG components routine to reduce the cost at the minimum level. This minimization problem's objective function can be presented, as shown in equation (3.2):

$$\min_x F = \sum_{t=1}^{T} w_1\left(Cost_{DG_t} + Cost_{grid_t}\right) + w_2\, Cost_{deg,ess_t} \qquad (3.2)$$

Here, F represents the overall CMG energy cost over the time of T. $Cost_{DG_t}$ consists of both maintenance and operational cost of all DG units at t^{th} time (in an hour). $Cost_{grid_t}$ and $Cost_{deg,ess_t}$ are utility grid power exchange and ESS degradation costs, respectively. Conversely, w_1 and w_2 are weighting factors for two different types of cost factor. For simplicity of application, this chapter assumes both $w_1 = w_2 = 1$. However, the costs of equation (3.2) can be further expressed as:

$$Cost_{DG_t} = \sum_{i=1}^{N_{DG}} \left\{ \frac{P_{DG_{i,t}}}{\eta_{DG_i}} Cost_{DG_i}^{oper} + P_{DG_{i,t}}\, Cost_{DG_i}^{maint} \right\} \qquad (3.3)$$

$$Cost_{grid_t} = \begin{cases} P_{grid_t} e_{price_t}^{buy} & \text{if } P_{grid_t} > 0 \\ P_{grid_t} e_{price_t}^{sell} & \text{if } P_{grid_t} < 0 \\ 0 & \text{else} \end{cases} \tag{3.4}$$

$$Cost_{deg,ess_t} = (Cost_{invest,ess} - C_{resd,ess})L_{deg} \tag{3.5}$$

The total cost of the DG units comprises of both maintenance and operational costs at t^{th} time. Here, $P_{DG_{i,t}}$ is the generated power (active) of ith distributed generator at t^{th} hour; operational and maintenance costs of generated power of the ith distributed generator is represented as $Cost_{DG_i}^{oper}$ and $Cost_{DG_i}^{maint}$, respectively; η_{DG_i} is the efficiency of the ith distributed generator; and N_{DG} is the total number of distributed generators.

Energy transaction is assumed as positive when energy is taken from the utility grid and negative when energy is given to the utility grid. Amount of transacted energy, per units of buying energy and selling energy costs, are represented by P_{grid_t}, $e_{price_t}^{buy}$, and $e_{price_t}^{sell}$, respectively. Although the microgrid buys energy when the price is low and sells energy when the price is high, this chapter assumes the same price for both the buying and selling purpose for simplicity. This simplification does not have noticeable effects on the problem formulation, and diversified rates of energy can readily be included whenever needed. $C_{invest,ess}$, $C_{resd,ess}$, and L_{deg} represent the cost of investment, life degradation factor of ESS, and residual value, respectively, in equation (3.6). The residual value is the final amount of a product after its useful life. Degradation cost and life calculation of Energy storage system start by measuring its life degradation; this degradation comprises two-steps: (a) static degradation (L_{static}), which is defined as the reduction of subjective functional properties of the ESS and (b) dynamic degradation ($L_{dynamic}$), which is defined as the operational degradation of the ESS. The static degradation enhances the impedance of ESS that ultimately lessens the output [31].

$$L_{deg} = L_{static} + L_{dynamic} \tag{3.6}$$

This static degradation is independent of the ESS operating modes and connected to shelf life ($T_{shelf-life}$) in years, which is shown in equation (3.7):

$$L_{static} = 1/T_{shelf-life} \tag{3.7}$$

However, the dynamic degradation depends on ESS operating modes. It particularly relies on the depth of operating cycles. It increases with the increase of cycle depths and can be determined using a well-known rain flow cycle counting algorithm [32]. This algorithm measures the asymmetric charge or discharge cycles of an energy storage system together with corresponding depths of the cycles [26], [33]. Using cycle depths $L_{dynamic}$ can be measured using equation (3.8):

$$L_{dynamic} = \sum_{n=1} \frac{p_n}{N_{cycle}} c_{d_n}^q \qquad (3.8)$$

Here, p_n is the amount cycle (full and a half) measured by rain flow algorithm at c_{d_n} cycle depth. q and N_{cycle} represent the ESS degradation curve's slope and the total cycle life of ESS, respectively. With the degradation, the expected life of ESS can be gained using a one-day simulation implemented in equation (3.9) [34]:

$$ESS_{life} = \frac{1}{L_{static} + L_{dynamic}} \qquad (3.9)$$

3.2.3 CONSTRAINTS

To work better, this model needs to satisfy both individual components and overall system constraints. All dispatchable DG units have active power upper and lower constraints that are given by following equation (3.10) [35]:

$$P_{DG}^{lb} \leq P_{DG_t} \leq P_{DG}^{ub} \qquad (3.10)$$

Here, P_{DG}^{lb} and P_{DG}^{ub} represent the minimum and maximum operating power of a single DG unit, respectively. All DG units also have ramp rate constraints, which can be expressed as follows:

$$\Delta P_{DG}^{down} \leq P_{DG_t} - P_{DG_{t-1}} \leq \Delta P_{DG}^{up} \qquad (3.11)$$

Here, ΔP_{DG}^{down} and ΔP_{DG}^{up} represent ramp down and up limits of a distributed generator, respectively, between time instants "t" and "$t - 1$."

The overall life of ESS increases when it is operated inside a safe limit [36]. ESS power is considered positive in times of discharge, and it is considered negative in times of charging. Safe limit along with other constraints of ESS are shown below:

$$P_{ess}^{Dmax} \leq P_{ess_t} \leq P_{ess}^{Cmax} \qquad (3.12)$$

$$E_{ess_{t+1}} = \begin{cases} E_{ess_t} + P_{ess_t} \eta^C \Delta t & if \ P_{ess_t} < 0 \\ E_{ess_t} + \frac{P_{ess_t}}{\eta^D} \Delta t & if \ P_{ess_t} > 0 \\ 0 & else \end{cases} \qquad (3.13)$$

$$E_{ess}^{lb} \leq E_{ess_t} \leq E_{ess}^{ub} \qquad (3.14)$$

Here, P_{ess}^{Dmax} and P_{ess}^{Cmax} represent the maximum allowable amount of power discharge and charge, respectively. η^C and η^D represent charging and discharging efficiency. E_{ess}^{lb} and E_{ess}^{ub} represent the minimum and maximum energy storage level.

The energy produced in a microgrid must be equal to the energy consumed here. All kinds of DG units, ESS discharging, and utility grid energy selling to the CMG are considered a source of energy production. While different kinds of loads, ESS charging, and utility grid energy buying from the CMG are considered as the source of energy consumption. Equality constraint at instant "t" in a CMG can be represented as:

$$P_{load_t} - \left(P_{grid_t} + P_{ess_t} + \sum_{i=1}^{N_{DG}} P_{DG_{i,t}} \right) = 0 \qquad (3.15)$$

3.3   GREY WOLF OPTIMIZATION ALGORITHM

Grey wolves (GWs) prefer to stay mostly in a pack where an average pack consists of 5–12 of GWs. They possess an incredibly strict hierarchy of social dominance [37]. The heads are female and a male is termed the alpha. Mostly the alpha is accountable for making decisions/choices about hunting, wake-up time, sleeping, and so on. The second level in their hierarchy is termed the beta. The betas are subservient wolves that assist the alpha in making decisions or executing the actions of the other packs. The lowest grey wolf ranking is known as the omega. The omegas perform the scapegoat role, and they must always submit to all other prevailing wolves. However, the fourth category, the delta, or subordinate wolf, is referenced to a wolf other than alpha, beta, or omega. They dominate the omegas while surrendering to the alphas and the betas. This category includes elders, sentinels, scouts, caretakers, and hunters. Aside from the wolves' social hierarchy, another fascinating social behavior of GWs is group hunting. According to Ref. [38], the main hunting stages of the GWs are:

- Tracking, chasing, and getting closer to the target.
- Pursuing, encircling, and harassing the target until it ceases to move.
- Assaulting prey.

3.3.1   GW HIERARCHY

The alphas (α) are considered the fittest solutions while mathematically defining the wolves' social hierarchy in the construction of the GWO algorithm. The second and the third levels of the finest solutions are therefore labeled the betas (β) and the deltas (δ), respectively. The remaining portions of the GWs are the omegas (ω). In a GWO, the hunting of the prey is dictated by the alpha, beta, and delta and the omega wolves are the followers of these three types.

3.3.2 PREY ENCIRCLING

As suggested by Mirjalili *et al.* [37], the GWs surround the prey during the hunting processes. In the GWO algorithm, the grey wolves change their positions by using equation (3.17) within the area around the prey at any arbitrary spot.

$$\vec{X}(i + 1) = \vec{X}_p(i) - \vec{A} . \vec{D} \tag{3.16}$$

Here, $\vec{D} = \left| \vec{C} . \vec{X}_P(i) - \vec{X}(i) \right|$; "$i$" shows the present iteration; $\vec{X}_p$ and $\vec{X}$ are lo-cation vectors of the targets and the GW; $\vec{A}$ and $\vec{C}$ are coefficient vectors and they are calculated using the following equations:

$$\vec{A} = 2 .\vec{a}. \vec{r_1} - \vec{a} \tag{3.17}$$

$$\vec{C} = 2. \vec{r_2} \tag{3.18}$$

Here, the elements of $\vec{a}$ are linearly reduced from 2 to 0, where $\vec{r_1}$ and $\vec{r_2}$ are the arbitrary vectors in the range of [0, 1].

3.3.3 HUNTING

As stated earlier, the alpha wolf leads the hunt. The beta and delta wolves also engage in hunting processing occasionally. For a mathematical depiction of the hunting activities of the GWs, α, β, and δ wolves are considered to have the updated information on the possible locations of the prey. Therefore, their positions are considered the fittest solutions while the positions of the other GWs are updated accordingly using the following equations [37][39]:

$$\vec{D}_\alpha = \left| \vec{C_1} . \vec{X}_\alpha - \vec{X} \right| \tag{3.19}$$

$$\vec{D}_\beta = \left| \vec{C_2} . \vec{X}_\beta - \vec{X} \right| \tag{3.20}$$

$$\vec{D}_\delta = \left| \vec{C_3} . \vec{X}_\delta - \vec{X} \right| \tag{3.21}$$

$$\vec{X_1} = \vec{X}_\alpha - \vec{A_1} . \vec{D}_\alpha \tag{3.22}$$

$$\vec{X_2} = \vec{X}_\beta - \vec{A_2} . \vec{D}_\beta \tag{3.23}$$

$$\vec{X_3} = \vec{X_\delta} - \vec{A_3} \cdot \vec{D_\delta} \tag{3.24}$$

$$\vec{X}(i+1) = \frac{\vec{X_1} + \vec{X_2} + \vec{X_3}}{3} \tag{3.25}$$

3.3.4 PREY ATTACKING (EXPLOITATION)

The GWs update their positions using equations (3.17) to (3.26) in several iterations and then come to an end of their hunting process by attacking the prey when they stop moving. To reflect the notion mathematically, the value of $\vec{a}$ is slowly decreased from 2 to 0. Thus, the scale of fluctuations of $\vec{A}$ is also lessened. If $\vec{A}$ has arbitrary value in the range of [−1, 1], the subsequent position of the agent will be anywhere in between its present position, and the position of the prey [37]. Finally, the grey wolves attack the prey when $|\vec{A}|$ is less than one.

3.3.5 PREY SEARCHING (EXPLORATION)

Like attacking the prey, the grey wolves also diverge from each other to locate the fitter prey mostly based on the positions of the alphas, betas, and deltas. To model their divergence characteristics mathematically, a random value of $\vec{A}$ outside of the [−1, 1] domain is utilized. It puts stress on the exploration characteristics of the GWO and makes the global exploration possible by diversifying the search agents' positions [37].

3.3.6 GWO APPLICATION FOR CMG ENERGY SCHEDULING

The GWO is one of the widely used meta-heuristic techniques in solving many engineering complex optimization problems due to its impressive characteristics over other peer algorithms. Besides, the simplicity, easiness of implementation, flexibility, scalability, and capability to balance between the exploration and exploitation also the major reasons for the widespread adoption of the algorithm [40]. Therefore, the GWO is employed in CMG energy scheduling in this chapter, where Figure 3.2 illustrates the flow chart of the suggested approach for better visualization. Besides, Algorithm A illustrates the pseudocode of the proposed approach. The GWO has the following steps while solving the CMG energy scheduling problem:

Step 1. **CMG data input:** All the required input data including the upper and lower limits of the DG and ESS units, PV and WT systems predicted output power, load requirement, energy prices (grid and utility bid rates), and operating and maintenance costs of the DG units.

Step 2. **Initialization:** The number of search agents or grey wolves (N) and the maximum number of iterations are determined. Based on the

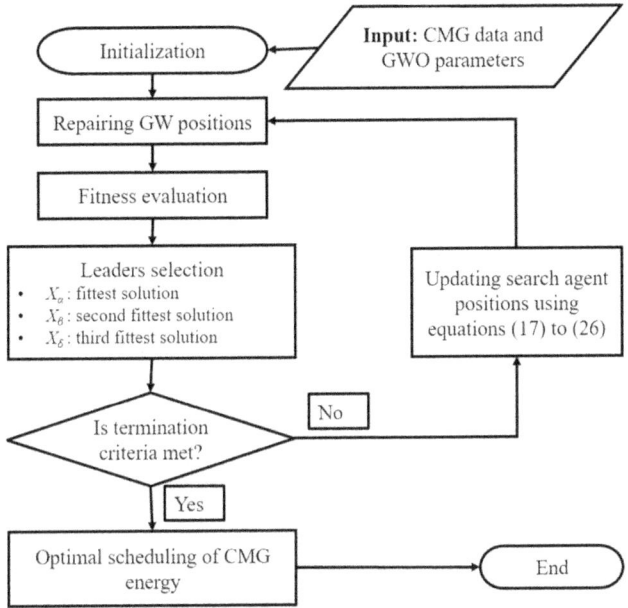

FIGURE 3.2 Flow chart of the GWO algorithm for CMG energy scheduling.

determined numbers, the positions of the grey wolves are initialized
in the X matrix according to equation (3.26):

$$X_m = X_{m,i}^{min} + rand * (X_{m,i}^{max} - X_{m,i}^{min}) \qquad (3.26)$$

Here, X_m is the position of the mth search agent (grey wolf) and $m = 1$,
2, 3, ..., N. Besides, $i = 1, 2, 3, ... D$, and D is the dimension of grey
wolves positions. In this chapter, D = 96 as the power output from the
MT, FC, ESS, and grid are considered to vary in each hour of the day.
Therefore, the position matrix of the grey wolves can be represented
as follows [39]:

$$X = \begin{bmatrix} X_1 \\ X_2 \\ \cdots \\ X_m \\ \cdots \\ X_N \end{bmatrix} = \begin{bmatrix} X_{1,1} & X_{1,2} \cdots \cdots \cdots \cdots & X_{1,D} \\ X_{2,1} & X_{2,2} \cdots \cdots \cdots & X_{2,D} \\ & \cdots & \\ X_{m,1} & X_{m,2} \cdots & X_{m,D} \\ & \cdots & \\ X_{N,1} & X_{N,2} \cdots \cdots \cdots & X_{N,D} \end{bmatrix} \qquad (3.27)$$

Step 3. **Repairing the solutions:** The positions of the search agents signify
potential solutions to the CMG energy scheduling problem. However,
the obtained numbers of the position matrix cannot be sent directly to

the objective function equation for calculation of their finesses as the ESS power exchange depends on its state of charge (SoC). Hence, such numbers associated with ESS are repaired considering the mentioned issue. Then, to avoid load-generation mismatch, the numbers related to grid power exchange are also updated.

Step 4. **Evaluating the fitness and selection of the alphas, betas, deltas, and omegas:** After repairing and updating the positions of the GWs, the fitness of each position is evaluated. The position associated with the fittest solution is assigned to X_α. Likewise, the positions of the second- and third-best solutions are assigned to X_β and X_δ, respectively. The remaining positions are for the omega grey wolves.

Step 5. **Updating GW positions and checking the constraints:** This step updates the positions of the GWs using equations (3.17) to (3.26). Then, the GWO algorithm for the CMG energy scheduling problem checks the constraints and updates the positions of the GWs if there is a violation.

ALGORITHM A: PSEUDOCODE OF THE PROPOSED GWO APPROACH IN CMG ENERGY SCHEDULING

Input: **CMG data and GWO parameters**

1. Initialization of the GW positions (X)
2. for *iteration = 1:IterMax* do
3. Repairing of the position matrix (X), especially, the items related to ESS and Grid
4. Evaluating the fitness of the GW
5. Selecting the alphas, betas, deltas, and omegas
6. Updating the position matrix (X) using equations (3.17) to (3.26)
7. Checking the constraints and updating the position matrix (X)
8. if termination criteria met then
9. STOP the iterative process
10. end if
11. end for

Output: **Scheduled 24-hour energy of the investigated CMG**

Step 6. **Checking the termination criteria:** The step checks the convergence principle and terminates the iterative process if it reaches to the pre-specified number of iterations. Otherwise, this step sends back the process to Step 3. However, after the termination of the iterative process, the position associated with the best fittest wolf is stored as the best solution (X_α).

3.4 RESULTS AND DISCUSSIONS

The proposed meta-heuristic technique, namely the GWO, is implemented to sche-dule CMG energy. Table 3.1 displays the day-to-day hourly forecasted for CMG data. The data shows that the PV system generates power generation from hours 8.00 to 17.00; on the other hand, the WT system generates power in whole days, but these data are fluctuating hour by hour. Consumer demand is also varying throughout the day ranging from 50 kW to 90 kW. Besides, the electricity price is quite high at noon almost 4.0 €¢/kWh compare to other times in a day. During this time, CMG targets to export the electricity to the grid so that CMG costs can be reduced and more efficient compare to the conventional system. Table 3.2 displays the rating of various DG units and their per-unit costs. The nominal (rated), minimum, and maximum energy ca-pacities of the ESS are set at 222 kWh, 40 kWh, and 200 kWh, respectively. This chapter used the data information of the Ref. [27,41]. At the beginning of the day

TABLE 3.1
Day-ahead Hourly Predicted CMG Data

Time (Hour)	PV Generation (kW)	WT Generation (kW)	Load Demand (kW)	Electricity Price (€¢/kWh)
1	0.0	1.785	52.0	0.23
2	0.0	1.785	50.0	0.19
3	0.0	1.785	50.0	0.14
4	0.0	1.785	50.0	0.12
5	0.0	1.785	56.0	0.12
6	0.0	0.915	63.0	0.20
7	0.0	1.785	70.0	0.23
8	0.20	1.305	75.0	0.38
9	3.75	1.785	76.0	1.50
10	7.525	3.09	80.0	4.00
11	10.45	8.775	78.0	4.00
12	11.95	10.41	74.0	4.00
13	23.90	3.915	72.0	1.50
14	21.05	2.37	72.0	4.00
15	7.875	1.785	76.0	2.00
16	4.225	1.305	80.0	1.95
17	0.55	1.785	85.0	0.60
18	0.0	1.785	88.0	0.41
19	0.0	1.302	90.0	0.35
20	0.0	1.785	87.0	0.43
21	0.0	1.3005	78.0	1.17
22	0.0	1.3005	71.0	0.54
23	0.0	0.915	65.0	0.30
24	0.0	0.615	56.0	0.26

TABLE 3.2

Specifications of the Distributed Generations in the Targeted Microgrid

DG Type	Fuel Cell	Microturbine	ESS	PV System	WT System
Minimum output (kW)	3.0	6.0	−30.0	0.0	0.0
Maximum output (kW)	30.0	30.0	30.0	25.0	15.0
Running cost (€¢/kWh)	0.20	0.40	–	–	–
Maintenance cost (€¢/kWh)	0.04	0.12	0.02	0.08	0.11

before scheduling, ESS was supposed to be drained; therefore, the initial energy level was set to 40 kWh.

This chapter employed one of the widely used meta-heuristic approaches, namely the GWO, to optimize CMG operating costs with consideration of ESS degradation costs. It set 384 search agents (grey wolves) that are four times the total number of decision variables. Besides, the maximum number of iterations was set to 2,000. Then, this chapter ran the formulated optimization problem for 31 times. It is worth mentioning that the formulated optimization problem was implemented in the MATLAB® (2018b) platform on a desktop computer with specifications of 3.5 GHz processor, 8 GB RAM, and Intel Core i5. Table 3.3 compares the operating computational platforms and meta-heuristic techniques parameters of this chapter and Ref. [27]. As can be seen, the computational platforms of this chapter and the referenced work are compatible with each other. Table 3.4 presents the computational time analysis of the employed GWO and GA techniques where the required mean times for 2,000 iterations were 408.40 and 403.04 seconds, respectively.

TABLE 3.3

Comparisons of Employed Computational Platforms and Meta-heuristic Technique Parameters

Items	Specifications	
	Proposed	Ref. [27]
Number of grey wolves/population size	384	384
Number of iterations/generations	2,000	1,200
Stopping criteria (iterations/ generations)	2,000	1,200
Number of runs	31	31
Computational system configuration	Intel Core i5 (3.50 GHz, 8 GB RAM)	Intel Core i7 (3.20 GHz, 16 GB RAM)
Software platform	MATLAB 2018b	MATLAB 2018a

TABLE 3.4
Computational Time Analysis for 31 Runs

Algorithm	Time (seconds)			Skewness	Kurtosis	Time required for each generation (seconds)
	Minimum	Maximum	Mean			
Proposed (GWO)	374.98	442.62	408.40	0.0825	1.8142	0.2042
Compared (GA)	381.97	449.49	403.04	2.0040	5.6195	0.2015
Ref. [27]	–	–	310.55	–	–	0.2588
Ref. [27]	–	–	246.96	–	–	0.2058

Conversely, the mean times of 1,200 generations for two different techniques of Ref. [27] are 310.55 and 246.96 seconds, respectively. As can be seen from the last column of Table 3.4, the computational time required for each generation/iteration with 384 individuals/grey wolves employing the proposed techniques and the techniques of the referenced works are quite compatible. More specifically, the employed GWO requires less average time per iteration than that of the referenced techniques.

Table 3.5 illustrates the analysis of objective function values obtained from three selected runs (best, median, and worst) and comparison with the values of the referenced techniques. It depicts that the proposed GWO algorithm significantly reduced the best, worst, mean, and median values compared to the GA, DE-NH, and DE-H. Besides, the standard deviation is also less than that of the values of the other techniques. Therefore, it can be concluded that the proposed technique outperformed other techniques. However, the convergence plot of three selected runs (best, median, and worst) of the proposed GWO algorithm is shown in Figure 3.3(a), whereas Figure 3.3(b) compares the objective values of the best runs of GWO and GA techniques. Objective function values for the best runs for both GWO and GA converged to 570.77 € and 573.58 €, respectively, before 1,000 iterations/generations. However, this chapter checked any further improvement

TABLE 3.5
Objective Function Values (€) Analysis for 31 Runs

Algorithm	Best	Worst	Mean	Median	Standard Deviation	Skewness	Kurtosis	Coefficient of Variation
Proposed (GWO)	570.77	592.16	578.75	576.09	7.59	0.46554	1.603394	0.013118
Compared (GA)	573.58	603.50	589.63	592.79	10.31	−0.32532	1.568618	0.017478
Ref. [27]	591.31	642.51	625.91	627.78	16.43	–	–	–
Ref. [27]	602.21	656.03	620.47	620.27	11.39	–	–	–

(a)

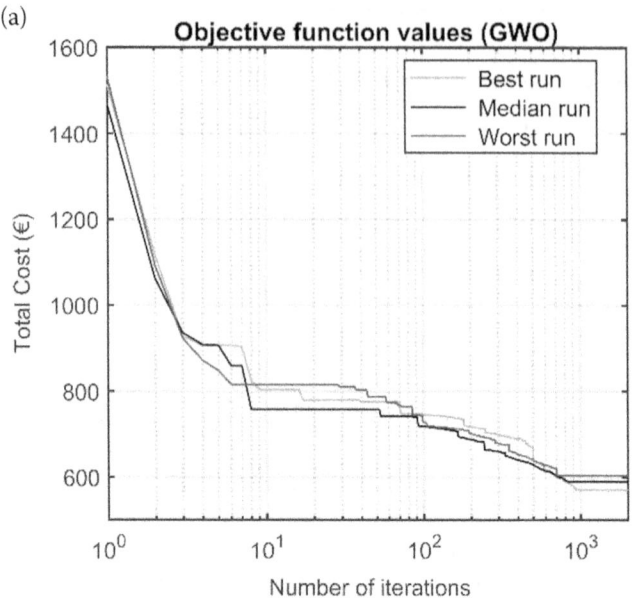

(b)

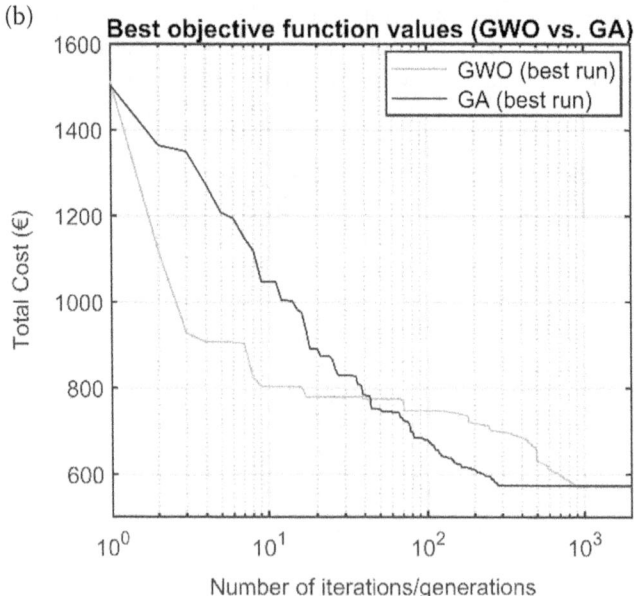

FIGURE 3.3 (a) objective function values for GWO (b) Comparative analysis of Best objective function values for GWO and GA.

until 2,000 iterations/generations. Other selected and non-selected runs also followed a similar trend where the total costs (objective function values) maintained constant values after the convergence points.

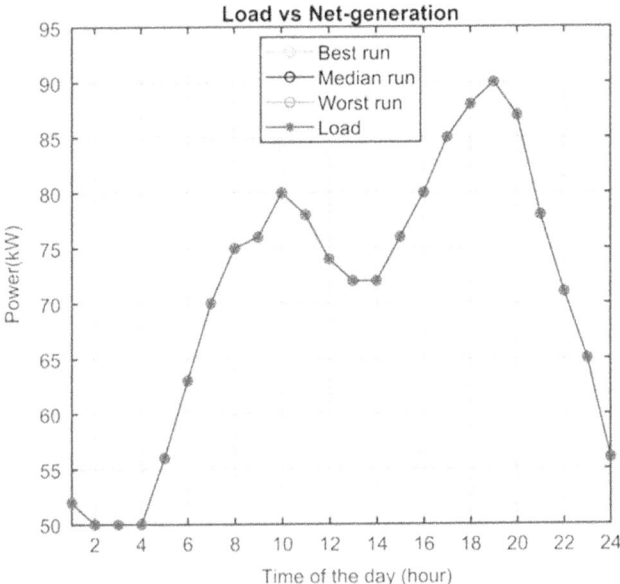

FIGURE 3.4 Net contribution of generation during three different runs to meet the community microgrid load.

Figure 3.4 depicts the net contribution of generation throughout the day during three different runs (best, median, and worst) to meet the CMG load. As can be observed, the net generation curves merged with the consumer demand curve for each selected run that satisfies the equality constraint of equation (3.16) of sub-Section 3.3.2. Generation schedules of MT and FC for 24 hours obtained using a GWO for three selected runs (best, median, and worst) are illustrated in Figure 3.5 and Figure 3.6, respectively. Both MT and FC schedule productions have been varied according to load demand and energy price. For the first couple of hours, MT production was around 6 kW up to hour 08:00 of the day owing to lower energy prices from the grid. Its production raised around 30 kW due to the high price of energy from hour 10:00 to 16:00, followed by another downtrend of energy production from hour 17:00 to 24:00. Likewise, the FC followed a similar trend of MT but the later one being more committed due to cheaper energy production cost and produced around 30 kW from hour 09:00 to 16:00. Then, the FC followed a downward trend of energy production from hour 17:00 to 24:00 except for hour 21:00 as the electricity price went high at that hour. All selected runs for both MT and FC energy scheduling followed almost similar patterns.

Energy exchange profiles of ESS and the utility grid are shown in Figure 3.7 and Figure 3.8, respectively, for 24 hours obtained using a GWO for three selected runs. From Figure 3.7, it can be observed that the ESS started charging for the period of the lower energy price from the grid at hour 01:00 to 06:00 until it was fully charged (maximum level of state of charge). Then, at hour 07:00, it did not charge or discharge. It delivered energy to the CMG from hour 08:00 to 14:00 except for

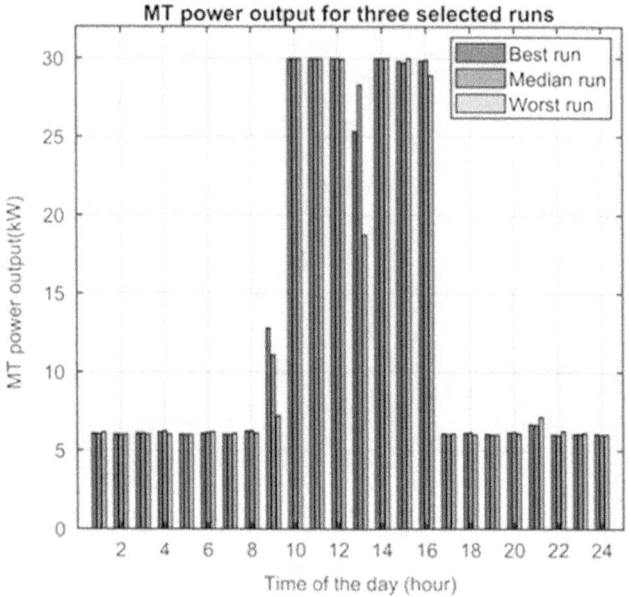

FIGURE 3.5 Output of MT for three selected runs.

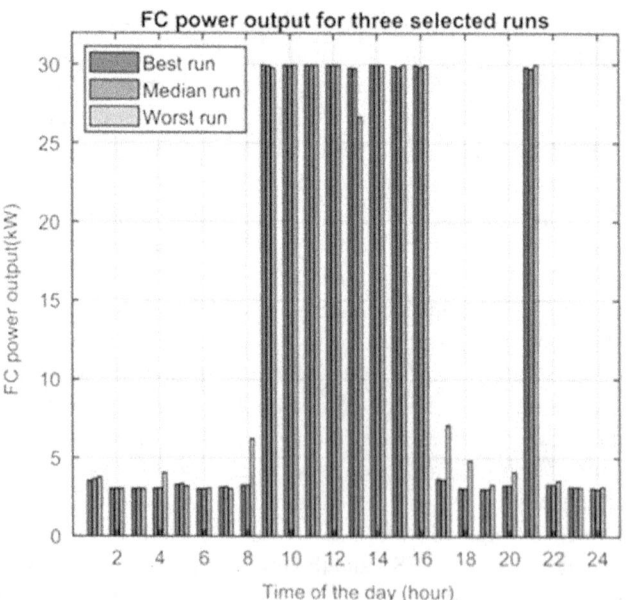

FIGURE 3.6 Output of FC for three selected runs.

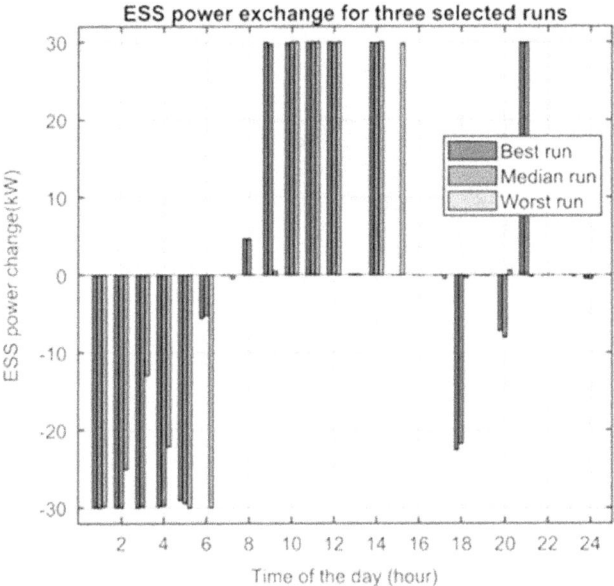

FIGURE 3.7   ESS power exchanges for three selected runs.

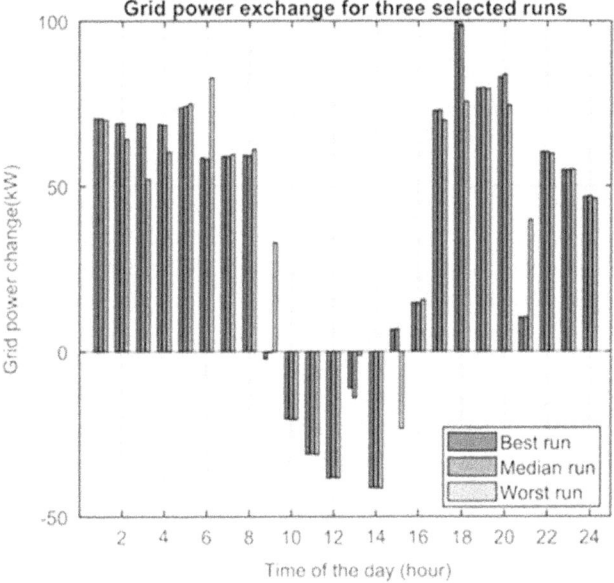

FIGURE 3.8   Grid power exchange for three selected runs.

the hour 13:00 where the energy price is lower compared to the previous and the next hour. Then, for the remaining hours of the day, the ESS system either took energy or acted as neutral except for the hour 21:00 where the energy price again went high. All selected runs for the ESS energy exchange profile followed almost

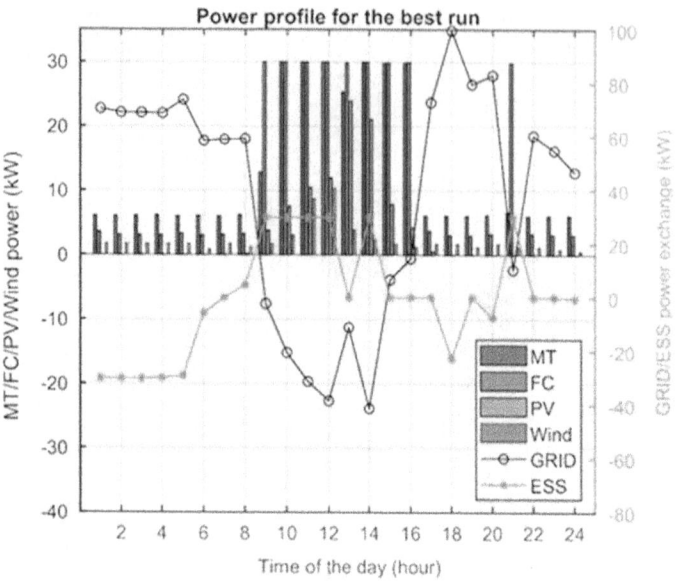

FIGURE 3.9 Power profile for the best run of the GWO technique.

similar patterns. From Figure 3.8, it is understood that the CMG bought energy from the grid at the early hour s of the day from 01:00 to 08:00 and the late hours from 16:00 to 24:00 due to lower energy prices.

Conversely, from hour 10:00 to 14:00, the CMG sold energy to the utility grid as the energy prices were higher during that period. During a few hours, the CMG exchanged a lower or minimal amount of energy with the grid considering its demand and energy prices of those hours. However, for the best run, maximum amount of energy (~100 kW) was taken from the grid at hour 18:00 and the maximum amount of energy was delivered to the grid at hour 14:00. All selected runs for utility grid energy exchange profiles followed almost similar patterns.

Figure 3.9 to Figure 3.11 show the power output of each source (MT, FC, ESS, and utility grid) to satisfy CMG load demand employing the GWO algorithm for three selected (best, median, and worst) runs. For the best run (Figure 3.9), it is seen that ESS was charged between hour 01:00 to 08:00 and discharged between hour 09:00–12:00. Then, the ESS got charged for the remaining hours of the day except for hour 14:00 and hour 21:00 where the energy prices were higher compared to the previous and later hours. The ESS charging periods of the coincide with the energy buying periods from the utility grid, whereas the discharging periods coincide with the energy selling periods to the utility grid. This charging and discharging of ESS make the system economically beneficial because the system can sell at a higher price and buy at a lower price. Besides, as can be seen from Figure 3.9, the generation of the dispatchable sources (MT and FC) varied according to CMG load demand and energy prices. As discussed earlier, the FC contributed more energy than that of the MT because of comparatively cheaper energy costs. For the availability and cost issue, WT has a greater effect on supplying energy compared to PV. The other two runs also followed similar patterns.

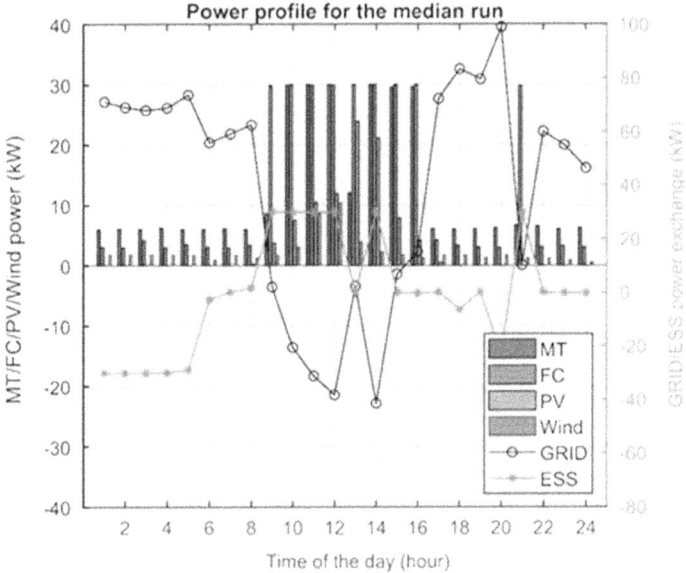

FIGURE 3.10 Power profile for the median run of the GWO technique.

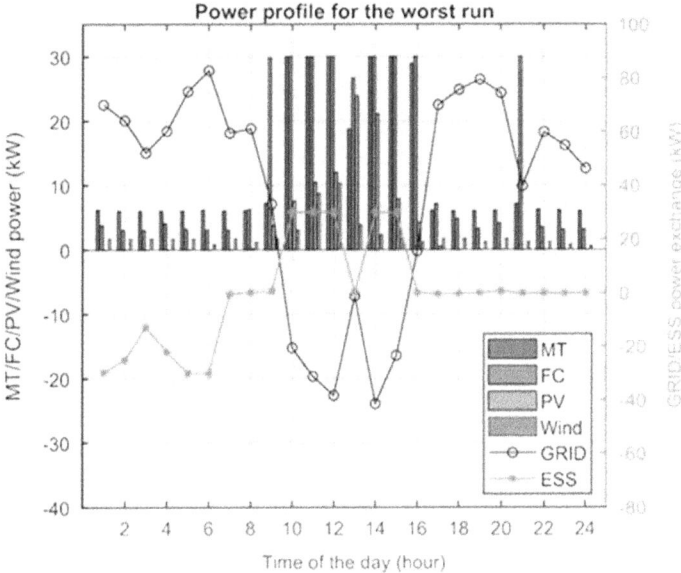

FIGURE 3.11 Power profile for the worst run of the GWO technique.

Figure 3.12(a) illustrates the ESS energy level (kWh), whereas Figure 3.12(b) illustrates ESS SoC in percentage for the best, median, and worst runs. As stated earlier, it was assumed that the ESS was fully exhausted at the beginning of the day, and its initial value was set to its lower limit. However, as can be seen from Figure 3.12, the ESS

(a)

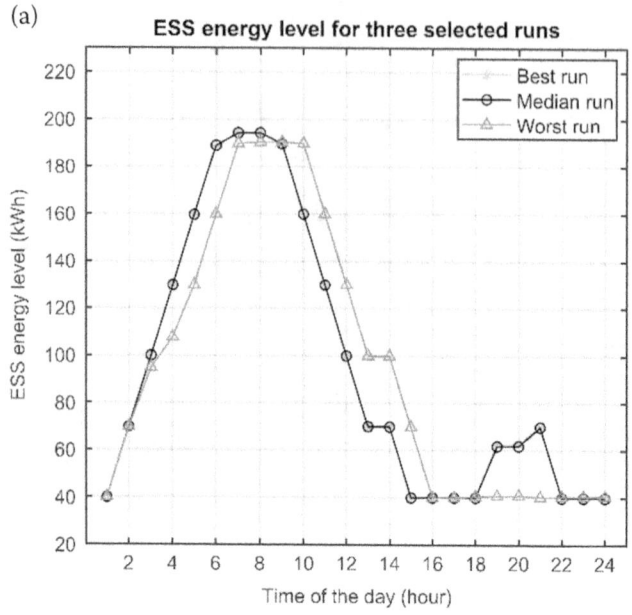

(b)

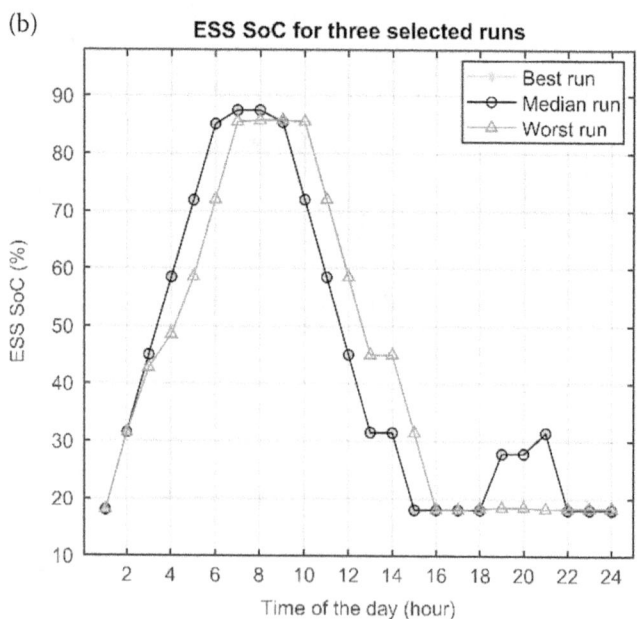

FIGURE 3.12 (a) ESS energy level (kWh) for three selected runs; (b) ESS SoC (in percentage) for three selected runs.

energy level varied in between the upper and lower limits. Finally, Figure 3.13 illustrates MT and FC power profiles, whereas Figure 3.14 presents ESS and utility grid power exchange profiles for the best runs of the employed techniques (GWO and GA). As can be seen from both figures, power profiles of the energy sources followed similar patterns.

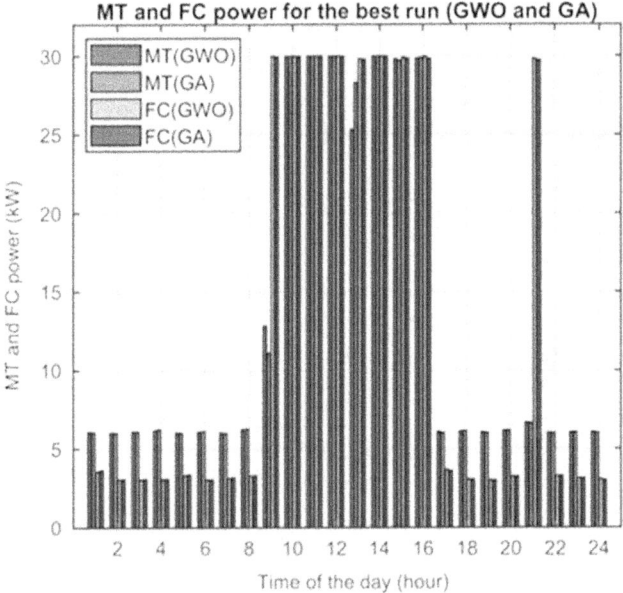

FIGURE 3.13 CMG energy generation comparisons of the DGs for the best runs of GWO and GA.

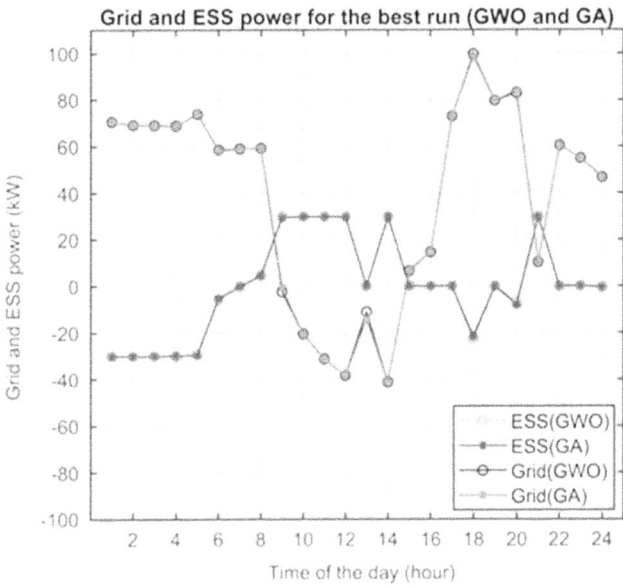

FIGURE 3.14 CMG energy exchange comparisons of the ESS and grid for the best runs of GWO and GA.

3.5 CONCLUSIONS

This chapter formulated a multi-objective constrained optimization problem for a CMG energy scheduling considering ESS degradation costs. Most of the available energy scheduling formulations rarely incorporate all complex equality and inequality constraints. However, this chapter considered most of the dominant equality and inequality constraints in the problem formulation. Then, it employed a newly developed meta-heuristic optimization technique; namely the GWO was employed here to solve the formulated optimization model. Comprehensive result analysis proved the efficacy and competitiveness of the employed technique in terms of CMG energy scheduling. Besides, the statistical analysis of the obtained results from different runs proved the robustness of the employed algorithm. Comparative results analysis showed that the proposed approach outperformed the genetic algorithm and algorithms of the reported literature in achieving better objective function values (energy cost of the CMG). Moreover, the proposed approach was found to be competitive with other peer algorithms in terms of the average computational cost. However, one of the future concerns of the topic under investigation could be the development of self-adaptive algorithms for CMG energy scheduling.

ACKNOWLEDGMENTS

The authors acknowledge the support received from King Fahd University of Petroleum & Minerals' Center of Research Excellence in Renewable Energy.The authors declare no conflict of interest.

REFERENCES

[1] M. T. Al-Nory, "Optimal decision guidance for the electricity supply chain integration with renewable energy: Aligning smart cities research with sustainable development goals," *IEEE Access*, vol. 7, pp. 74996–75006, 2019.
[2] N. K. Meena, J. Yang, and E. Zacharis, "Optimisation framework for the design and operation of open-market urban and remote community microgrids," *Appl. Energy*, vol. 252, p. 113399, Oct. 2019.
[3] R. Lasseter et al., "The CERTS microgrid concept, white paper on integration of distributed energy resources," *Calif. Energy Comm. Off. Power Technol. Dep. Energy, LBNL-50829*, http://certs.lbl.gov, p. 29, 2002.
[4] B. Khan and P. Singh, "Optimal power flow techniques under characterization of conventional and renewable energy sources: A comprehensive analysis," *J. Eng.*, vol. 2017, Article ID 9539506, 16 pages, 2017.
[5] P. Singh and B. Khan, "Smart microgrid energy management using a novel artificial shark optimization," *Complexity*, vol. 2017, Article ID 2158926, 22 pages, 2017.
[6] T. Molla, B. Khan, B. Moges, H. H. Alhelou, R. Zamani and P. Siano, "Integrated optimization of smart home appliances with cost-effective energy management system," *CSEE J. Power Energy Syst.*, vol. 5, no. 2, pp. 249–258, June 2019.
[7] Z. Tang, Y. Lin, M. Vosoogh, N. Parsa, A. Baziar and B. Khan, "Securing microgrid optimal energy management using deep generative model," *IEEE Access*, vol. 9, pp. 63377–63387, 2021.

[8] S. P. Bihari et al., "A comprehensive review of microgrid control mechanism and impact assessment for hybrid renewable energy integration," *IEEE Access*, vol. 9, pp. 88942–88958, 2021, doi: 10.1109/ACCESS.2021.3090266.

[9] B. Khan and P. Singh, "Selecting a meta-heuristic technique for smart micro-grid optimization problem: A comprehensive analysis," *IEEE Access*, vol. 5, pp. 13951–13977, 2017.

[10] O. P. Mahela et al., "Comprehensive overview of multi-agent systems for controlling smart grids," *CSEE J. Power Energy Syst.*, vol. 8, no. 1, pp. 115–131, Jan. 2022, doi: 10.17775/CSEEJPES.2020.03390.

[11] Z. Wang, B. Zhang, M. Mobtahej, A. Baziar and B. Khan, "Advanced reactive power compensation of wind power plant using PMU data," *IEEE Access*, vol. 9, pp. 67006–67014, 2021.

[12] S. Padmanaban, C. Dhanamjayulu and B. Khan, "Artificial Neural Network and Newton Raphson (ANN-NR) algorithm based selective harmonic elimination in cascaded multilevel inverter for PV applications," *IEEE Access*, vol. 9, pp. 75058–75070, 2021.

[13] M. S. Fakhar et al., "Implementation of APSO and improved APSO on non-cascaded and cascaded short term hydrothermal scheduling," *IEEE Access*, vol. 9, pp. 77784–77797, 2021.

[14] M. Shafiullah, A. T. Al-Awami, and I. M. ElAmin, "Profit maximization planning of a load aggregator using electric vehicles through optimal scheduling of day ahead load," *2015 18th International Conference on Intelligent System Application to Power Systems (ISAP)*. IEEE, pp. 1–6, Sep-2015.

[15] A. M. Bouzid, J. M. Guerrero, A. Cheriti, M. Bouhamida, P. Sicard, and M. Benghanem, "A survey on control of electric power distributed generation systems for microgrid applications," *Renew. Sustain. Energy Rev.*, vol. 44, pp. 751–766, Apr. 2015.

[16] M. Ahmed, S. Kuriry, M. Shafiullah, and M. A. Abido, "DC microgrid energy management with hybrid energy storage systems," in *2019 23rd International Conference on Mechatronics Technology (ICMT)*, 2019, pp. 1–6.

[17] C. A. Macana, E. Mojica-Nava, H. R. Pota, J. Guerrero, and J. C. Vasquez, "Accurate proportional power sharing with minimum communication requirements for inverter-based islanded microgrids," *Int. J. Electr. Power Energy Syst.*, vol. 121, p. 106036, Oct. 2020.

[18] E. Rokrok, M. Shafie-Khah, and J. P. S. Catalão, "Review of primary voltage and frequency control methods for inverter-based islanded microgrids with distributed generation," *Renew. Sust. Energ. Rev.*, vol. 82, pp. 3225–3235, Feb 2018.

[19] F. Moazeni and J. Khazaei, "Optimal operation of water-energy microgrids: A mixed integer linear programming formulation," *J. Clean. Prod.*, vol. 275, pp. 122776, Jul. 2020.

[20] M. Shafiullah and A. T. Al-Awami, "Maximizing the profit of a load aggregator by optimal scheduling of day ahead load with EVs," *2015 IEEE International Conference on Industrial Technology (ICIT)*. IEEE, pp. 1342–1347, Mar 2015.

[21] E. C. Umeozor and M. Trifkovic, "Operational scheduling of microgrids via parametric programming," *Appl. Energy*, vol. 180, pp. 672–681, Oct. 2016.

[22] M. Strelec and J. Berka, "Microgrid energy management based on approximate dynamic programming," in *IEEE PES ISGT Europe 2013*, vol. 2013, pp. 1–5, 2013.

[23] S. Mandal and K. K. Mandal, "Optimal energy management of microgrids under environmental constraints using chaos enhanced differential evolution," *Renew. Energy Focus*, vol. 34, pp. 129–141, Sep. 2020.

[24] Z. Xin-Gang, Z. Ze-Qi, X. Yi-Min, and M. Jin, "Economic-environmental dispatch of microgrid based on improved quantum particle swarm optimization," *Energy*, vol. 195, p. 117014, Mar. 2020.

[25] C. Li, X. Jia, Y. Zhou, and X. Li, "A microgrids energy management model based on multi-agent system using adaptive weight and chaotic search particle swarm optimization considering demand response," *J. Clean. Prod.*, Vol. 262, p. 121247, Jul. 2020.

[26] M. A. Hossain, H. R. Pota, S. Squartini, F. Zaman, and J. M. Guerrero, "Energy scheduling of community microgrid with battery cost using particle swarm optimisation," *Appl. Energy*, vol. 254, p. 113723, Nov. 2019.

[27] M. J. Rana, F. Zaman, T. Ray, and R. Sarker, "Heuristic enhanced evolutionary algorithm for community microgrid scheduling," *IEEE Access*, vol. 8, pp. 76500–76515, 2020.

[28] A. Askarzadeh, "A memory-based genetic algorithm for optimization of power generation in a microgrid," *IEEE Trans. Sustain. Energy*, vol. 9, no. 3, pp. 1081–1089, Jul. 2018.

[29] S. Leonori, M. Paschero, F. M. Frattale Mascioli, and A. Rizzi, "Optimization strategies for microgrid energy management systems by Genetic Algorithms," *Appl. Soft Comput. J.*, vol. 86, pp. 105903, Jan. 2020.

[30] M. Hemmati, B. Mohammadi-Ivatloo, M. Abapour, and A. Anvari-Moghaddam, "Day-ahead profit-based reconfigurable microgrid scheduling considering uncertain renewable generation and load demand in the presence of energy storage," *J. Energy Storage*, vol. 28, pp. 101161, Apr. 2020.

[31] G. Yan, D. Liu, J. Li, and G. Mu, "A cost accounting method of the Li-ion battery energy storage system for frequency regulation considering the effect of life degradation," *Prot. Control Mod. Power Syst.*, vol. 3, no. 1, p. 4, Dec. 2018.

[32] S. D. Downing and D. F. Socie, "Simple rainflow counting algorithms," *Int. J. Fatigue*, vol. 4, no. 1, pp. 31–40, 1982.

[33] X. Ke, N. Lu, and C. Jin, "Control and size energy storage systems for managing energy imbalance of variable generation resources," *IEEE Trans. Sustain. Energy*, vol. 6, no. 1, pp. 70–78, 2015.

[34] A. M. Gee, F. V. P. Robinson, and R. W. Dunn, "Analysis of battery lifetime extension in a small-scale wind-energy system using supercapacitors," *IEEE Trans. Energy Convers.*, vol. 28, no. 1, pp. 24–33, Mar. 2013.

[35] J. Qi, C. Lai, B. Xu, Y. Sun, and K. S. Leung, "Collaborative energy management optimization toward a green energy local area network," *IEEE Trans. Ind. Informatics*, vol. 14, no. 12, pp. 5410–5418, Dec. 2018.

[36] S. Zhou, Z. Hu, W. Gu, M. Jiang, and X.-P. Zhang, "Artificial intelligence based smart energy community management: A reinforcement learning approach," *CSEE J. Power Energy Syst.*, vol. 5, no. 1, pp. 1–10, 2019.

[37] S. Mirjalili, S. Mohammad, and A. Lewis, "Grey wolf optimizer," *Adv. Eng. Softw.*, vol. 69, pp. 46–61, 2014.

[38] C. Muro, R. Escobedo, L. Spector, and R. P. Coppinger, "Wolf-pack (Canis lupus) hunting strategies emerge from simple rules in computational simulations," *Behav. Processes*, vol. 88, no. 3, pp. 192–197, 2011.

[39] S. Sharma, S. Bhattacharjee, and A. Bhattacharya, "Grey wolf optimisation for optimal sizing of battery energy storage device to minimise operation cost of microgrid," *IET Gener. Transm. Distrib.*, vol. 10, no. 3, pp. 625–637, 2016.

[40] H. Faris, I. Aljarah, M. A. Al-Betar, and S. Mirjalili, "Grey wolf optimizer: A review of recent variants and applications," *Neural Comput. Appl.*, vol. 30, no. 2, pp. 413–435, Jul. 2018.

[41] A. Anvari-Moghaddam, A. R. Seifi, T. Niknam, and M. Alizadeh Pahlavani, "Multi-objective operation management of a renewable MG (micro-grid) with back-up micro-turbine/fuel cell/battery hybrid power source," *Fuel Energy Abstr.*, vol. 36, pp. 6490–6507, Nov. 2011.

NOMENCLATURE (ABBREVIATION)

CMG   Community Microgrid
DE   Differential Evolution
DE-H   Heuristic Enhanced Differential Evolution
DE-NH   Non-heuristic Differential Evolution
DG   Distributed Generator
DP   Dynamic Programming
ESS   Energy Storage Systems
FC   Fuel Cell
GA   Genetic Algorithm
GW   Grey Wolves
GWO   Grey Wolf Optimization
HP   Hydraulic Powerplants
MT   Microturbine
MILP   Mixed-Integer Linear Programming
PCC   Point of Common Coupling
PSO   Particle Swarm Optimization
PV   Photovoltaic
RER   Renewable Energy Resources
WT   Wind Turbine

4 Different Optimization Algorithms for Optimal Coordination of Directional Overcurrent Relays

Ahmed Korashy and Salah Kamel
Department of Electrical Engineering, Faculty of
Engineering, Aswan University, Aswan, Egypt

Abdel-Raheem Youssef
Department of Electrical Engineering, Faculty of
Engineering, South Valley University, Qena, Egypt

Francisco Jurado
Department of Electrical Engineering, University of Jaén,
EPS Linares, Jaén, Spain

CONTENTS

DOI: 10.1201/b22884-4

4.1   INTRODUCTION

Present day, the stability of the power network is becoming more challenging because of the growing size of the electric network. Protective relays are used in the electrical system at different voltage levels [1]. The main role of relays is to detect defected component of the electric network and initiate proper control circuit and maintain healthy part in service to keep electric system reliability [2–11]. The electrical network is planned to be as faultless as possible through careful planning, proper installation, and regular maintenance of electrical equipment [12]–[13]. As the electric system becomes complex, the coordination of relays becomes a very hard task [14]. Directional overcurrent relays (DOCRs) are known as directional units with over current relay (OCR), where this relay initiated when current flows in the same direction and exceeds [15]. Due to simplicity and low cost, the DOCRs are usually used in protecting sub-transmission and distribution systems [16].

DOCRs operating time is based on two settings, which are considered a decision variable (DV). The correctly chosen relay settings (time dial setting (TDS) and pick-up current (Ip)) are important to solve the problem of coordination [17]. The target of solving the problem of coordination is to find the right settings that decrease the operation time of DOCR in the case of faults within its zone and at the same time give a time delay to maintain operation sequential for relays [18]–[19]. The backup relay shall be initiated to disconnect the faulty part after a time delay if the main relay is unsuccessful initiating [20]. In other words, the main relay shall be initiated firstly to minimize outage of the electric system to the smallest zone [21].

4.1.1   LITERATURE REVIEW

In this chapter, a comprehensive survey including traditional optimization methods and recent optimization methods for the problem of coordination are presented. Several algorithms are implemented to deal with such problems, such as graph theory technique and manual methods, where these methods successfully solved the coordination problem in the case of a small system. However, such methods were very time consuming in the large network. The chosen DOCRs settings were performed manually [22]. Also, the trial-and-error method was suggested in the 1960s in order to solve the problem and get DOCRs settings [23]. The linear programming was proposed to solve the coordination problem in the 1980s [24–28], where the TDS is optimally calculated. However, the Ip in this method is assumed to be predefined [29]. DOCRs settings are simultaneously calculated using nonlinear programming (NLP) such as Sequential Quadratic Programming (SQP) [30]. The solution obtained from conventional methods is far away from a globally optimal solution.

Recently, meta-heuristic and hybrid optimization methods have been developed and most widely used in the coordination of relay to get the globally optimal solution and able to escape from local minima solutions [31] such as:

- Particle Swarm Optimization (PSO) that simulated hunting mechanism for birds in nature [32–35].
- Genetic Algorithm (GA) that inspired from concepts of Darwinian evolution [34–36].
- Seeker algorithm that simulated memory of human and experience of social learning [19].
- Ant Colony Optimization (ACO) simulate the searching mechanism of an ant for the shortest path between a source of food and their home colony [37], [38].
- Harmony Search (HS) inspired from searching for a perfect state of harmony during the process of getting music composition [15], [33].

Other techniques have been suggested to find a solution for the problem of coordination problem as Teaching-Learning-Based Optimization (TLBO) [31], Biogeography-Based Optimization (BBO) [23], MEFO [39], FFA [40], BH [41], and GSA [14]. Many hybrid techniques were suggested to catch the benefits of multiple techniques and to produce and achieve superior results compared to the original algorithm, such as CSA-FFA [42], BBO-LP [16], GA-NLP [43], GSA-SQP [44], and BBO-DE [45]. These methods use the advantages of both approaches and are extensively used to solve a relay coordination problem.

4.1.2 MAIN GOALS OF THIS CHAPTER

The main goals are to present a study of different methodologies for the coordination problem in electric systems. These goals are achieved through applying mathematical optimization methods and a computational tool to determine the best settings for the relays been selective. In order to reach the main objective, the above specific objectives are defined:

- Review of specialized methodologies used to get a solution to the problem of coordination.
- Formulate the problem of relay coordination like a mathematical optimization problem.
- Develop metaheuristic algorithms to solve the problem of coordination.
- Assess the performance of the suggested techniques using different standard test networks.
- Compare and evaluate the results applying the proposed methodology to the literature reviewed.

This chapter presents a general overview, the background, and the motivation of the study. Also, this chapter gives a comprehensive survey of the suggested solution for solving the coordination optimal. The OF and constraints for the problem of coordination are described in this chapter. Many methods are suggested to solve the problem of coordination. The results of solving a coordination problem using different techniques are presented in this chapter. The results are compared with different techniques. A summary and highlights of the main achievement are given in this chapter.

4.2 DOCRS' COORDINATION PROBLEM

Keeping electrical system stability is an important issue for getting the solution the coordination problem. The problem of coordination was described as a non-linear problem of optimization with high boundaries. This problem deals with calculating the relay operations sequence without excessive time delay and gives sufficient margins. All protection relays in the electric system shall be set correctly to ensure good coordination among protective devices [17]. The problem coordination in this chapter is mathematically expressed as an optimization problem. Goal function for this problem can be expressed as [46]:

$$OF = Minimize \sum_{d=1}^{P} Tpri_d \tag{4.1}$$

where P is the number of DOCRs and Tpri is the primary relay operating time, which this time can be determined as [47]:

$$T_{pri} = \frac{A \times TDS^d}{\left(\frac{I_f^d}{I_p^d}\right)^B - 1} \tag{4.2}$$

$$I_p^i = CT^i \times PS^i \tag{4.3}$$

where If is the short circuit magnitude and B and A have constant values. These constant values represent the relay characteristics, which can be summarized as follows [48]:

- In the case of standard IC, the values for A and B are equal to 0.14 and 0.02.
- In the case of very IC, the A and B values are equal to 13.5 and 1.
- In the case of extremely IC, the A and B values A and B are equal to 80 and 2.
- In the case of long time IC, the A and B values A and B are equal to 120 and 1.

In this chapter, the normal inverse is chosen and the value of the relay parameters B and A in that scenario are set as 0.02 and 0.14, respectively [49].

4.2.1 BOUNDARIES OF THE COORDINATION PROBLEM

The OF shall be accomplished within two types of constraints. These constraints are coordination constraints and constraints of DOCR characteristics.

4.2.1.1 Limits on Relay Characteristics

4.2.1.1.1 Limits on Pickup Current Setting

The constraint on Ip can be described as:

$$Ip_{Mini}^{d} \le Ip \le Ip_{Maxi}^{d} \tag{4.4}$$

$$PS_{\min}^{i} \le PS^{i} \le PS_{\max}^{i} \tag{4.5}$$

where Ip_{Maxi} I and p_{Mini} are the maximum and minimum limits of Ip, respectively. These limits are dependent on the maximum loading and minimum short circuit current; also, to ensure that at the smallest short circuit current the relay will be sensitive [43], [50].

4.2.1.1.2 Limits on TDS

TDS boundaries can be described as:

$$TDS_{Mini}^{d} \le TDS^{d} \le TDS_{Maxi}^{d} \tag{4.6}$$

where TDS_{Maxi} and TDS_{Mini} are maximum and minimum values of TDS, respectively. These limits depend on the manufacturer of relay [51].

4.2.1.1.3 Boundaries on DOCRs' Coordination

The avoidance of mal-operation of DOCRs is the main target of the boundaries on coordination. This goal could be achieved by the right sequence of operation between P/B pairs. The main relays are considered as the first defenses to isolate the faults, while the main and backup relays simultaneously sense the fault. In order to prevent mal-operation of DOCRs, a specified margin between relays is required, and this margin is called the CTI [22]. This delay is a very important issue that guarantees the backup relays will be initiated after this delay if the main relay is to initiate. The CTI can be calculated as [52]:

$$T_{backup} - T_{pri} \ge CTI \tag{4.7}$$

4.3   OPTIMIZATION TECHNIQUES

In mathematics and computer science, the optimization process is called mathematical programming, as it is related to computer programming. The optimization process is used in all different fields, especially engineering systems. It is applyied to get the best solution (minimum/maximum) of the single OF or multi-objective function between numbers of variables under desired constraints [52]. In this chapter, two algorithms are suggested:

1. GWO and EGWO
2. HWGO and WOA

4.3.1 GWO AND EGWO

4.3.1.1 Conventional GWO

This algorithm simulates the attacking method and the hierarchy of leadership for wolves in nature [53]. The first and second stages in the hierarchy are *Alp* and *Bet*. The third stage in the group is known as Del [54]. The social hierarchy can be described as follows:

1. Encircling prey: Wolves during a hunt encircle prey. This concept is described as [54]:

$$\overrightarrow{P}(h+1) = \overrightarrow{P}(h) - \overrightarrow{E} \cdot \overrightarrow{S} \tag{4.8}$$

$$\overrightarrow{S} = \left| \overrightarrow{F} \cdot \overrightarrow{P_p}(h) - \overrightarrow{P}(h) \right| \tag{4.9}$$

where P refers to a grey wolf location, P_p refers to the prey location, h is the current iteration, and E and F are coefficients that can be expressed as:

$$\overrightarrow{E} = 2\overrightarrow{g} \cdot \overrightarrow{rand_1} - \overrightarrow{g} \tag{4.10}$$

$$\overrightarrow{F} = 2 \cdot \overrightarrow{rand_2} \tag{4.11}$$

where the component g is decreased linearly from two to zero [54].

2. Hunting manner: Alp, Bet, and Del are considered the best candidate solutions. Regarding the Alp, Bet, and Del, other solutions update their location [53–55]. The hunting process can be described as follows:

$$\overrightarrow{S_{Alp}} = \left| \overrightarrow{F_1} \cdot \overrightarrow{P_{Alp}} - \overrightarrow{P} \right| \tag{4.12}$$

$$\overrightarrow{S_{Bet}} = \left| \overrightarrow{F_2} \cdot \overrightarrow{P_{Bet}} - \overrightarrow{P} \right| \tag{4.13}$$

$$\overrightarrow{S_{Del}} = \left| \overrightarrow{F_3} \cdot \overrightarrow{P_{Del}} - \overrightarrow{P} \right| \tag{4.14}$$

$$\overrightarrow{P_1} = \overrightarrow{P_{Alp}} - \overrightarrow{E_1} \cdot (\overrightarrow{S_{Alp}}) \tag{4.15}$$

$$\vec{P_2} = \vec{P}_{Bet} - \vec{E_2} \cdot (\vec{S}_{Bet}) \qquad (4.16)$$

$$\vec{P_3} = \vec{P}_{Del} - \vec{E_3} \cdot (\vec{S}_{Del}) \qquad (4.17)$$

$$\vec{P}(h+1) = \frac{\vec{P_1} + \vec{P_2} + \vec{P_3}}{3} \qquad (4.18)$$

3. Attacking process: When E < 1, wolves move to the prey [54].
4. Search process: The wolves search for prey when E > 1 [55–58].

4.3.1.2 EGWO Algorithm

The performance of metaheuristic techniques can be improved through the right balance between the two conflicting elements. The first element that aims to search locally is known as exploitation. The second element that aims to search globally is known as exploration. Global minima can be guaranteed and the search space is reduced by those elements and prevents the technique from being stuck in local minima [59]. The balance between these elements can be performed in a GWO by the value of parameter g, where this parameter is linearly decreased over the iterations cycle from two to zero [33]. In the EGWO, the component that balances between the exploration and exploitation, to decrease exponentially rather than decrease linearly over the iteration process to converge quickly to the best global solution. The suggested parameter can be described as follows [60]:

$$q = 2 \times e^{-\left(\frac{4 \times h}{Maxi.It}\right)^2} \qquad (4.19)$$

As mentioned before, the component E pushes the algorithm to move toward prey when E < 1 and to search for prey when E > 1. In the EGWO, this component can be determined as [60]:

$$\vec{E} = 2q \cdot \vec{rand_1} - \vec{q} \qquad (4.20)$$

The EGWO process is shown in Figure 4.1.

4.3.2 WOA AND HWGO

4.3.2.1 WOA Technique

The WOA simulated humpback whales social behavior in nature. These whales' hunting technique is called bubble-net foraging. This technique generates bubbles along 9' shaped path or circle form [61]. The hunting mechanism for whales can be explained as:

Step 1) Generate an initial population of the position candidate solutions and initialize the iteration counter $h=1$;

Step 2) Initialize control components (E, F, and q);

Step 3) Sort the search agents as follows: P_{Alp}, P_{Beta}, and P_{Del};

Step 4) Generate the components E, F, and q using (20), (11), and (19), respectively;

Step 5) Update the location of the current search agents using (18);

Step 6) Update E, F, and q;

Step 7) Assess the OF value for search agents;

Step 8) Update the search agents as follows: the best search agent is P_{Alp}, the second search agent is P_{Beta}, and the third search agent is P_{Del};

Step 9) Update the iteration $h=h+1$

Step 10) Repeat the process from Step 5 to Step 9 until the convergence criteria are r reached ($h \geq$ Maxi.It);

Step 10) The optimal solution (P_{Alp}).

FIGURE 4.1   The overall EGWO process.

1. Encircling prey: Whales update their location regarding the best agents; this process can be described as [61], [62]:

$$\vec{S} = \left| \vec{F} \cdot \vec{P}_p(h) - \vec{P}(h) \right| \tag{4.21}$$

$$\vec{P}(h + 1) = \vec{P}_p(h) - \vec{E} \cdot \vec{S} \tag{4.22}$$

2. Bubble-net attacking technique

As mentioned before, whales use the bubble-net strategy to attack the prey, where these techniques can be described as follows [61]:

1. Shrinking encircling: This method is accomplished using (10) and (11).
2. Spiral updating location: The whales' helix movement can be expressed:

$$\vec{P}(h + 1) = \vec{S}^* \cdot e^{m*rand} \cdot \cos(2\pi * rand) + \vec{P}^*(h). \tag{4.23}$$

$$\vec{S}^* = \left| \vec{P}^*(h) - \vec{P}(h) \right| \tag{4.24}$$

where rand is a random number in (-1) and (1), $\vec{P}$ is the whale location, $\vec{P}^*$ is the prey location, and m is a constant number [62]. The whales move toward prey within a spiral shape and a shrinking circle. The probability is 50 percent to select between these methods, which these mechanisms can be described as [61]:

$$\vec{P}(h+1) = \begin{cases} \vec{P^*}(h) - \vec{E} \cdot \vec{S} & \text{if } rand\ 1 < 0.5 \\ \vec{S^*} \cdot e^{m*rand} \cdot \cos(2\pi * rand) + \vec{P^*}(p) & \text{if } rand\ 1 > 0.5 \end{cases} \quad (4.25)$$

3. Search for prey: when $|E| > 1$, whales search for prey, where this process can be described [61]:

$$\vec{S} = \left| \vec{F} \cdot \vec{P}_{rand} - \vec{P} \right| \quad (4.26)$$

$$\vec{P}(h+1) = \vec{P}_{rand} - \vec{E} \cdot \vec{S} \quad (4.27)$$

4.3.2.2   HWGO Algorithm

The HWGO enhance the traditional WOA method. This enhancement used the GWO leadership hierarchy, where this hierarchy is utilized in the bubble-net attacking strategy. The HWGO choose the three best search agents (Alp, Bet, and Del) from the search space. The other solution will update their location regarding the position of Alp, Bet, and Del [63]. Within two methods, the whales move around the prey. In the HWGO, whales utilize the GWO leadership hierarchy for updating their position. This suggested technique is described as:

- Shrinking encircling method: Whales update their location using equations (4.28)–(4.32).

$$\vec{S_\alpha^*} = \left| \vec{P}_{Alp}(h) - \vec{P} \right|, \quad \vec{S_\beta^*} = \left| \vec{P}_{Bet}(h) - \vec{P}(h) \right|, \quad \vec{S_\delta^*} = \left| \vec{P}_{Del}(h) - \vec{P}(h) \right| \quad (4.28)$$

$$\vec{P_1}(h) = \vec{P}_{Alp}(h) + \vec{S}_{Alp}^* \cdot e^{m*rand} \cdot \cos(2\pi * rand) \quad (4.29)$$

$$\vec{P_2}(h) = \vec{P}_{Bet} + \vec{S}_{Bet}^* \cdot e^{m*rand} \cdot \cos(2\pi * rand) \quad (4.30)$$

$$\vec{P_3}(h) = \vec{P}_{Del}(h) + \vec{S}_{Del}^* \cdot e^{m*rand} \cdot \cos(2\pi * rand) \quad (4.31)$$

Step 1: Initialize an initial population for candidate solutions.

Step 2: Assess the OF value for each candidate solutions.

Step 3: Sort the search agents as follows: P_{Alp}, P_{Beta}, and P_{Del};

Step :6 *While*) h >Maxi.It)

Step 8:   Update control parameter (E, F, g, and x).

Step 9:   If$_1$ (x<0.5)

Step 10:   If$_2$ (|E|<1)

Step 11:   Update the position of the candidate solutions

Step 12:   else If$_2$ (|A| ≥1)

Step 13:   Select a random agent ( ).

Step 14:   Update the position of the candidate solutions

Step 15:   end If$_2$

Step 16:   else If$_1$ (x ≥0.5)

Step 17:   Update the position of the candidate solutions

Step 18:   end If$_1$

Step 21: Determine the OF value for each candidate solution.

Step 22: Update the position of P_{Alp}, P_{Beta}, and P_{Del}.

Step 23: h=h+1

Step 24: end while

Step 25: Return P_{Alp}

FIGURE 4.2   The overall process for the HWGO algorithm.

$$\vec{P}(h + 1) = \frac{\vec{P_1} + \vec{P_2} + \vec{P_3}}{3} \qquad (4.32)$$

- Spiral updating position: whales update their location using a spiral shape according to the following [63]:

The HWGO algorithm process is shown in Figure 4.2.

4.4   RESULTS AND DISCUSSION

The suggested techniques are applied in different systems to get a solution to the problem of coordination in this section. The suggested methods are compared with recent and other optimization methods (EFO [39], Modified Electromagnetic Field Optimization (MEFO) [39], differential evolution (DE) [33], HS [22], Black Hole (BH) [39], and GSA-SQP [44], Gravitational Search Algorithm (GSA) [44], PSO [33], GA [33], and SQP [44]) to prove their superiority to solve the coordination problem. The proposed techniques are implemented using the MATLAB® environment and the short circuit and load flow are accomplished using DIgSILENT Power Factory.

4.4.1 Description of Test System

In this chapter, a different network system has been used to verify the ability of the suggested methods. These test networks are listed below:

- Eight-bus network system
- IEEE 30-bus network system

4.4.1.1 The Eight-Bus Network

The eight-bus system is presented in Figure 4.3 [64–66]. There are seven lines in this system. Each line has DOCRs in its end. There are 28 DVs for this system. The constraint limits for TDS in this system are 0.05 and 1.1, respectively. The co-ordination margin is equal to 0.3 s. Other data such as Ip and If are found in [67].

4.4.1.2 IEEE 30-Bus Test System

The 30-bus network is presented in Figure 4.4 [51]. There are 20 lines in this system. Each line has DOCRs in its end. There are 62 DVs for this system. The constraint limits for TDS in this system are 0.1 and 1.1, respectively. The range for plug setting (PS) is considered 1.5 to 6. The coordination margin is equal to 0.3 s. Other data such as the Ip and If are found in [51].

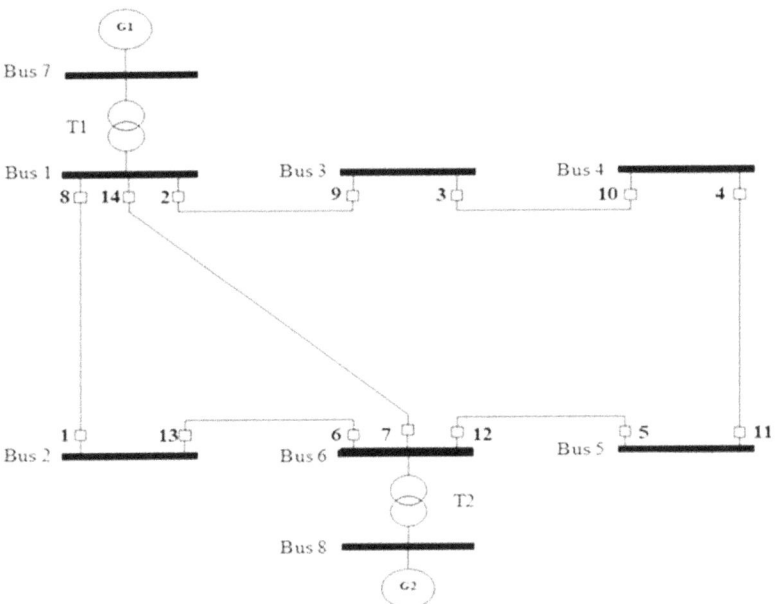

FIGURE 4.3   The network of the eight-bus network.

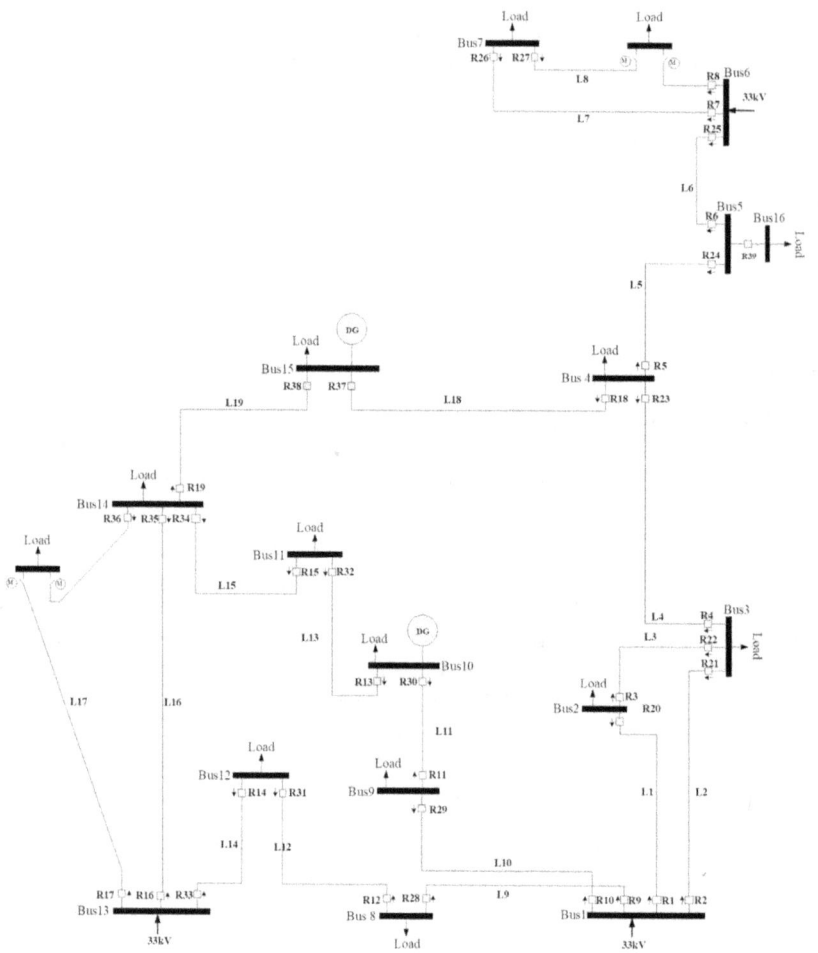

FIGURE 4.4   The network of the 30-bus network.

4.4.2   Using the EGWO for Solving the Coordination Problem

CASE 1   EIGHT-BUS NETWORK

In this case, the GWO and Enhanced Version of Grey Wolf Optimizer (EGWO) are tested using the eight-bus, which is presented in Figure 4.1. The operating time (OT) of the main and backup relays is presented in Figure 4.5 and Figure 4.6. It can be noticed from these figures that the backup relays will be initiated if the main relays fail to operate. It can be said that both techniques (GWO and EGWO) successes to find optimal relay setting that maintains the sequence of operation between relay pairs.

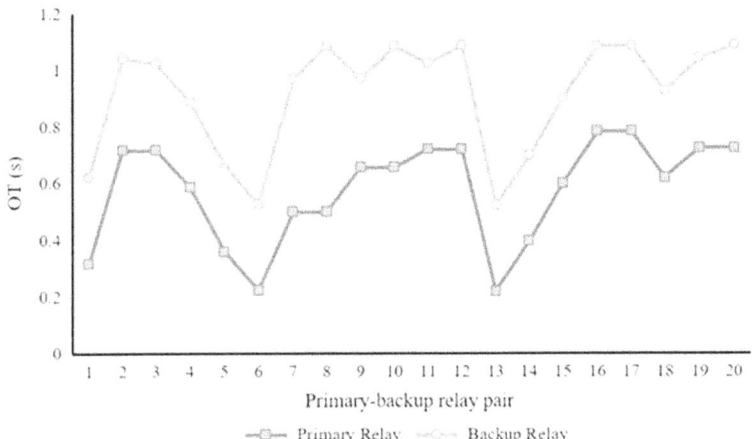

FIGURE 4.5   Relay pairs OT of the eight-bus network using GWO.

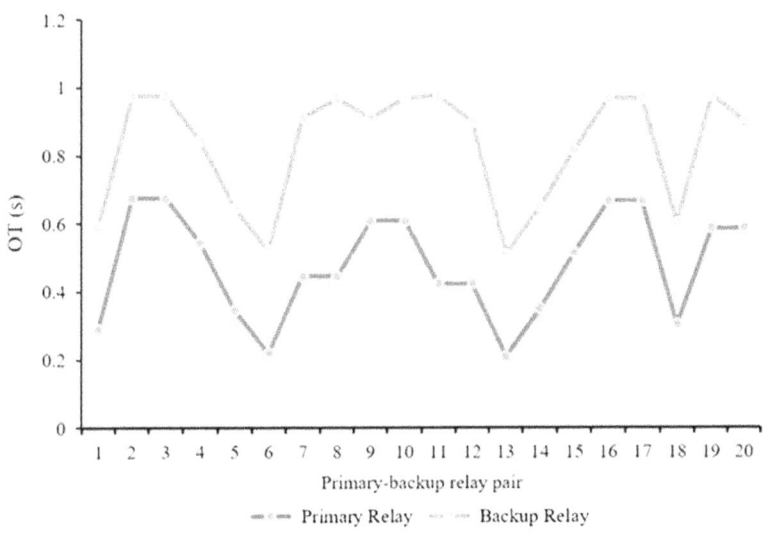

FIGURE 4.6   Relay pairs OT of the eight-bus network using EGWO.

The convergence charecteristics (CCs) of GWO and EGWO techniques is presented in Figure 4.7. It can be noticed from this figure that the EGWO converge to the promising solution faster than Water cycle algorithm (WCA). Also, the Objective Function (OF) value that given from the MWCA is better than OF value that given from WCA, where the reduction on OF reaches to 16.64%.

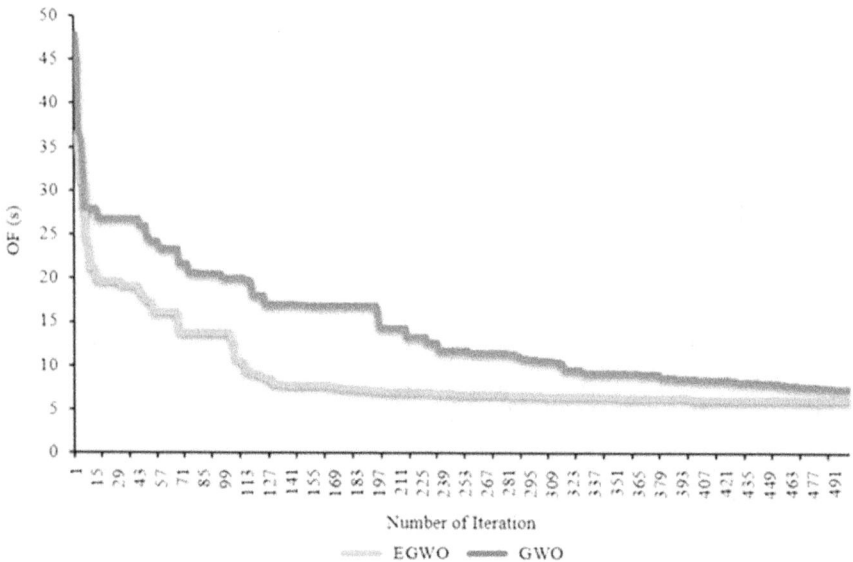

FIGURE 4.7   The EGWO and GWO fitness function for the eight-bus network.

The comparison between well-known methods and GWO and EGWO is presented graphically in Figure 4.8. The EGWO technique gives the least OF as noticed in Figure 4.6. That proves the EGWO superiority to find the best relay settings and maintain the relay coordination between P/B pairs.

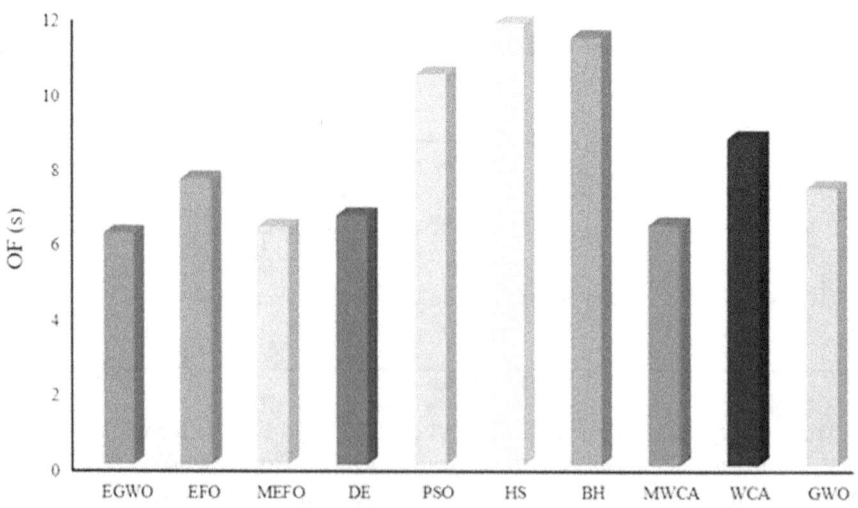

FIGURE 4.8   Comparing the EGWO with other optimization algorithms.

CASE 2   IEEE 30-BUS SYSTEM

The GWO and EGWO are assessed using the 30-bus system, which this system is presented in Figure 4.3. The main and backup relays OT using GWO and EGWO are shown in Figure 4.9 and Figure 4.10. In the event of the main relays fail to operate, the backup relays will be initiated after the specified margin as noticed in these figures. It can be said that both techniques successes to get the best setting that keeps the operation sequential between DOCRs pairs.

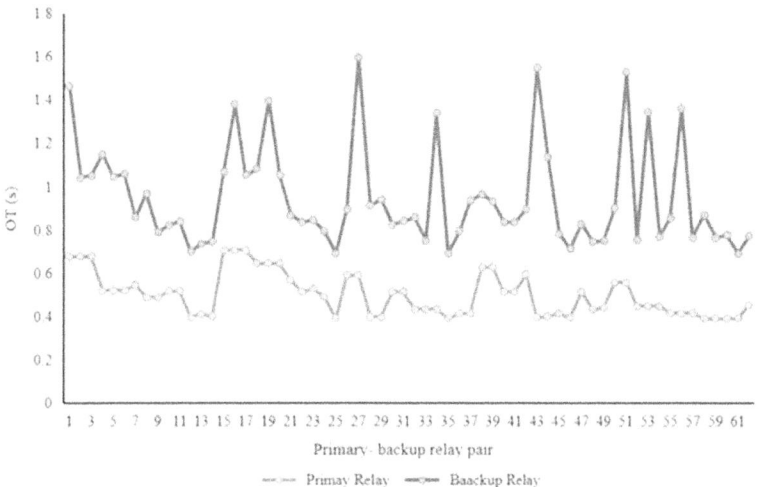

FIGURE 4.9   Relay pairs OT of the 30-bus network using GWO.

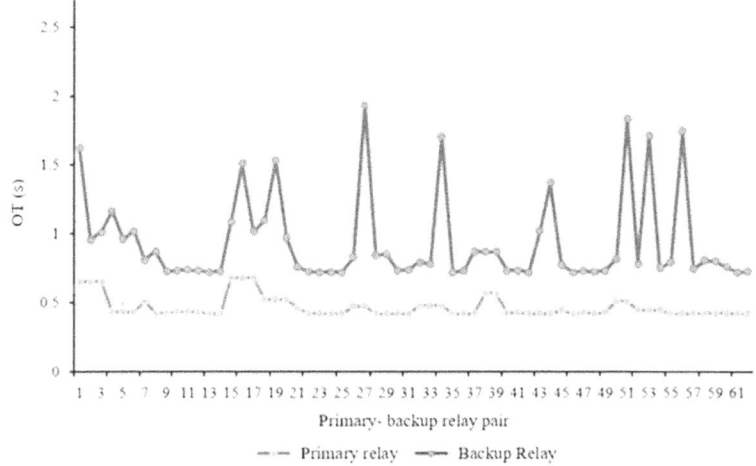

FIGURE 4.10   Relay pairs OT of the 30-bus network using EGWO.

The CC of GWO and EGWO techniques is presented in Figure 4.11. The EGWO converge to the optimal solution faster than GWO as shown this figure. Also, the OF value given by the EGWO is better than OF value that is given from GWO, where the reduction on OF reaches 8.4%.

The comparison between well-known methods and GWO and EGWO is presented graphically in Figure 4.12. The EGWO technique gives the least OF, as noticed in Figure 4.10. That proves the EGWO superiority to find the best relay settings and maintain the relay coordination between P/B pairs.

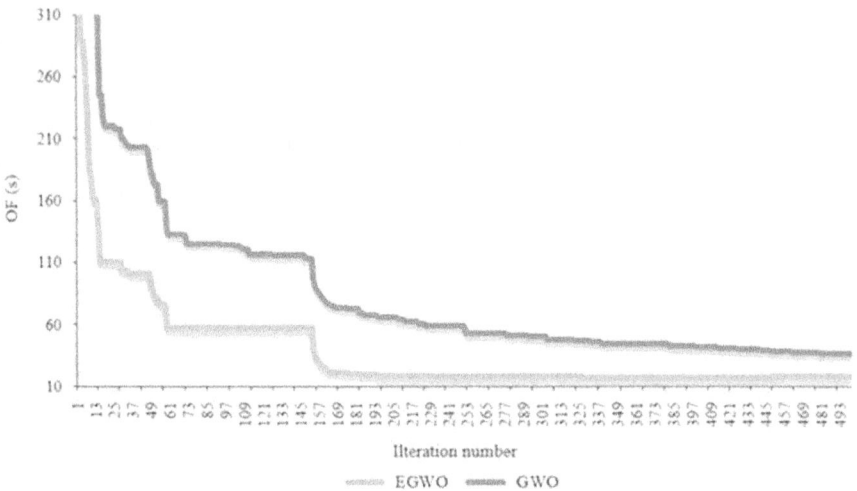

FIGURE 4.11 The EGWO and GWO fitness function for the 30-bus network.

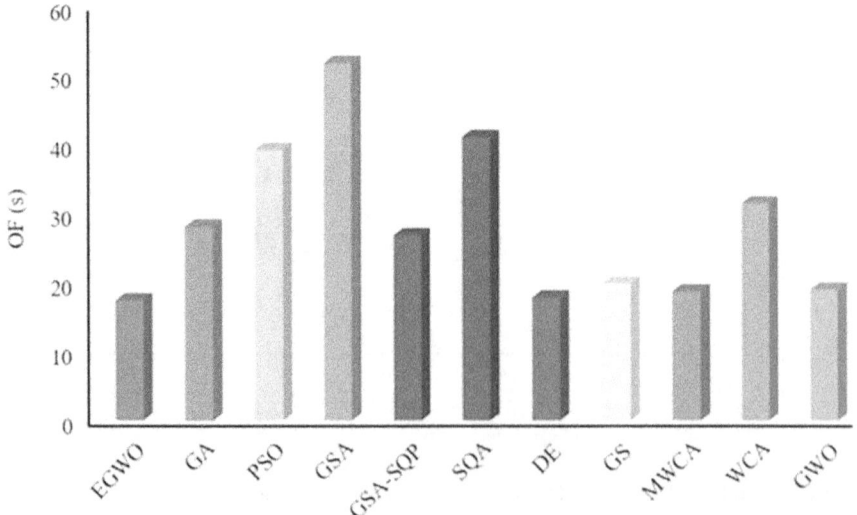

FIGURE 4.12 Comparing the EGWO with other optimization algorithms.

4.4.3   USING THE HWGO FOR SOLVING THE COORDINATION PROBLEM

CASE 1   EIGHT-BUS NETWORK

In this case, the WOA and Hybrid Whale Optimization Algorithm (HWGO) are tested using the eight-bus, which this system is presented in Figure 4.1. The OT of the main and backup relays is presented in Figure 4.13 and Figure 4.14. It can be noticed from these figures that the backup relays will be initiated if the main relays fail to operate. It can be said that both techniques (WOA and HWGO) successes to find optimal relay setting that maintains the sequence of operation between relay pairs.

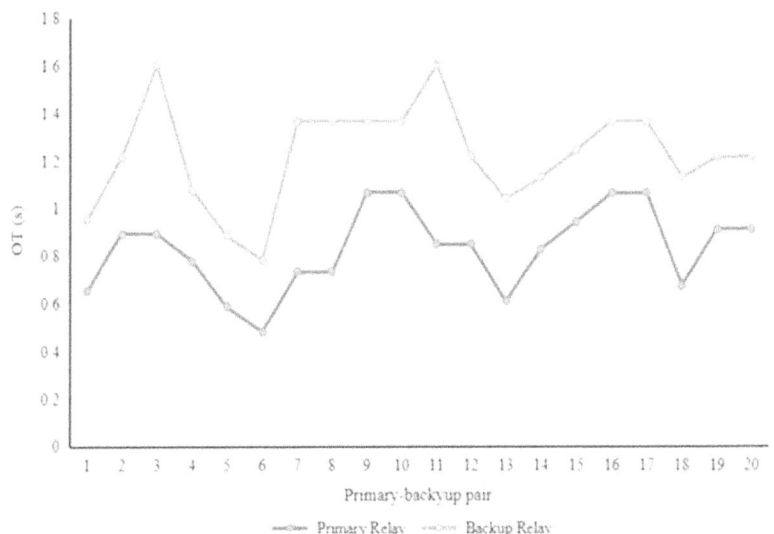

FIGURE 4.13   Relay pairs OT of the eight-bus network using WOA.

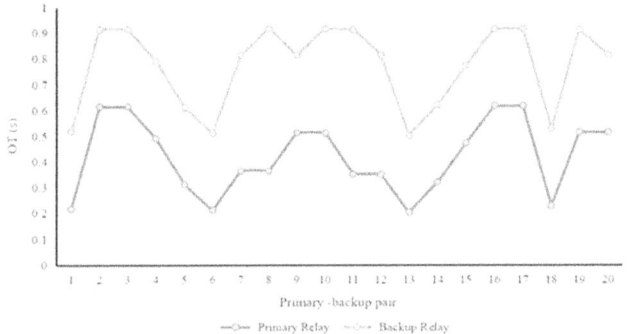

FIGURE 4.14   Relay pairs OT of the eight-bus network using HWGO.

The CC of WOA and HWGO techniques is presented in Figure 4.15. It can be noticed from this figure that the HWGO converge to the promising solution faster than WOA. Also, the OF value that gives the HWGO is better than the OF value that is given from WOA, where the reduction on OF reaches 53.5%.

The comparison between well-known methods and WOA and HWGO is presented graphically in Figure 4.16. The HWGO technique gives the least OF, as noticed in Figure 4.16. That proves the HWGO superiority to find the best relay settings and maintain the relay coordination between P/B pairs.

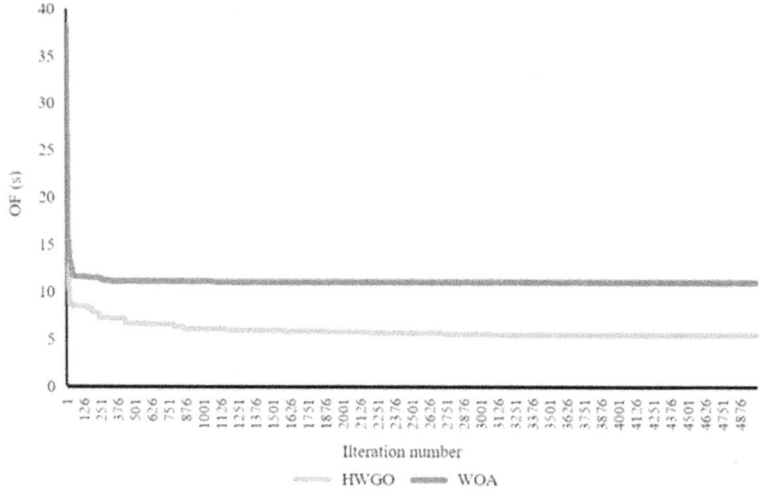

FIGURE 4.15   The fitness function for the nine-bus network using HWGO and WOA.

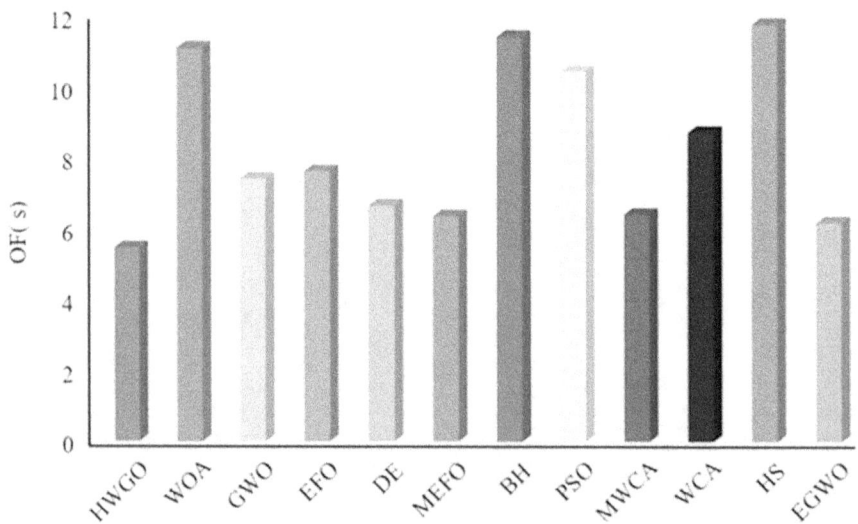

FIGURE 4.16   Comparing the HWGO with other optimization algorithms.

CASE 2   IEEE 30-BUS SYSTEM

In this case, the WOA and HWGO are tested using the 30-bus system, which is presented in Figure 4.4. The main relay OT and the backup relay OT using HWGO and WOA are shown in Figure 4.17 and Figure 4.18. In the event the main relays fail to operate, the backup relays will be initiated after the specified margin, as noticed in these figures. It can be said that both technique successes to get the best setting keeps the operation sequential between DOCRs pairs.

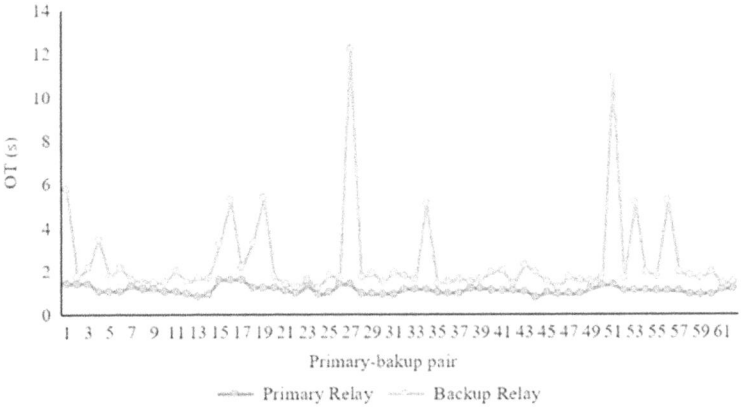

FIGURE 4.17   Relay pairs OT of the 30-bus network using WOA.

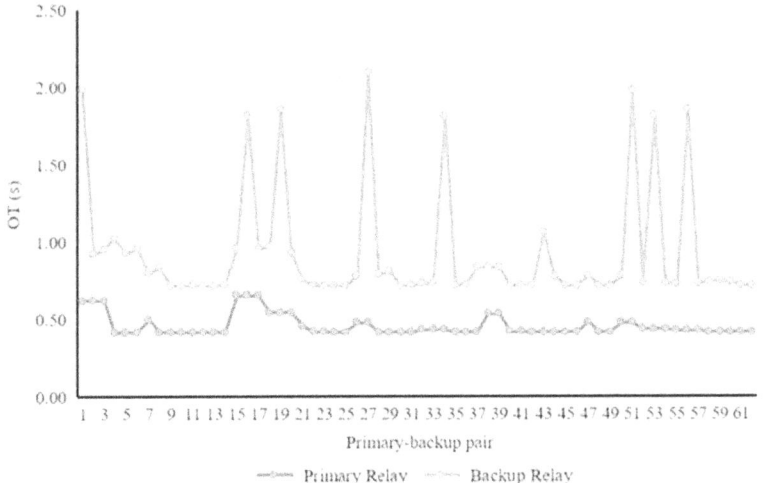

FIGURE 4.18   Relay pairs OT of the 30-bus network using HWGO.

The CC of WOA and HWGO techniques is presented in Figure 4.19. The HWGO converge to the optimal solution faster than WOA, as shown in this figure. Also, the OF value given by the EGWO is better than the OF value that is given from WOA, where the reduction on OF reaches 44.9%.

The comparison between well-known methods and WOA and HWGO is presented graphically in Figure 4.20. The HWGO technique given the least OF as

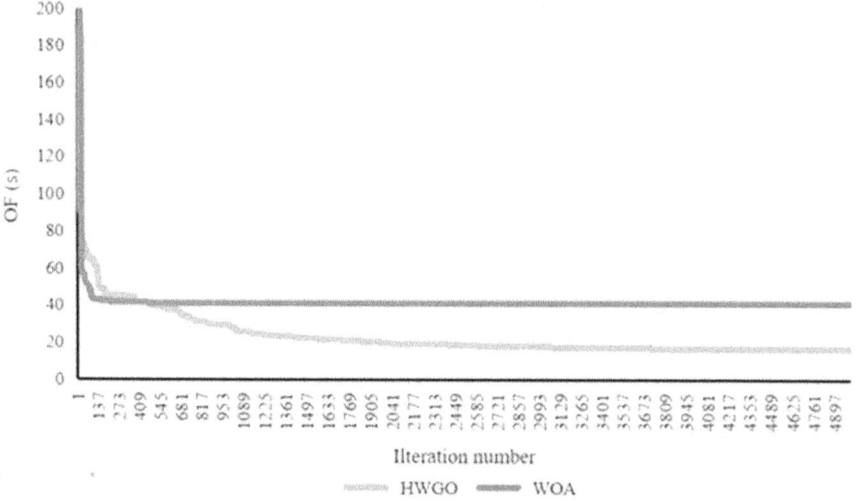

FIGURE 4.19 The fitness function for the 30-bus network using HWGO and WOA.

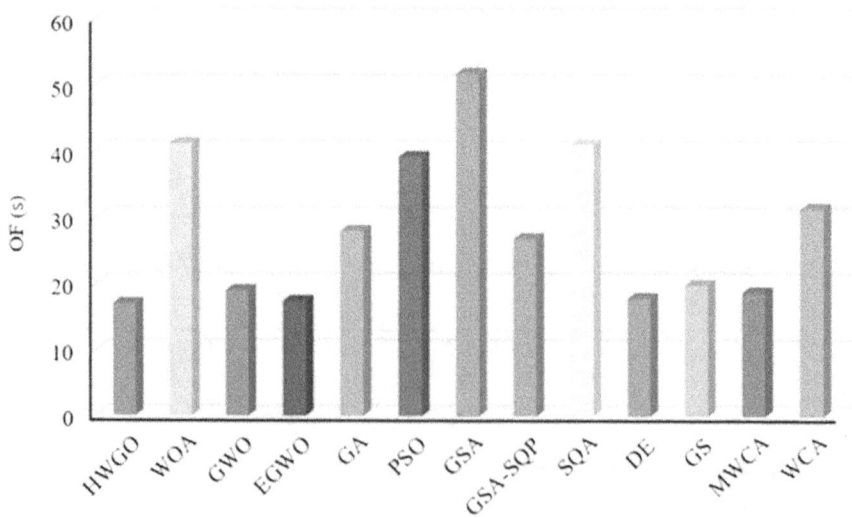

FIGURE 4.20 Comparing the HWGO with other optimization algorithms.

noticed in Figure 4.20. That proves the EGWO superiority to find the solution of the coordination problem.

4.5 CONCLUSIONS

Find correct relay settings that reduce the DOCRs operating time is the main objective from solving the problem of coordination. In this chapter, different techniques have been suggested to find the solution for coordination problems. Comparing well-known techniques with methods have been accomplished to prove the effectiveness of the suggested techniques to deal with such a problem. The main achievement can be concluded as follows:

- Solving the problem of coordination using GWO and EGWO has been presented. The EGWO has enhanced the traditional technique performance through the right balance between globally and locally searching. That improves the ability of the technique during the search to find the global minimum. This balance has been reached by decreasing K operator during the iterative cycle. The results from using EGWO prove that the proposed technique finds the best relay setting that reduced the OT for all primary relays. Also, the suggested technique maintained the sequence of operation between P/B pairs. The EGWO reaches the optimal solution faster than the conventional algorithm. The reduction in OF reaches about 19.39% compared to the GWO. The EGWO also has been compared with recent techniques. The results of the comparison show the ability of the EGWO in solving the coordination problem.
- Solving the problem of coordination using WOA and HGWO has been performed. The HWGO has enhanced the traditional technique performance. This enhancement is achieved through utilized leadership hierarchy of GWO that is applied in WOA algorithm. The suggested enhancement improved the exploitative phase of the WOA. The suggested technique chooses the three best search agents. The other agents modify their location regarding the positions of the three best agents. The results from using HWGO prove that the suggested technique finds the optimal relay setting that reduces the summation OT. Also, the suggested technique maintains the sequence of operation between P/B pair. The HWGO reaches the optimal solution faster than the conventional technique. The reduction in OF reaches about 44.9% compared to the WOA. The HWGO also has been compared with other techniques. The results of the comparison show the ability of the EGWO in solving the coordination problem.

REFERENCES

[1] Yen Shih, M., "Real time coordination of overcurrent relays by means of optimization algorithm," Master thesis, *Universidad Autónoma de Nuevo León*, 2013.
[2] Khan, B. and Singh, P., "Optimal power flow techniques under characterization of conventional and renewable energy sources: a comprehensive analysis," *Journal of Engineering*, Vol. 2017, Article ID 9539506, 16 pages, 2017.

[3] Singh, P. and Khan B., "Smart microgrid energy management using a novel artificial shark optimization," *Complexity*, Vol. 2017, Article ID 2158926, 22 pages, 2017.

[4] Molla, T., Khan B., Moges B., Alhelou H. H., Zamani R., and Siano P., "Integrated optimization of smart home appliances with cost-effective energy management system," in *CSEE Journal of Power and Energy Systems*, Vol. 5, No. 2, pp. 249–258, June, 2019.

[5] Tang, Z., Lin Y., Vosoogh M., Parsa N., Baziar A., and Khan B., "Securing microgrid optimal energy management using deep generative model," in *IEEE Access*, Vol. 9, pp. 63377–63387, 2021.

[6] Bihari, S. P. et al., "A comprehensive review of microgrid control mechanism and impact assessment for hybrid renewable energy integration," in *IEEE Access*, Vol. 9, pp. 88942–88958, 2021, doi: 10.1109/ACCESS.2021.3090266

[7] Khan, B. and Singh P., "Selecting a meta-heuristic technique for smart micro-grid optimization problem: a comprehensive analysis," in *IEEE ACCESS*, Vol. 5, pp. 13951–13977, 2017.

[8] Mahela, O. P. et al., "Comprehensive overview of multi-agent systems for controlling smart grids," in *CSEE Journal of Power and Energy Systems*, doi: 10.17775/CSEEJPES.2020.03390

[9] Wang, Z., Zhang B., Mobtahej M., Baziar A., and Khan B., "Advanced reactive power compensation of wind power plant using PMU data," in *IEEE Access*, Vol. 9, pp. 67006–67014, 2021.

[10] Padmanaban, S., Dhanamjayulu C., and Khan B., "Artificial Neural Network and Newton Raphson (ANN-NR) algorithm based selective harmonic elimination in cascaded multilevel inverter for PV applications," in *IEEE Access*, Vol. 9, pp. 75058–75070, 2021.

[11] Fakhar, M. S. et al., "Implementation of APSO and improved APSO on non-cascaded and cascaded short term hydrothermal scheduling," in *IEEE Access*, Vol. 9, pp. 77784–77797, 2021.

[12] Breitfelder, K. and Messina D., "IEEE 100: the authoritative dictionary of IEEE standards terms," *Standards Information Network IEEE Press*. Vol. 879, 2000.

[13] Phadke, A. G. and Thorp J. S., *Computer relaying for power systems*, 2009: John Wiley & Sons.

[14] Rashedi, E., Nezamabadi-Pour H., and Saryazdi S., "GSA: a gravitational search algorithm," *Information Sciences*, Vol. 179, pp. 2232–2248, 2009.

[15] Hussain, M., Rahim S., and Musirin I., "Optimal overcurrent relay coordination: a review," *Procedia Engineering*, Vol. 53, pp. 332–336, 2013.

[16] Al-Roomi, A. R. and El-Hawary M. E., 'Optimal coordination of directional overcurrent relays using hybrid BBO-LP algorithm with the best extracted time-current characteristic curve,' Electrical and Computer Engineering, IEEE 30th Canadian Conference on, 2017, pp. 1–6.

[17] Korashy, A., Kamel S., Youssef A.-R., and Jurado F., "Solving optimal coordination of direction overcurrent relays problem using grey wolf optimization (GWO) algorithm," in *2018 Twentieth International Middle East Power Systems Conference (MEPCON)*, 2018, pp. 621–625.

[18] Zellagui, M. and Abdelaziz A. Y., "Optimal coordination of directional overcurrent relays using hybrid PSO-DE algorithm," *International Electrical Engineering Journal (IEEJ)*, Vol. 6, pp. 1841–1849, 2015.

[19] Amraee, T., "Coordination of directional overcurrent relays using seeker algorithm," *IEEE Transactions on Power Delivery*, Vol. 27, pp. 1415–1422, 2012.

[20] Blackburn, J. L. and Domin T. J., *Protective relaying: principles and applications.* 2015: CRC Press.

[21] Chelliah, T. R., et al. "Coordination of directional overcurrent relays using opposition based chaotic differential evolution algorithm," *International Journal of Electrical Power & Energy Systems*, Vol. 55, pp. 341–350, 2014.

[22] Rajput, V. N. and Pandya K. S., "Coordination of directional overcurrent relays in the interconnected power systems using effective tuning of harmony search algorithm," *Sustainable Computing: Informatics and Systems*, Vol. 15, pp. 1–15, 2017.

[23] Albasri, F. A., Alroomi A. R., and Talaq J. H., "Optimal coordination of directional overcurrent relays using biogeography-based optimization algorithms," *IEEE Transactions on Power Delivery*, Vol. 30, pp. 1810–1820, 2015.

[24] Noghabi, A. S., Sadeh J., and Mashhadi H. R., "Considering different network topologies in optimal overcurrent relay coordination using a hybrid GA," *IEEE Transactions on Power Delivery*, Vol. 24, pp. 1857–1863, 2009.

[25] Rashtchi, V., Gholinezhad, J., Farhang, P., "Optimal coordination of overcurrent relays using Honey Bee Algorithm," in 2010 International Congress on Ultra Modern Telecommunications and Control Systems and Workshops, ICUMT 2010, 2010, pp. 401–405.

[26] Bedekar, P. P., Bhide S. R., and Kale V. S., "Optimum time coordination of overcurrent relays in distribution system using big-M (penalty) method," *WSEAS Transactions on Power System*, Vol. 4, pp. 341–350, 2009.

[27] Birla, D., Maheshwari, R. P., and Gupta, H. O., "Time-overcurrent relay coordination: a review," *International Journal Emerging Electrical Power System*, vol. 2, 2005.

[28] Bedekar, P. P., Bhide, S. R., and Kale, V. S., "Optimum time coordination of overcurrent relays using two phase simplex method," *World Academy of Science, Engineering and Technology*, Vol. 52, pp. 1110–1114, 2009.

[29] Urdaneta, A. J., Nadira R., and Jimenezssss L. P., "Optimal coordination of directional overcurrent relays in interconnected power systems," *IEEE Transactions on Power Delivery*, Vol. 3, pp. 903–911, 1988.

[30] Birla, D., Maheshwari R. P., and Gupta H., "A new nonlinear directional overcurrent relay coordination technique, and banes and boons of near-end faults-based approach," *IEEE Transactions on Power Delivery*, Vol. 21, pp. 1176–1182, 2006.

[31] Singh, M., Panigrahi B., and Abhyankar A., "Optimal coordination of directional over-current relays using Teaching Learning-Based Optimization (TLBO) algorithm," *International Journal of Electrical Power & Energy Systems*, Vol. 50, pp. 33–34, 2013.

[32] Zeineldin, H., El-Saadany E., and Salama M., "Optimal coordination of overcurrent relays using a modified particle swarm optimization," *Electric Power Systems Research*, Vol. 76, pp. 988–995, 2006.

[33] Alam, M. N., Das B., and Pant V., "A comparative study of metaheuristic optimization approaches for directional overcurrent relays coordination," *Electric Power Systems Research*, Vol. 128, pp. 39–52, 2015.

[34] Davis, L., Handbook of genetic algorithms. 1991: Van Nostrand Reinhold Company.

[35] Mirjalili, S., Gandomi A. H., Mirjalili S. Z., Saremi S., Faris H., and Mirjalili S. M., "Salp swarm algorithm: a bio-inspired optimizer for engineering design problems," *Advances in Engineering Software*, Vol. 114, pp. 163–191, 2017.

[36] Singh, D. K. and Gupta S., "Use of genetic algorithms (GA) for optimal co-ordination of directional over current relays," in Engineering and Systems (SCES), 2012 Students Conference on, 2012, pp. 1–5.

[37] Shih, M. Y., et al., "On-line coordination of directional overcurrent relays: Performance evaluation among optimization algorithms." *Electric Power Systems Research*, Vol. 110, pp. 122–132, 2014.

[38] Rivas, A. E. L. and Pareja L. A. G., "Coordination of directional overcurrent relays that uses an ant colony optimization algorithm for mixed-variable optimization problems," Environment and Electrical Engineering Industrial and Commercial Power Systems Europe (EEEIC/I&CPS Europe), IEEE International Conference, 2017, pp. 1–6.

[39] Bouchekara, H., Zellagui M., and Abido M. A., "Optimal coordination of directional overcurrent relays using a modified electromagnetic field optimization algorithm," *Applied Soft Computing*, Vol. 54, pp. 267–283, 2017.

[40] Tjahjono, A., Anggriawan D. O., Faizin A. K., Priyadi A., Pujiantara M., Taufik T., et al., "Adaptive modified firefly algorithm for optimal coordination of overcurrent relays," *IET Generation, Transmission & Distribution*, Vol. 11, pp. 2575–2585, 2017.

[41] Hatamlou, A., "Black hole: A new heuristic optimization approach for data clustering," *Information Sciences*, Vol. 222, pp. 175–184, 2013.

[42] Rajput, V., Pandya K., and Joshi K., "Optimal coordination of Directional Overcurrent Relays using hybrid CSA-FFA method," Electrical Engineering/ Electronics, Computer, Telecommunications and Information Technology, 12th International Conference, 2015, pp. 1–6.

[43] Bedekar, P. P. and Bhide S. R., "Optimum coordination of directional overcurrent relays using the hybrid GA-NLP approach," *IEEE Transactions on Power Delivery*, Vol. 26, pp. 109–119, 2011.

[44] Radosavljević, J. and Jevtić M., Hybrid GSA-SQP algorithm for optimal co-ordination of directional overcurrent relays. *IET Generation, Transmission & Distribution*, Vol. 10, No. 8, pp. 1928–1937, 2016.

[45] Al-Roomi, A. R. and El-Hawary M. E., "Optimal coordination of directional overcurrent relays using hybrid BBO/DE algorithm and considering double primary relays strategy," Electrical Power and Energy Conference (EPEC), IEEE, 2016, pp. 1–7.

[46] Korashy, A., Kamel S., Youssef A.-R., and Jurado F., "Evaporation rate water cycle algorithm for optimal coordination of direction overcurrent relays," in *2018 Twentieth International Middle East Power Systems Conference (MEPCON)*, 2018, pp. 643–648.

[47] Rao, T. M., *Power system protection: static relays*. 1989: Tata McGraw-Hill Education.

[48] Sharaf, H. M., Zeineldin H., Ibrahim D. K., and Essam E., "A proposed co-ordination strategy for meshed distribution systems with DG considering user-defined characteristics of directional inverse time overcurrent relays," *International Journal of Electrical Power & Energy Systems*, Vol. 65, pp. 49–58, 2015.

[49] El-Fergany, A. and Hasanien H. M., "Optimized settings of directional over-current relays in meshed power networks using stochastic fractal search algorithm," *International Transactions on Electrical Energy Systems*, Vol. 27, 2017.

[50] Singh, M., Panigrahi B. K., Abhyankar A. R., and Das S., "Optimal coordination of directional over-current relays using informative differential evolution technique," *Journal of Computational Science*, Vol. 5, pp. 269–276, 2014.

[51] Mohammadi, R., Abyaneh H. A., Rudsari H. M., "Fathi S. H., and Rastegar H., "Overcurrent relays coordination considering the priority of constraints," *IEEE Transactions on Power Delivery*, Vol. 26, pp. 1927–1938, 2011.

[52] Kiranyaz, S., Ince T., and Gabbouj M., *Multidimensional particle swarm optimization for machine learning and pattern recognition*. 2014: Springer.

[53] Kishor, A. and Singh P. K., "Empirical study of grey wolf optimizer," in Proceedings of Fifth International Conference on Soft Computing for Problem Solving, 2016, pp. 1037–1049.

[54] Mirjalili, S., Mirjalili S. M., and Lewis A., "Grey wolf optimizer," *Advances in Engineering Software*, Vol. 69, pp. 46–61, 2014.

[55] Mittal, N., Singh U., and Sohi B. S., "Modified grey wolf optimizer for global engineering optimization," *Applied Computational Intelligence and Soft Computing*, Vol. 2016, pp. 8, 2016.

[56] Long, W., et al., "Inspired grey wolf optimizer for solving large-scale function optimization problems," *Applied Mathematical Modelling*, Vol. 60, pp. 112–126, 2018.

[57] Long, W. and Xu S., "A novel grey wolf optimizer for global optimization problems," in Advanced Information Management, Communicates, Electronic and Automation Control Conference (IMCEC), 2016 IEEE, 2016, pp. 1266–1270.

[58] Faris, H., et al., "Grey wolf optimizer: a review of recent variants and applications," *Neural Computing and Applications*, Vol. 30, No. 2, pp. 413–435, 2018.

[59] Črepinšek, M., Liu S.-H., and Mernik M., "Exploration and exploitation in evolutionary techniques: A survey," *ACM Computing Surveys (CSUR)*, Vol. 45, pp. 35, 2013.

[60] Kamel, S., et al., "Development and application of an efficient optimizer for optimal coordination of directional overcurrent relays," *Neural Computing and Applications*, Volume 32, No. 12, pp. 8561–8583, 2019.

[61] Mirjalili, S. and Lewis, A., "The whale optimization algorithm," *Advances in Engineering Software*, Vol. 95, pp. 51–67, 2016.

[62] Kaur, G. and Arora S., "Chaotic whale optimization algorithm," *Journal of Computational Design and Engineering*, Vol. 5, No. 3, pp. 275–284, 2017.

[63] Korashy, A., et al., "Hybrid whale optimization algorithm and Grey Wolf optimizer algorithm for optimal coordination of direction overcurrent relays," *Electric Power Components and Systems*, Vol. 47, No. 6–7, pp. 644–658, 2019.

[64] Braga, A. and Saraiva J. T., "Coordination of overcurrent directional relays in meshed networks using the Simplex method," in Electrotechnical Conference, 1996. MELECON'96., 8th Mediterranean, 1996, pp. 1535–1538.

[65] Ezzeddine, M., et al., "Coordination of directional overcurrent relays using a novel method to select their settings." *IET Generation, Transmission & Distribution*, Vol. 5, No. 7, pp. 743–750, 2011.

[66] Razavi, F., Abyaneh H. A., Al-Dabbagh M., Mohammadi R., and Torkaman H., "A new comprehensive genetic algorithm method for optimal overcurrent relays coordination," *Electric Power Systems Research*, Vol. 78, pp. 713–720, 2008.

[67] Rajput, V. N. and Pandya K. S., "On 8-bus test system for solving challenges in relay coordination," in Power Systems (ICPS), 2016 IEEE 6th International Conference on, 2016, pp. 1–5.

NOMENCLATURE

A	Constant Values for The Characteristic of The Relay
ACO	Ant Colony Optimization
B	Constant Values for The Characteristic of The Relay
BH	Black Hole
BBO	Biogeography-Based Optimization
CSA	Cuckoo Search Algorithm
CT	Current Transformer
CTI	Coordination Time Interval
DOCRs	Directional Overcurrent Relays
DV	Decision Variables
E	Number of Primary Relay
EGWO	Enhanced Version of Grey Wolf Optimizer
FFA	Firefly Algorithm
GA	Genetic Algorithm
GSA	Gravitational Search Algorithm
GWO	Grey Wolf Optimizer
HWGO	Hybrid Whale Optimization Algorithm
HS	Harmony Search
IC	Inverse characteristics
I_p	Pickup Current
I_f	Fault Current
Iter	Current Iteration value
LP	Linear Programming
LoBo	Lower Ranges of Decision Variables
MEFO	Modified Electromagnetic Field Optimization
Maxi.It	Number of Maximum Iteration
Nvars	Number of Decision Variables
Npop	Number of Population
NLP	Nonlinear Programming
OCR	Over-Current Relay
OF	Objective Function
OT	Operating Time
P	Number of Primary Relay
P/B	Primary and backup
PSO	Particle Swarm Optimization
rand	Random Number
random1	Random Number between (0) and (1)
random2	Random Number between (0) and (1)
rand	Random value
SA	Seeker algorithm
SQP	Sequential Quadratic Programming
T_{backup}	Backup Relay Operating Time

TDS	Time Dial Setting
TLBO	Teaching-Learning-Based Optimization
T_{pri}	Primary Relay Operating Time
UpBo	Upper Ranges of Decision Variables
WOA	Whale Optimization Algorithm
Xp	Position Vector of The Prey

5 Microgrids—A Future Perspective

Akhil Gupta
Electrical Engineering Department, I. K. Gujral Punjab
Technical University Batala Campus, District Gurdaspur,
Punjab, India

Kamal Kant Sharma
Electrical Engineering Department (UIE), Chandigarh
University, Mohali Punjab, India

Gagandeep Kaur
Electrical Engineering Department, I. K. Gujral Punjab
Technical University Main Campus, Kapurthala Punjab, India

CONTENTS

DOI: 10.1201/b22884-5

5.1 INTRODUCTION

A microgrid (MG) is a kind of system in which various sources of energy are connected together with different characteristics and operating principles with variable controlling strategies. The concept of a MG has been developed with an objective to reduce the gap between demand and supply in which conventional and non-conventional sources are connected together to cater a basic load and interment load, which has peak characteristics along with less load factor. Peak demand is catered by the use of non-conventional energy sources mainly solar, wind, tidal, and mini-hydropower plants [1–5]. These resources have the capability of connecting with a utility grid and can provide additional amounts of energy when required and also takes away energy from the grid. Excessive energy is produced, termed distributed energy sources (DERs). A MG is mainly designed for a local load connected, subjected to control by a certain mechanism that allows the flow of real and reactive between source and load and integrated together. There are various kinds of generators available although a specific type of distributed generator specifies the particular kind of operation of a MG. Every MG is unique in their constitution of components and their configuration. A MG is a kind of grid mechanism that is controlled by a single unit comprising individual controllers and compensation techniques associated with different configurations of plants. Across the globe, two different types of systems prevail and are categorized as centralized power systems and decentralized power systems. A MG is a small formation of centralized power system with its localized loads and can be controlled through a pre-defined mechanism. A MG evolved with a period of time and the grid at a micro-scale can be used for a load that is present locally; or in other words, the load is present at a considerably significant small distance so that feeder losses and lowest distribution losses are high in the case of a conventional power system [6–10]. In a conventional power system, the system is designed with a number of sources; in India, there are five grids and sources of energy are far away and power is transported from various distances with different transformation of voltage and current with the help of different auxiliaries present, but in case of a MG, more reliability indices can be achieved as a MG is present at a small distance and its efficiency can be monitored. Various case studies have been formulated to study the behavior of MGs and have been found that more than 40 percent of the losses can be reduced with the combination of an available technological framework [11]. A MG is able to reduce carbon footprints as more than 95 percent of MGs employ renewable energy sources of energy and reduce a risk of integrating renewable sources of energy with conventional energy sources. MGs can have different individual sources and can be integrated with different sources to make them a bigger size, which can have different sizes and different component structures to deliver in a better way. A MG is a two-way approach with increased participation of consumers and gives a transparent mechanism of generation companies (GENCOs) so that the energy market can be sustained with different terminologies like cogeneration and deregulated environment.

There are different ways of connecting MGs and their operation methodology. MGs can be sufficiently large to cater a load of a connected system and not need to deliver their amount of energy to other systems and owned by private owners in their respective area; this type of configuration system is a standalone MG, which is almost preferred. On the other hand, MGs can also be designed and operate in a grid-connected mode in which the flow of power takes place and the state utility pays the money to a MG system if excessive power is taken from the MG and vice versa. This kind of system incorporates a deregulated environment and can be applied to a P2P framework in which parties can be private or government, but in India, P2P is only limited to the government and private; this defined that MGs can be owned by a private or government company, but the amount of energy can only be transferred to the government or state utility. The connection of MGs is augmented through power electric interfaces as some of the MGs do not have reactive power inheritance, so reactive power compensation is being provided by power electronic devices and being followed in the European Union and many developing countries [12,13].

Certain rules and regulations are framed by various agencies in designing a MG and their evaluation in terms of efficiency and reliability. In India, deregulated environment approval is one of the hurdles in making a MG a big success, as government agencies decide on an energy market in consideration of different climate conditions and availability of systems. Many new technologies have been looked into the making of MGs like fuel cell, solar and wind integration, and collaboration of thermal systems, which help different customers to work on a model of cogeneration. The term cogeneration deals with a generation of electricity on-site; this signifies the kind of land and available source of energy to be delivered. The main problem with MGs is constraints in designing and drawing a formulated strategy to deliver maximum efficiency. DERs have opened a new space for GENCOs and customers to have active participation and different optimization and prediction algorithms have been designed to meet the required deliverables, which involves a reduction in the gap of supply and demand. A DER provides better reliability indices and comprises different types of generators and storage mechanisms with power electronic interfaces that control the power flow between the utility and provider of MG output. A power electronic interface has been provided on the basis of the type of source used in making a MG, like solar energy incorporates shunt compensation techniques and wind employs a series compensation technique. Nowadays, a MG employs a hybrid configuration of interfaces for MGs of different sizes and structures.

A MG is defined by various groups of researchers as "a group of interconnected loads and distributed energy resources within clearly defined electrical boundaries that act as a single controllable entity with respect to the grid [14]. A MG can connect and disconnect from the grid to enable it to operate in both grid-connected or island mode." In addition to existing terminology, a MG can be operated in isolated mode or has an islanding structure with the same configuration of grid but islanded with a specific type of load that might not have a relation with a load connected with a conventional grid. A MG is connected in parallel with an existing grid to keep the voltage profile the same and allow a wide range of frameworks in deciding on different levels of voltage that need to considered with intermittent load mechanisms.

The Consortium for Electric Reliability Technology Solutions (CERTS) defines a MG as a combination and places all sources and loads connected under a defined cluster in such a way that a MG acts as a medium for delivering CHP and other applications for sustainable use. It has been studied that MGs are static and dynamic and requires equal balance of real and reactive power; however, reactive power devices can't be incorporated for small voltage index, and therefore, MGs are used with power electronic interface to increase the security and quality indices in consideration with stability. A very important feature of MGs defined by CERTS is an ability to be self-sufficient and controlled in such a manner that it is presented as a grid instead of a different entity and considered as a grid at a customer location. The main advantage of MGs is to incorporate every auxiliary placed on a load or generation side as per requirements and load consideration at the customer side [14].

5.2 A NOTE FROM NREL

Various investigations have been carried out and the National Renewable Energy Laboratory (NREL) proposes a definition of MGs similar to CERTS with a difference in terms of technological advancements and methods proposed that make DERs act as full-fledged sources to be utilized as similar with the conventional grid. NREL states that MGs can be incorporated for different infrastructures that comprise buildings, campuses, and homes and can be customized for varying types of loads with better reliability and dynamic assessment. NREL also defined MGs as an alternative in the case of emergency requirements or failures of the main grid in which outages or blackout-like conditions occur, which also empower different DERs to incorporate various devices to make grids self-sufficient and allow for the sale and purchase of power with P2P agreements in which government agencies can purchase power in consideration with seasonal requirements. It has been investigated that with the proper location of devices required to maintain power flow, MGs are able to provide better dynamic analysis with different interconnect arrangements giving high output and better regulation indices keeping sending and receiving end voltage the same so that transmission losses can be minimized and can be operated for an emergency line in use whenever required in stand-alone or grid-connected modes. As depicted in Figure 5.1, the main components of a typical MG are distributed generators, distribution feeder from substation, interconnection switch, control system, load, and distributed storage (DS) [15,16].

5.3 WORKINGS OF A MICROGRID

Various operating modes have been investigated and demonstrated for MGs as a MG is a small analogy depicting the configuration of a main grid but needs to be connected for delivering power to consumers. Some connections are being recommended from the point of stability like islanded and utility connected modes. The islanded mode is a new approach in which sources are disconnected from the

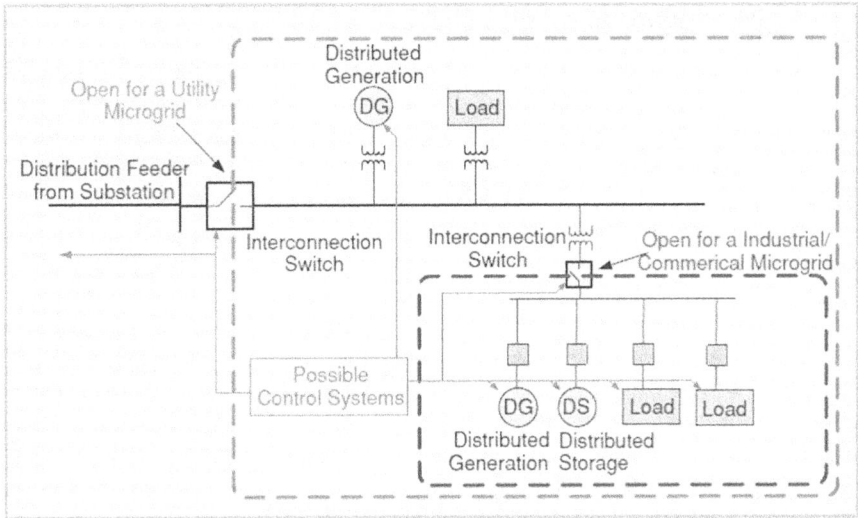

FIGURE 5.1   Schematic of a typical microgrid and its components.

utility supply to the check stability of a system in the case of an emergency. The islanding mode is important to check the dynamic stability of a system and is classified as intentional and unintentional. In the intentional mode, the system is trained and sources are being disconnected, whereas in the unintentional mode, sources are disconnected in the case of fault, an outage occurs, and the system collapses, which results in the disruption of the power system and the system gets halted. The importance of islanding is also important in that the system needs to be checked at various instances as a MG deals with intermittent loads and peaks have a system where connectivity is of prime importance as a conventional grid is not able to fulfill the requisite demand.

The other important connection topology for a MG is a utility-connected MG in which the conventional grid is connected with the main grid with a switch that enables a switching operation between both grids. This kind of operation with the suggested connection caters a load of emergency systems and is required to maintain reliability. For example, if some fault occurs in a conventional grid system, then a MG is allowed to deliver sufficient power to the conventional grid so that the main utility grid shouldn't be disrupted and the localized load connected with grid must be considered if it falls into the emergency category that provides a P2P connection between two government parties. A P2P arrangement also can be modified depending upon the mutual consent of two parties, enabled to cater a demand of their respective loads. Therefore, a MG can be considered a system with a single utility having all controlled parameters including generation and load parameters.

5.3.1 GRID-CONNECTED MODE

Every renewable source of energy is environmentally friendly and behaves on the nature and environment relationship, although various forms of renewable energy are by-products of solar energy due to the sun; various factors are being controlled and maneuvered. These sources of energy are unpredictable as some of them are seasonal and depend upon natural conditions, depicting uncertainty in their output. It has been investigated that sources like wind and solar requires additional types of generators instead of conventional generators; and their output is fluctuating. Various other forms of renewable energy sources like micro-turbines and fuel cells are independent of natural conditions, but their response towards changes occurs in a system are quite slow compared to changes occurred in a system that enables a system not fit to be incorporated for managing changing loads. Therefore, a need is being felt to integrate these sources of energy with conventional energy sources in order to achieve reliability and to be able to maintain stability with respect to dynamic changes that occur in the system. The main reason of absence of dynamic responses in renewable energy sources is the inability to generate reactive power to control real power flow. However, faults are not frequent in their occurrence but whenever severe faults occur in a system, the system collapses and is not able to sustain stability for a significant period of time.

Stability is a major concern of developing different topologies and different configurations of its kind as various resources are being integrated and their characteristics and voltage profile are independent of controlled parameters in a system, but various reactive power compensation and reduction in number of source connections and making a MG helps a system to improve efficiency and reliability. These kinds of topologies are being addressed with the consideration of various components used and different strategies like islanded and grid-connected systems to maintain voltage profile and maintaining static and dynamic stability in large.

- **Structure and main components selection in distributed hybrid energy system (DHES):** A hybrid energy system comprises different renewable energy systems and is also termed dispersed or distributed energy sources are elaborated as DHES and incorporates a dual structure of AC-DC systems as many loads are connected together near the supply system and some loads are connected at very significant distances with respect to the supply system. It has been seen that power driven for localized loads controlled by power electronic devices used that are available to convert AC to DC supply, but in the case of a bulk nature of load, the devices that can be driven by a supply are preferred. In order to maintain a balance, a converter configuration is used in a hybrid system depending upon the type of sources connected so that desirable deliverables can be achieved. DHES also behaves in a stable manner considering different converter topologies are used in a hybrid manner with a reduction in harmonics and other disturbance components.

- **Design of the circuitry and modeling of various components:** Various components are involved in developing a hybrid system with an approach that every component has been studied and a reliable strategy has been examined, considering energy storage mechanisms and various interconnection configurations. It has been seen that micro-sources and incorporating renewable energy sources have been explained through different non-linear electrical and mechanical components and having nonlinear characteristics and forming a differential equation of a certain higher order that needs to be linearized in order to identify the results and components to be integrated. On the basis of parameters involved, various compensation devices have been discussed and examined at the micro- and macro-level to obtain desired results. Every result needs to be validated and modeled on a simulation platform and if all the results are satisfactory, then designing a hybrid system has progressed to the next stage of evolution. A parametric approach and continuous evolution of a system is a key feature of developing a sustained hybrid power system.
- **Control coordination problem of the micro-sources:** The coordination of various micro-sources connected together is a challenging task; therefore, various benchmark models need to be validated and other models are required to follow a same structure. It has been examined that various micro-sources are connected to DC or AC line of system possessing different characteristics in the case of static performance with a relationship between voltage or current or having switching characteristics in the form of a derived relation between power and frequency. Both characteristics are important and need to be evaluated before validation and constitution of a model with a different time constant. On the basis of different selected parameters, a model is constituted in a proficient manner.

Due to variable features of renewable energy sources, a constituent of a MG with a different interconnection of a MG such that reliability and load demand requirements must be satisfied considering environmental safety, revenue, clean mechanism of energy generation, and electricity energy market forecasting. It has been seen that various advantages of hybrid energy systems have been considered over conventional—power system and models have been developed considering various parameters with respect to control, generation, and load strategies. The importance of integrating of MG into a conventional grid is presented below:

- Burden of load of conventional grids in terms of meeting the requirement of a peak load has been reduced and behaves as a back-up strategy for MGs with enhanced parameterization.
- Direct connections between a MG and a conventional grid help to make a system more reliable and efficient in maintaining a constant power supply and minimum losses.
- Various modes of connection of a MG with a conventional grid has been evaluated, but a grid-connected mode has been preferred, considering more revenue and reliable operation.

- Various strategies are being followed to meet the requisite demand and load structured in a conventional system. During the extra amount of energy generated by a MG in the case of less load requirements, it can add to the revenue of a system by helping the conventional grid to operate in a better way by keeping less reactive power compensation techniques.
- Various types of connections are being evaluated in the consideration of a conventional and an MG with the reduction in fuel costs. It provides flexibility on the customer side to select a particular kind of connection from any generation source available so that cost-effective operation can be achieved along with optimised operation on the load side, keeping customer benefits at large.

5.3.2 ISLAND MODE

A MG possesses an important characteristic of working independently in every condition like a conventional grid. A MG is connected with a grid and can be separated and works properly without the integration of a conventional grid and is able to cater a need of localized loads. This whole mechanism and mode of operation is termed an island mode. It has been seen in various studies that the island mode is also suggested to check the stability of a system at the occurrence of fault and a MG also works as a small function of a grid with the same characteristics and relationship between voltage and current. The only difference between a MG and conventional grid in the island mode is the operation that only limits to localized loads instead of the bulk of load connection available.

Other major uses of a MG structure is the availability of regulated power supply with constant properties such that dynamic responses can be controlled and loads that are subjected to transients and variations of voltage and current, can be protected with a change in voltage and current at switching instants. It has seen that a MG is dominant in the case of localized loads and control the dynamic behavior of loads and controlling reactive power efficiently. The island mode of a MG is also used to operate repair and maintenance task on conventional grid and enhance value of available transfer capability (ATC) in order to increase the transmission capability of the line which needs to be ascertained and maintained properly at every expansion of power system with the same or more voltage profile with optimization.

5.4 MICROGRID CONTROL

MGs can operate in two different modes considering the grid connected or island mode but different control strategies need to be employed to make the system secure and friendly in nature. Control strategies of MGs are also classified on the basis of different micro-sources connected and their feasibility structure with continuous output in terms of constant power supply. There are various controllers associated with MGs considering individual operation of controller embedded for all types of sources or every source will have separate controllers for controlling and maintaining compensation for different types of sources of energy. The main objective of this controller is to provide a constant frequency at a localized end along with maintaining an independent operation. However, if operated in the island mode, such a difference between real and reactive power instantaneous difference is kept in limits considering generation value at the source end and demand value at the load end, in order to mange sustainability in the internal structure of a MG. It plays an important role in developing a constant supply operation in grid-connected or islanded mode with different types of embedded controllers.

A MG operated in the island mode must be analyzed properly as controlling a frequency at localized end is not easy as mass of system is not significant. It can be understood from the fact that conventional power system has huge masses connected to each other and their respective inertia maintained a stability of system in defined limits and synchronism is achieved with constant value of frequency. On the other hand, distributed energy sources employed dispersed generation stability depends upon power electronic interfaces connected and real and reactive power relationship which has no direct connection with mass of system and cannot be controlled by such operation and frequency is somehow unstable. Many micro-sources are generating DC value at output and frequency control is subjected to instability in the island mode. Some micro-sources like fuel cell and micro-turbine have weak responses towards dynamic behavior of load and not be able to track the frequency stability and small signal stability issues in a system. However, a control strategy in MG operating in island mode must be defined, considering converter operational characteristics and maintain a relation with respect to masses of a system so that a likewise conventional power system, these MG operations can also be controlled with integration of different micro-sources and frequency can be controlled, which is essential requirement.

Another problem with the operation of a MG to maintain a voltage profile at sending at the receiving end; as a MG is employed for localized loads, therefore dynamic responses are quite high, which leads to fluctuations and uncontrolled reactive power also leads to uncontrolled real power which makes a system oscillated at a certain frequency. This is a problem with voltage magnitude in both grid-connected and island modes. Various researchers recommend voltage and frequency instability in a MG system considering dynamic responses of a system at the localized end and need to improve the voltage and frequency profile in a system in which different micro-sources are connected.

5.4.1 CENTRALIZED MICROGRID CONTROL

Centralized control of a MG is useful and helpful in optimizing productivity at localized loads along with different security mechanisms for exchanging maximum power from the MG to the central grid or between two different MGs of the same and different configurations. This strategy is very efficient for the internal operation of a MG by optimizing all control set points considering all dispersed energy sources and loads in controllable limits so that maximum power can be exchanged with minimal losses. Figure 5.2 depicts the two-way communications setup between the microgrid central controller (MCC) and local controller (LC) in order to exchange information of centralized control strategy. Mainly, the communication channel is through the telephonic lines, power line communication (PLC), or some

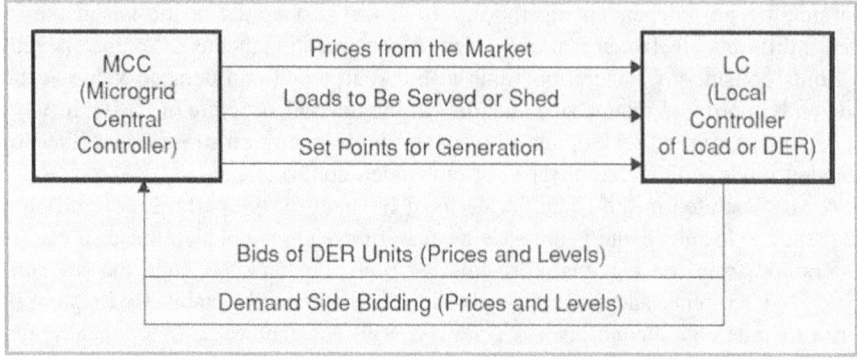

FIGURE 5.2 Flow of information in a centralized controlled microgrid.

kind of wireless source. The MG control mechanism comprises two controllers, like central controller and local controller, which work independently but a local controller behaves like a predictable controller to provide important information about load changes, variable market conditions, and different types of capacities of different units of MGs to provide sustainability. On the other hand, the central controller decides on every information of local controller associated with localized loads and fixed time intervals for predefined time intervals on the basis of priority and takes every requisite decision on the basis of requirement. It can be understood that the central controller behaves as the central brain of a system and the local controller provides information about the nervous system and other organ of the body to decide what decision is appropriate or not and divide every decision on a priority index. The main advantage of this control strategy is to obtain two controls and feedback mechanisms in which a local controller is a part of a control mechanism and controls the parameters of outer disturbances and provide information to a central controller to modify a strategy or improve the existing one for better sustainability indices [17].

A central controller decides a number of bids and a local controller develops a load and demand strategy to balance and bids are being generated at the generation side. Various functions have been performed by a central controller, such as demand generation balance, heat forecasting, unit commitment, economic dispatch, and security constraints in consideration with a local controller. Furthermore, it has the responsibility of managing load and demand in real time, which is dependent upon localised loads.

5.4.2 DECENTRALIZED MICROGRID CONTROL

Different terminologies are being used in this control strategy with a difference from centralized MG control that all decisions are to be taken by local controller and every local controller must be intelligent enough with security mechanism that stability should be enhanced. The main objective is to incorporate a local controller as a decisive component of a system to enhance controller parameters instead of generating revenue by balancing load and demand. Revenue is also a main parameter for establishing a control strategy but environmental conditions and unpredictable features that require a control approach are to be considered in this control strategy. Various intermediate systems have been used between a local controller and central controller, which initialize on certain weights decides on making the priority and objective to improve overall performance of a system in difficult conditions of mismatched load demand and other variable factors. This type of control strategy is termed decentralized and used for larger systems in which multiple numbers of MGs are connected together.

As illustrated in Figure 5.3, the multi agent system (MAS) is evolved from the classical theory of distributed control system that has features to control the system of large sizes and non-linear parameters, making them complex and driven by multiple order differential equations. The main feature of the MAS is to have independent software that uses a classical system and drives through a unique approach with different framework strategies. There can be multiple units in MGs where every unit represents a type of resource and every unit has their own intelligent framework with their driven system that helps them to target unpredicted parameters like environmental and weather conditions. A multi-agent system can use crisp and fuzzy logic along with different optimization techniques so that effective communication can be established with most energy-efficient mechanisms. The objective of an introduction of a MAS is to keep a system stable and intelligent and reduces a burden of predicting value of infinite value at large. A system comprises a distribution network operator (DNO), market operator (MO), MCC, and LC [17].

The industrial setups and commercial ones are now in search of a high level of power quality (PQ) and efficient and reliable power, due to the increase of advanced

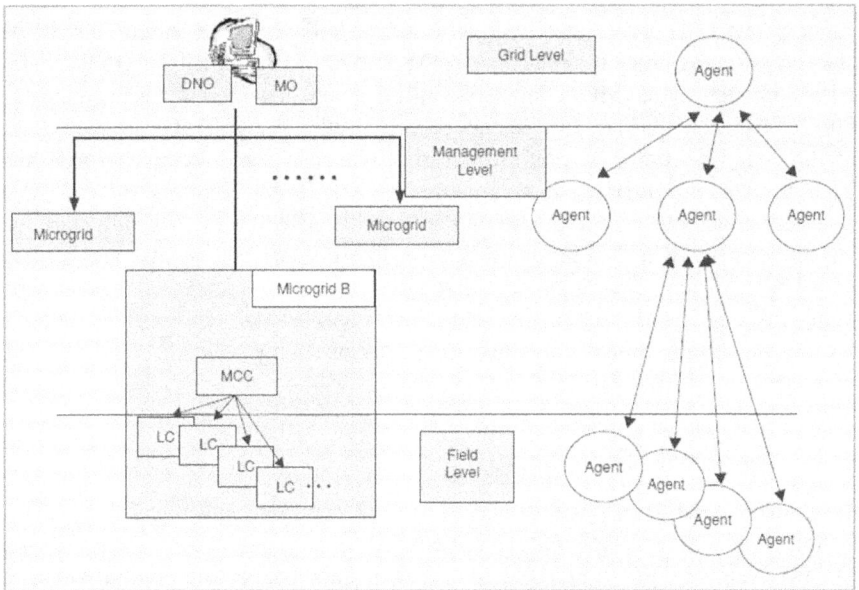

FIGURE 5.3 Schematic architecture of MAS in decentralized controlled microgrid.

control techniques. These customers are especially sensitive to momentary voltage sags caused by remote faults [18]. Various studies have been studied to understand the problem of frequency in standalone systems and it's been summarized that stand-alone systems are smaller in size and don't have big rotating parts and also don't have flexibility of having energy storage devices; therefore, larger systems have less problems in frequency control as compared to smaller systems and smaller systems must contain more MGs. It has also been emphasized that larger systems also have a frequency control approach on the basis of inertial constant and flywheel mechanism for storing additional amounts of energy. Various studies also show that small systems must incorporate converter topology to overcome frequency problems as improvement in real power can counter the frequency change and control the deviations produced due to the absence of large masses in a system [16].

Voltage control at both sending and receiving ends of a system is a dominant factor in both modes of operation considering grid-connected or island modes. Th main problem of voltage control is the large difference between the sending and receiving end, which are mainly termed the generation end and load end. It has been investigated that voltage regulation should be small so that an efficient topology can be achieved. Voltage regulation can be controlled by reducing a circulating current driven by reactive power and power electronic interface is required, which can be controlled by using a strategy of control of voltage versus reactive current droop so that voltage and current characteristics can be maintained in order to main the stability of a system.

With exponential growth in dispersed generation units, more complexity has been introduced in distribution networks lead to more congestion in an existing network and poor regulation. This kind of growth is appreciable and environmentally friendly and also requires a coordinated approach of controlling the voltage profile and maintaining the frequency range in limits so that the power flow can be maintained appropriately without a sudden collapse of a system. This can be achieved by integrating a small type of MGs with a predefined controlled parametric approach so that MGs can operate in full swing, considering a grid connected with a conventional grid or work independently. Both modes of MGs are acceptable but the island operation is preferred as it works in a stand-alone approach and different levels of integrity can be achieved, making every component responsible for a particular kind of operation and can be utilized while maintaining stability. Various controls of MGs are also available, like primary, secondary, and tertiary control with different parameters and segregated on different voltage levels and complexity of network architecture and allow a user-friendly approach to define a network in certain limits of voltage and current along with real and reactive power [19].

5.4.3 PRIMARY CONTROL STRATEGY

The primary control for MGs consists of the power electronic interfaces and the distributed generators, depicted in Figure 5.4. In this strategy, a MG is connected with a conventional grid with the help of a switch and connecting point driven by a point of common coupling. A DG comprises various components like a source of

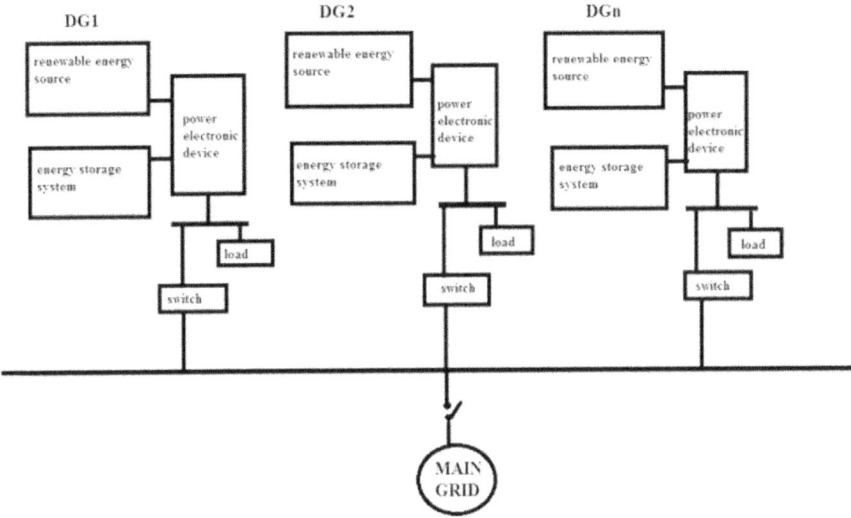

FIGURE 5.4   A general structure of a microgrid.

energy in the form of micro-sources and a storing mechanism that can also incorporate a flywheel, converter topology working in a dual AC-DC structure, and load and interconnection mechanism for DC and AC architecture [20]. Converter topology incorporates different types of inverters segregated as a current source inverter (CSI) or voltage source inverter (VSI). A CSI is defined as a current source inverter employed in the case of grid intrgration and is helpful in injecting current whenever there is a requirement for reactive power. A VSI is defined as a voltage source inverter that is used in grid integration and island mode and is helpful in controlling voltage profiles and improving small signal stability in terms of frequency and voltage limits. Because of their multi-level approach, VSIs are most commonly used in island mode without any hybrid configuration.

Converter topology is very essential as they provide reactive power compensation techniques and inverters are designed in such a way that they act as reactive power generators and provide virtual inertial through negative characteristics of voltage driven droop method in such a way that power flow is efficiently shared without increased power sharing of real and reactive power among them. It has been studied and shown by various studies that a connection of MGs with conventional grid is an important aspect as a mismatch of connection results in transients that can damage a system at large. A MG operating in island mode is a must when resorted with a conventional grid and must be supervised; a person or a control signal must be provided with a small value of bandwidth and through a regular communication channel. This kind of mechanism is required to maintain synchronism with a conventional grid and make safe transitions so that a MG, which doesn't carry large masses, should not be affected with large masses present in a conventional grid and collapsed. The other main objective is to maintain a constant power pool with two energy grids and prevent localized effect on grids while making a connection with the same frequency and intermittent voltage profile as a MG associated with DG sources and can operate in parallel with a conventional grid [21].

The primary control strategy deals with voltage and current loops, virtual impedance loop, and droop control strategy. A droop controller and virtual impedance loop act as reference values of voltage and current loops in a system. A droop controller adjusts the frequency and amplitude of voltage reference, whereas the virtual impedance loop and virtual voltage reference control the output impedance of VSI so that the output remains independent and accurate parameterization can be obtained.

5.4.4 SECONDARY CONTROL STRATEGY

Although the primary strategy is used to control output parameters of converter topology, this kind of strategy with a secondary approach is also interactive with DG units so that controlled parameterization can be formed effectively and improve the overall performance of a MG. The basic scheme is shown in Figure 5.5. It is of a different structure and placed between primary control and communication channel and collects data from each unit with an objective to quantify the values of parameter voltage, frequency, active power, and reactive power. It follows a step-by-step procedure and uses all data for measurement and validated with reference value and removes steady-state errors so that voltage and frequency deviation from their reference values can be seen and placed accurately and a mechanism can be made to achieve in different operation conditions of MGs. This approach of secondary control is a fast process and simultaneously reduces the deviation of voltage reference and frequency reference to a minimum value, even if a system is changing and a strategy is to evolve with a dynamic change in a system and calculate values of every parameter at every stage and make them accurate without making it so large as to create a disturbance or instability in a system [22].

The main function of a central controller is to make frequency and voltage deviation a minimum value so that the system remains stable and parameters defined

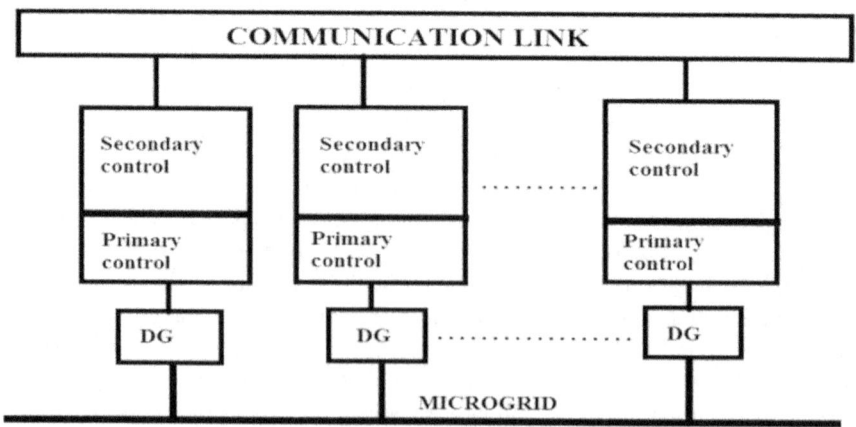

FIGURE 5.5 Network-controlled MG system.

for restoration should be in limits. Various network systems are being introduced with an objective to replace a central controller with an additional secondary control system so that the burden on the central system can be reduced with a security mechanism. This strategy has two major functions comprising equal sharing control of flow among different devices and making the energy pool constant with respect to deviations in frequency and voltage, which can be present with changes in load characteristics. This secondary strategy also gives flexibility to a secondary controller to work independently and have no relation with a primary or central controller in such a way that decision making is efficient without much use of resources and less time consuming and network efficiently with multiple units present in a MG. This strategy is used in larger systems of MGs with equal penetration of static and dynamic loads with a load factor nearly equal to unity and interconnecting strategy of both conventional and DGs [23].

5.4.5 TERTIARY CONTROL STRATEGY

Various strategies have been incorporated for evaluation of operation of MGs with a parameterization approach or making the system free from errors and improving sustainability indices. This layer of strategy deals with economic analysis and decision criterion totally dependent upon revenue and associated power flow in a system. Power flow is managed at localized spots or globally accepted points by the set points of the inverter and the tertiary control scheme allows the necessary changes to be made in the set points of the system comprises different units of MGs. Tertiary control evolves with converter topology with economic constraints [21].

5.5 NEED FOR MICROGRIDS

MGs are now seen as replacements of convention grids in terms of customer satisfaction, reliability, improved sustainable indices, economic benefits, and constant supply with efficient power flow in a system. Various benefits are also important by matching load and demand and determine the stability of a system and motivate different sections of customers to implement in a larger way of interest. Some of them are listed below:

- **Accelerates improvement:** It gives a flexibility of integrating different sources of energy available to customers with conventional grid and making it an intelligent decision-making grid which can improve structural and economic benefits accessed by a user.
- **Reliability:** MG-enhanced reliability indices of an existing system as customers have a privilege to add as many as sources connected to the grid or stand-alone system according to its capacity and requirement. In case of a sudden collapse of the central system, an additional system can save the time of a user and also provide security against blackouts and brownouts. Reactive power control has been provided by power electronic interfaces that overcome the problem of voltage fluctuations and MGs also give customers an additional advantage of storing excessive amounts of energy

generated at a localized point and can be sent back to the grid or, with the help of DERs, can be customized to their use also.

- **Helps consumers save money:** MGs also give a customer insight about their load requirements; with different types of sources connected, customers can choose an appropriate time of using a conventional grid at lower rates and also close down when energy is not required locally. This kind of operation saves customer time and money, which was wasted when operated with a conventional grid mechanism.
- **Generates revenue:** Customers can earn profit by implementing MG customized at their end and send surplus energy back to the grid and money can be adjusted in their electricity bills or in hand by which their capital cost on installation of a MG can be recovered.
- **Encourages economic growth:** A MG also creates job opportunities for people working in the power industry and different stakeholders are also investing money in creating grids and associated frameworks. Various developing and developed countries establish this kind of architecture and provide constructive mechanisms for the same.
- **Future-proof grid:** Various industrial setups and establishments require the highest load demand without interruption, which is quite difficult with a conventional grid. A MG also enhances the power capacity of a grid and can be locally connected in providing a constant supply without affecting productivity and matching the load and demand scenario in the peak season. MGs are real-time mechanisms without any delay in establishing a connection to deliver a sufficient amount of energy at any period of time.
- **Reduces carbon footprint:** MGs employ dispersed generation sources, which are renewable energy sources like solar, wind, and hydro, with properties of clean energy and environmentally friendly. Unlike conventional energy sources, they reduce carbon emissions and are also helpful in reducing carbon footprints by a significant amount and are easily to utilized with less payback period. Solar, wind, and hydro are available in plenty and can be integrated with a conventional grid at ease.

5.6   ISSUES RELATED TO MICROGRID INSTALLATIONS

Various issues pertain to utilizing technological features that are used at the front end and can be classified and mentioned as follows:

- **Power quality:** Maintaining PQ in terms of sustainability indices, certain parameters must be controlled in limits and must be reduced in order to have sinusoidal waveforms and uniform structure, whereas in the case of a MG, power electronic interfaces are being used in a form of various converters and produce harmonics along with compensation of reactive power. However, it has been seen that power electronics interface also inject sufficient amounts of current in the form of circulating current and create a disturbance in a system and must be shunted out to prevent a load from variations in voltages. This problem of power quality is more dominant in the distribution end and causes instability in a system.

- **Reactive power coordination:** DERs have acharacteristics to be connected near the load centers and share equally the amount of reactive power driven by sources and improve coordination among them but some of the micro-sources like wind energy have considerably significant distances from load centers and, due to huge masses present in a system, reactive power injects a large amount of current and causes instability.
- **Reliability and reserve margin:** Most of the DERs are dependent upon weather conditions and environmental conditions, due to this dependency; the value of reliable indices is not correct and requires advanced power electronic devices, which also results in providing distortion and instability in matching load and demand fractions. With the help of a power electronic interface, system properties have been compromised and issues of reserve margins exist.
- **Reliability and network redundancy:** DERs are incorporated in a system at distribution side; however, there are various types of networks available to implement but a radial system is preferred, which results in more power outages along with brownouts, as distributed energy sources cannot provide whole sufficient amounts of energy to load and mismatch can lead to failure of the system and network redundancy is a major issue of concern.
- **Safety:** MGs are more secure compared to conventional grids as it has been assumed while designing a MG that fault can be cleared in such a way that repair work can be carried out safely with an interconnected configuration of MG.

5.7 FAULTS AND THEIR CLASSIFICATION

Power is the most complex system across the globe and is used in different configurations involving many conventional and non-conventional sources and processed through generation, transmission, and distribution. Although designing a power system structure is followed by various steps, various faults also occur in a system due to various reasons like human intervention in the form of creating errors, natural disturbances like flashovers, and use of non-suitable materials that can cause insulation failures in a system. Various types of faults occur in a system and are classified as balanced faults and unbalanced faults. Unbalanced faults occur due to conductor connection or interconnection of conductors and termed series faults, whereas balanced faults are due to disconnecting different wires in terms of lines and grounds in a system and termed shunt faults and can be categorized as LL, LG, LLG, and LLL faults. In these notations used, L represents line and G represents ground terminal and are used in three-phase systems and lead to collapse of a system. Under this category of shunt faults, LG is frequent fault and LLL is the most severe fault and its severity also depends upon the location of fault that occurs in a system.

Various faults have been encountered and various unbalanced faults include opening of a conductor in a series of conductor configurations and involves symmetrical issues in one or two phases such that current distribution is non-uniform [24]. Various methods are used in identifying the location and type of faults but a single line diagram is used to solve LL and LLG faults comprising all phases in a system. Various asymmetrical faults have been a prominent effect in a system and symmetrical components are used for the calculation as it reduces the complexity of a system and also increases iteration values so that convergence can be easy and efficient. It has been seen that inductance of a line has been increased exponentially along with an AC component of current, but due to high inductance. Algebraic sum of AC and DC component after fault is the same as the AC component before the occurrence of fault and compromise with the high value of DC transient. It has been seen that in a three-phase system, all phases are not equally effected, except the three-phase fault in which a fault occurs in all phases but with different intensities. Balanced faults can be segregated as:

- **LL fault:** A short circuit occurs between two phases or lines because of the reason of ionization of air, or because of the physical contact between the two caused by the broken conductor. This kind of fault can be minimized by keeping a significant distance between two lines.
- **LG fault:** A short circuit between the phase or line and ground generally caused by the storm or some other natural reason. It is a very frequent fault that occurs normally with insulation failure at the ground level.
- **LLG fault:** Two lines or phases come in contact with the ground. This type of fault is an enlarged case of LL fault.

Various studies have confirmed that symmetrical and asymmetrical faults have different causes of occurrence and also their effects on a system can be categorized as linear and non-linear. Symmetrical faults are very rare in nature but all phases have an equal effect and their performance is equally rated; on the other hand, asymmetrical faults have different phases and all phases are affected in different proportions and easy to occur so that the system remains stable and can be restored effectively.

5.8 FACTS DEVICES

An electric power system is comprised of various elements connected together in different configurations and structures having non-linear characteristics; however, it behaves as a dominant system in certain cases and is considered a non-linear entity in which various security interfaces have been implemented to cater a need of reducing a gap between desired load requirements. A power system is considered stable but depends upon various types of disturbances in the form of faults encountered in such a way that their effect on a system with significant duration matters. It can be understood from the fact that various symmetric and unsymmetrical devices are connected and faults occur for a long duration and then interruption of a supply leads to blackouts, brownouts, or a sudden collapse of a system irrespective of the size of a system and class of disturbance. A power system comprises a generator, transmission lines, and loads in general, where loads are kept changing in a sustained environment and a system must withstand all conditions so that a parametric association should not be disturbed and attain adaptability. Various conditions can be ascertained in consideration with a load-changing environment and must efficiently deliver constant power to the load. It has been seen that due to various conditions like tripping a transmission line or non-synchronism of a generator leads to interruption of supply and inability to provide a constant amount of supply in emergency cases. Due to these problems, a power system deals in two topologies like stand-alone and grid-connected systems in order to fulfill the demand of localized loads and grid networks to feed the loads at far distance with ease [25].

A centralized power system has centralized structure of loads along with dynamic loads in which characteristics are not similar, although various systems in a generation system transmission and distribution incorporate different topologies. It has been seen that utilities are required to operate in conditions where occurrence of faults is frequent and a transmission system is adequate enough to support a system in a better way. However, the transmission structure is different and dependent on different power ratings and has inefficient indices due to various limitations like stability categorization, voltage profile, thermal resistance, non-uniform temperature, and stability at various loads and at different switching times. These limitations can also be classified into various categories as follows:

- Stability characteristics
- Voltage compensation
- Thermal sinks and temperature dependency

These limits are being constrained to areas of concern of additionally connected various equipment of different sizes and characteristics, keeping the system intact with an inherent mechanism and left no stone unturned to keep the power system in specific limits of stability with no change in transfer of power. In addition to this, it also governs and restricts a feature to connect new equipment and increase the capacity of existing transmission lines, but this problem has been overcome by the use of a new family of devices termed FACTS devices that can be employed to improve voltage and active power issues in a constrained infrastructure without any additional burden on a system with no relative interference. It has been seen that FACTS devices can be controlled and incorporated with a power electronic interface in which a triggering angle plays an important role and can smooth out the required deliverables with more degree of reliability. It also provides series and shunt compensation along with increasing static and dynamic stability of a system [26,27]. FACTS devices have been developed and validated in the late nineteenth century with an objective to increase voltage magnitude in consideration with stability indices and control the power flow from a voltage structured mechanism between two terminals of high terminal voltage end to lower end voltage end. These devices are quite costly when developed in the start but with relatively better efficiency and ergonomics, researchers developed different methodologies with different types of configurations keeping series, series parallel, parallel, dynamic, and static types to keep reactive power in limits [28,29]. The most popular are static VAR compensators (SVCs), thyristor controlled series compensators (TCSCs), static compensator (STATCOM), static synchronous series compensator (SSSC), unified power flow controllers (UPFCs), and interline power flow controller (IPFC) [30]. Various types of FACTS devices are briefly explained in the following sub-sections.

5.8.1 STATCOM

STATCOM is a static compensator that comprises invertors and a transformer with an ability of controlling dynamic stability. Its operation can be relatively compared with dynamic compensator as it is synchronous in nature and has a feature of generating reactive power in a sufficient way in the case of non-availability of moving parts. It has been seen that its usage is more dominant in renewable energy sources, especially solar power plants in which reactive power is required for active power to flow and controlled due to the absence of moving parts and output deliverable in constant power supply. Active and reactive power control can easily be addressed between two different voltage ends as it also gives a difference in critical angle stability and also stabilizes frequency variations in a system [31]. Reactive power can be controlled by changing the voltage profile or having a sufficient voltage difference but active power operation in a steady state is zero to keep inverter losses at a minimum and neglected in a operation where the condition is being maintained by other equipment of a device [32–34].

5.8.2 SVC

A SVC is a device that can control the voltage profile by improving at points in a network in which stability is being compromised due to insufficient reactive power. Its operation is static but connected in parallel and termed a shunt device. This kind of device can be connected at any point or at a point where a voltage dips by a significant amount, like the midpoint of a transmission network. A SVC controls the voltage profile by keeping reactive power in limits by either absorbing or generating it. The control of reactive power is driven by a power electronic interface in a form of anti-parallel SCR, which can be controlled with the help of a firing angle that can be determined between an interval of zero and waveform extinction. A coupling transformer is connected at a point of common coupling and improves the voltage at different buses, keeping the value of the voltage to its reference defined value in a network [34].

5.8.3 SSSC

FACTS devices are non-linear in nature due to the presence of power electronic interfaces such as a non-sinusoidal waveform may be injected into a system and leads to distortion and produces harmonics that induce instability in a system. A SSSC is developed with an objective to provide a sinusoidal waveform that has aligned a phase with a transmission network along with the ability to control the amplitude of a voltage profile. It has been seen that a SSSC behaves like a static voltage source in which injective voltage by generating reactive power into a system improves inductive and capacitive reactance in such a way that a line current acts in a quadrature with a voltage phasor. It is evident that converter operation controls the amplitude of voltage by adding inductive and capacitive effect in series with a transmission network in which a device is connected [35].

5.8.4 UPFC

A UPFC is a type of FACTS device that has a distinct feature of shunt, series compensation along with a shifting mechanism. This device is termed a unified power flow controller and comprises a combination of a static compensator that is connected in series and a shunt device that is used to regulate voltage at terminals due to insufficient reactive power. It provides two different kinds of mechanisms with series and parallel structures and is connected with a DC capacitor that actively controls the flow of active power into a transmission network. Every FACTS device is connected in a transmission network with an ability to control losses at a distribution side of network. It has been used extensively for shifting series voltage and controlling shunt voltage and can be classified into a category where dynamic stability is being controlled along with the dynamic voltage profile. It is quite expensive and used as a protection strategy in which the control mechanism is not proper and insufficient to keep voltage disturbances in defined limits. It has an ability to provide reactive power compensation by absorbing or injecting in a shunt mode and gives flexibility to a user to define constrained limits. It provides accurate amplitude and phase of voltage to terminals of networks to enhance the stability of a system [36].

5.8.5 IPFC

An IPFC has been developed as an extension of an SSSC with the objective of controlling the flow of reactive power in different transmission lines comprising the same or different configurations. An A IPFC is an interline power flow controller that maintains dynamic control of power flow across multiple transmission lines and is a powerful tool for reducing voltage fluctuations.It has been seen that transmission lines are also classified as overloaded and lightly loaded, and many effects have been analysed that lead to an inefficiency in a system. This type of FACTS device regulates reactive power by absorbing or generating it, keeping the transmission line under a normal load with no overheating and improved stability indices. It is not only used as a reactive power compensator but also improves voltage stabilisation with dynamic stability characteristics [37].

5.9 CONCLUSION

A MG is a concept introduced in the energy sector with an objective to improve reliability and power quality into a system with more compact and stability indices. It is also helpful in reducing the gap with demand and requisite load for domestic and industrial customers. This concept of stand-alone and grid-connected MGs has been accepted globally and many countries have started developing different models to improve the performance of hybrid systems, reducing carbon footprints, and catering to a need for localized loads. This chapter has presented an introduction of all the components of MGs. It deals with the introduction, basic operation, various control techniques, and protection schemes of its various components. Different control schemes utilized are explained and the study of different types of faults is presented. For mitigating various PQ issues, various possible controlling schemes using FACTS devices have been elaborated on with different configurations and structured formations.

REFERENCES

[1] Khan, B. and Singh P., "Optimal power flow techniques under characterization of conventional and renewable energy sources: a comprehensive analysis," *Journal of Engineering*, Vol. 2017, Article ID 9539506, 16 pages, 2017.

[2] Singh, P. and Khan B., "Smart microgrid energy management using a novel artificial shark optimization," *Complexity*, Vol. 2017, Article ID 2158926, 22 pages, 2017.

[3] Molla, T., Khan B., Moges B., Alhelou H. H., Zamani R. and Siano P., "Integrated optimization of smart home appliances with cost-effective energy management system," in *CSEE Journal of Power and Energy Systems*, Vol. 5, No. 2, pp. 249–258, June 2019.

[4] Tang, Z., Lin Y., Vosoogh M., Parsa N., Baziar A. and Khan B., "Securing microgrid optimal energy management using deep generative model," in *IEEE Access*, Vol. 9, pp. 63377–63387, 2021.

[5] Bihari, S. P. et al., "A comprehensive review of microgrid control mechanism and impact assessment for hybrid renewable energy integration," in *IEEE Access*, Vol. 9, pp. 88942–88958, 2021, doi: 10.1109/ACCESS.2021.3090266.

[6] Khan, B. and Singh P., "Selecting a meta-heuristic technique for smart micro-grid optimization problem: a comprehensive analysis," in *IEEE Access*, Vol. 5, no., pp. 13951–13977, 2017.

[7] Mahela, O. P. et al., "Comprehensive overview of multi-agent systems for controlling smart grids," in *CSEE Journal of Power and Energy Systems*, doi: 10.17775/CSEEJPES.2020.03390.

[8] Wang, Z., Zhang B., Mobtahej M., Baziar A. and Khan B., "Advanced reactive power compensation of wind power plant using PMU data," in *IEEE Access*, Vol. 9, pp. 67006–67014, 2021.

[9] Padmanaban, S., Dhanamjayulu C. and Khan B., "Artificial Neural Network and Newton Raphson (ANN-NR) algorithm based selective harmonic elimination in cascaded multilevel inverter for PV applications," in *IEEE Access*, Vol. 9, pp. 75058–75070, 2021.

[10] Fakhar, M. S. et al., "Implementation of APSO and improved APSO on non-cascaded and cascaded short term hydrothermal scheduling," in *IEEE Access*, Vol. 9, pp. 77784–77797, 2021

[11] Lasseter, R. H., "Microgrid" in *IEEE Power Engineering Society Winter Meeting*, pp. 305–308, 2002.

[12] Guerrero, J. M., Blaabjerg F., Mohamed Y. A. R. I. and Salama M. M., "Introduction to the special section on distributed generation and micro grids," *IEEE Transactions on Industrial Electronics*, Vol. 60, No. 4, pp. 1251–1253, April 2013.

[13] Guerrero, J. M., Blaabjerg F., Zhelev T., Hemmes K., Monmasson E., Jemei S. and Frau J. I., "Distributed generation: toward a new energy paradigm," *IEEE Industrial Electronics Magazine*, Vol. 4, No. 1, pp. 52–64, March 2010.

[14] "Integration of Distributed Energy Resources", The CERTS Microgrid Concept, California Energy Commission (Consultant Report), October 2003.

[15] NREL Distributed Grid Integration. [Online]. Available: www.nrel.gov/electricity/distribution/microgrids.html. Aug. 2012.

[16] Kroposki, B., Lasseter R., Ise T., Morozumi S., Papatlianassiou S. and Hatziargyriou N., "Making micro grids work," *IEEE Power and Energy Magazine*, Vol. 6, No. 3, pp. 40–53. May/Jun. 2008.

[17] Katiraei, F., Iravani R., Hatziargyriou N. and Dimeas A., "Microgrid management," *IEEE Power and Energy Magazine*, Vol. 6, No. 3, pp. 54–65, May/Jun. 2008.

[18] Lasseter, R. H., "Smart distribution: coupled microgrid," *In: Proceedings of the IEEE*, Vol. 99, No. 6, June 2011.

[19] Vandoorn, T. L., Guerrero J. M., De Kooning D. M., Vasquez J. and Vandevelde L., "Micro grids: hierarchical control and an overview of the control and reserve management strategies," *IEEE Industrial Electronics Magazine*, Vol. 7, No. 4, pp. 42–55, Dec. 2013.

[20] Shafiee, Q., Guerrero J. M. and Vasquez J. C., "Distributed secondary control for islanded micro grids- A novel approach," *IEEE Transactionson Power Electronics*, Vol. 29, No. 2, pp. 1018–1031, Feb. 2014.

[21] Guerrero, J. M., Vasquez J. C., Matas J., Castilla M. and De L. G. Vicuňa, "Control strategy for flexible Microgrid based on parallel line-interactive UPS systems," *IEEE Transactions on Industrial Electronics*, vol. 56, No. 3, pp. 726–736, March 2009.

[22] Guerrero, J. M., Chandorkar M., Lee T. L. and Loh P. C., "Advanced control architectures for intelligent micro grids, part I: decentralized and hierarchical control," *IEEE Transactions on Industrial Electronics*, Vol. 60, No. 4, pp. 1254–1262, April 2013.

[23] Shafiee, Q., Vasquez J. C. and Guerrero J. M., "Distributed secondary control for islanded micro grids-A novel approach," *IEEE Transactions on Power Electronics*, Vol. 29, No. 2, pp. 5637–5642, Feb. 2014.

[24] Wadhwa, C. L., *Electrical Power Systems*. New Age Int. ltd., 2010.

[25] Kundur, P., *Power System Stability and Control*. McGraw-Hill, 1994.

[26] Song, Y. H. and Johns A., *Flexible AC Transmission Systems (FACTS)*, no. 30, IET, London, 1999.

[27] Hingorani, N. G. and Gyugyi, L., *Understanding Flexible AC Transmission System: Concept and Technology of FACTS*. IEEE Press, 1999.

[28] Hingorani, N. G., "Flexible AC transmission," *IEEE Spectrum*, Vol. 30, No. 4, pp. 40–45, 1993.

[29] Mathur, R. Mohan and Varma Rajiv K., *Thyristor-Based FACTS Controllers for Electrical Transmission Systems*. Wiley, 2011.

[30] Georgilakis, P. S. and Peter G., "Flexible AC transmission controllers: an evaluation," *Material Science Forum*, Vol. 670, pp. 399–406, Dec. 2010.

[31] Rai, A., "Enhancement of voltage stability and reactive power control of multimachine power system using FACTS devices," *International Journal of Engineeringand Innovative Technology*, Vol. 3, No. 1, pp. 123–127, July 2013.

[32] Wang, H. F., "Interactions and multivariable design of STATCOM AC and DC voltage control," *InternationalJournal of Electrical Power and Energy Systems*, Vol. 25, No. 5, pp. 387–394, Jun. 2003.

[33] Canizares, C. A., Pozzi M., Corsi S. and Uzunovic E., "STATCOM modeling for voltage and angle stability studies," *InternationalJournal of Electrical Power and Energy Systems*, Vol. 25, No. 6, pp. 431–441, 2003.

[34] Murali, D. and Raja Ram M., "Active and reactive power flow control using FACTS devices," *International Journal of Computer Applications*, Vol. 9, No. 8, pp. 45–50, Nov. 2010.

[35] Rao, G. S. and Bodha V. R., "Improvement of transient stability in power systems with neuro fuzzy UPFC," *American Journal of Engineering Research*, Vol. 2, No. 11, pp. 48–60, 2013.

[36] Mishra, I. P. and Kumar S., " Control of active and reactive power flow in multiple lines through interline power flow controller," *InternationalJournal of Emerging Technology and AdvancedEngineering*, Vol. 2, No. 11, pp. 86–93, Nov. 2012.

[37] Musirin, I., Diamah N., Radzi M., Murtadhaothman M., Idris M. K. and Rahman T. K. A., "Voltage profile improvement using UPFC via artificial immune system," *WSEAS Transactions on Power Systems*, Vol. 3, No. 4, pp. 194–204, April 2008.

6 Control Techniques for the Operation and Power Management of Smart DC Microgrids

Mahesh Kumar
Department of Electrical & Electronics Engineering, Amity
School of Engineering & Technology, Amity University
Haryana, Gurugram, India

CONTENTS

6.1 INTRODUCTION

Nowadays, green energy sources such as wind, solar, fuel cells, geothermal, biomass, and tides are being used for power generation. Due to having several issues in the integration of green energy sources to the main grid, the microgrid (AC or DC) is the only promising option. The microgrid is being widely used to integrate the different distributed generators (DGs), energy storage systems (ESSs), and loads, using the pulse width modulation (PWM)–based voltage source converters (VSCs) [1–3]. Thus, the microgrid is defined as a small-scale power supply network, consisting of a cluster of DGs, loads, and ESSs, to provide the power to a small community. A microgrid can be operated in islanded and/or grid-connected modes.

DOI: 10.1201/b22884-6

127

A DC microgrid (DCMG) has several benefits over the AC microgrid, which are high current carrying, no need of controlling frequency, reactive power, and phase angle, reduced corona loss i.e., higher efficiency, and no need of frequency synchronization for interconnections of multiple DGs [1], [3], [4]. Therefore, the smart DCMG is a promising option for facilitating the connections of several DGs, ESSs, and loads [5–14]. Several DCMG architectures have been presented in the literature [4,15–26]. A ring-type DCMG with decentralized control, based on DC power pool, is presented and discussed in this chapter. By using the decentralized control, the reliability can be maintained in the DCMG. The decentralized control also eliminates the circulating currents among the DGs in the proposed DCMG.

A control algorithm for energy management and coordinated control of the DGs has been proposed for different operating conditions, including the fault scenario, in this chapter. The PWM-based single-phase voltage source inverters (VSIs) have been widely used in adjustable speed drives, uninterruptible power supplies, marine and military applications, power factor controllers, and utility interfaces with the non-conventional sources, such as solar photovoltaic (SPV), fuel cells (FCs), etc. The single-phase VSIs have been widely used to integrate the single-phase loads to the smart DCMG as in [1], [27–32]. With the DCMG under islanded mode, it is necessary to maintain the sinusoidal rated voltage across all the single-phase AC loads. Several control techniques of the one-phase full-bridge VSI in stand-alone mode are proposed in the literature as hysteresis band control, dead-beat control, proportional-resonant (PR) control, predictive control, PI control, and control techniques based on d-q synchronous reference frame (SRF), for maintaining rated sinusoidal load voltage. As mentioned in [29,31,33–44], all these existing techniques have the drawbacks such as hysteresis band control requires variable switching frequency for variable loads and high switching frequency for fast dynamic response, dead-beat control requires the load parameters because of more sensitivity to the variations of the load parameters, PR control suffers from the frequency variations and instability margins, predictive control provides poor transient performance and more complex to implement and requires high sampling rate for good performance, PI control provides slow transient response, and d-q–based control techniques provide the DC component of the reactive power and poor performance. The control scheme [28], with the voltage feedback, cannot maintain the rated sinusoidal voltage for non-linear loads or under the DCMG voltage fluctuations.

To overcome problems associated with the previous methods, this chapter has also developed a d-q SRF-based control technique of the one-phase full-bridge VSI, using the feedforward and feedback controls. The developed technique has been aimed to maintain the rated sinusoidal voltage with low total harmonic distortion (THD), and needed power to the single-phase AC loads under different conditions. The performance analyses of the proposed control techniques along with smart DCMG have been carried out for different conditions in this chapter.

6.1.1 ARCHITECTURE OF DC MICROGRIDS

The architecture of the smart DCMG with decentralized control is shown in Figure 6.1. Several green energy DGs, such as wind turbine (WT), micro-turbine

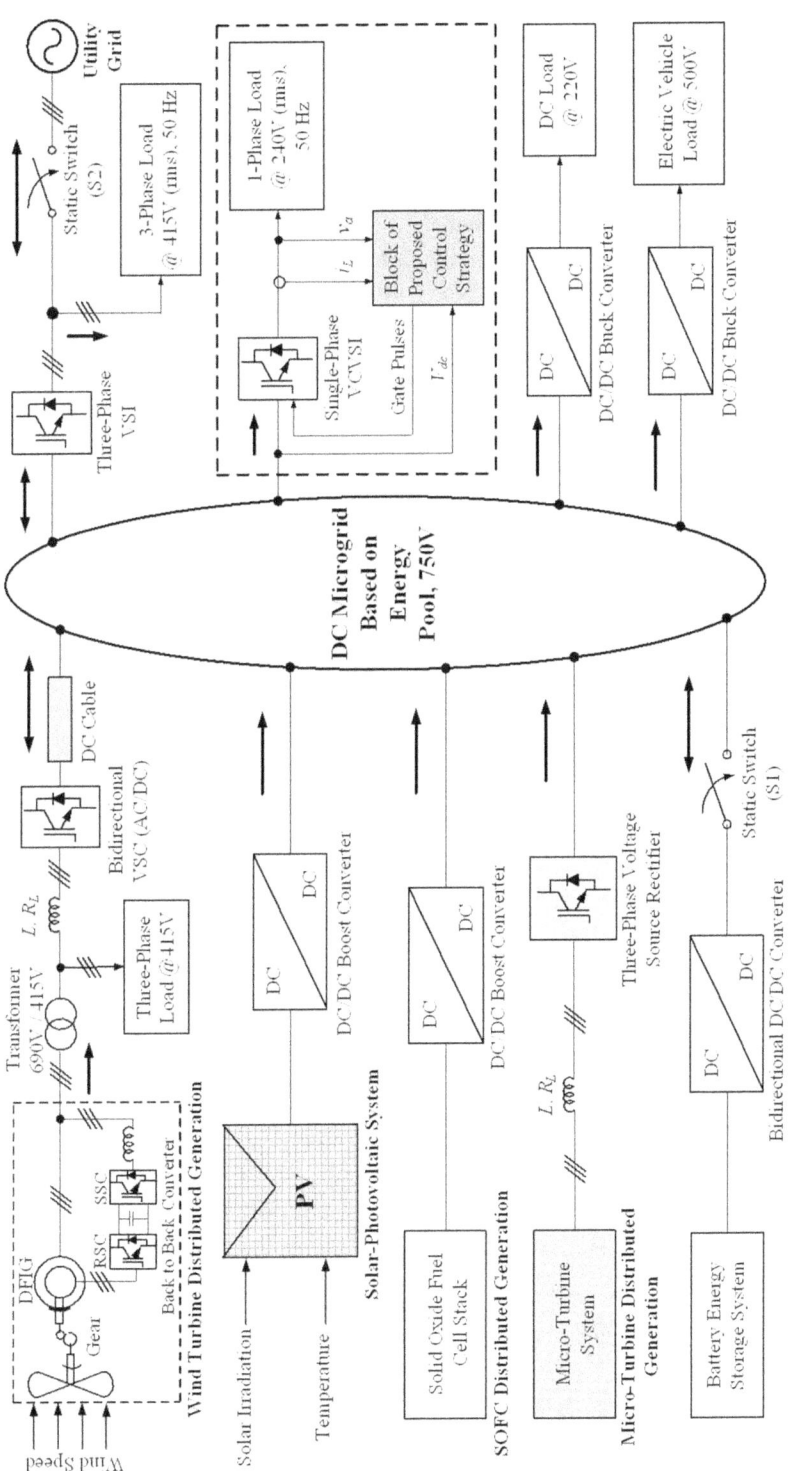

FIGURE 6.1 Architecture of smart DCMG for integrating the several DGs.

(MT), solid oxide fuel cells (SOFCs), solar photovoltaic (SPV), battery energy storage system (BESS), and various loads, have been integrated to the DCMG, and their interconnection to the main grid, as shown in Figure 6.1.

The proposed DCMG provides the reliable and high-quality power supply. This DCMG will supply the power to all loads (both DC and AC), simultaneously. The autonomous coordinated control or decentralized control can be obtained by considering the DCMG voltage as the reference signal to all DGs. In the decentralized method, no communication link is used with each other. In this DCMG, no transformer is required across three-phase VSI output. The operating DC voltage of the microgrid is considered 750 V, in this work.

A bidirectional three-phase VSC is used to connect the wind turbine with the doubly fed induction generator (DFIG) to the DCMG. The power reduction of the WT DG can be achieved either by controlling the DFIG back-to-back VSC, or by changing pitch angle. The SPV and SOFC DGs are connected to the DCMG through the DC-DC boost converters. The SOFC output changes with the variation of load. A voltage source rectifier has been used to connect the MT DG. In islanded mode, a BESS is used to balance the power mismatch in the DCMG. However, in grid-connected mode, the power mismatch is managed by the utility grid.

6.1.2 POWER CONTROL ALGORITHM FOR SMART DCMG

A control algorithm for coordinated operation of the DGs has been proposed for balancing the power mismatch under different operating conditions. The controllable DGs have been used when the WT and SPV DGs are unable to fulfill the load demand. A flow chart of the proposed control algorithm [1] has been mentioned in Figure 6.2. The following steps have been used in the power control algorithm, where P_{batt_avail} is power available in the BESS; P_{BESS} is output power of the BESS; P_{batt_max} is the maximum storage capacity of the BESS; P_{batt_min} is the minimum power in the BESS; $P_{G,MT}$ is the output power of the micro-turbine; $v_{wind,cut-in}$ and $v_{wind,cut-off}$ are the cut-in and cut-off wind speed, respectively; and v_{wind} is the available wind speed.

Step 1. Check the output power of the wind turbine and SPV DGs ($P_{G,WT}$ and $P_{G,SPV}$, respectively) with the intermittent and uncertainty nature of the renewable sources:

Power generation by WTG

$$= \begin{cases} P_{G,WT} & ; \text{when } v_{wind,cut-in} < v_{wind} \\ & \qquad\qquad\qquad \leq v_{wind,cut-off} \\ 0 & ; \text{otherwise} \end{cases}$$

Power generation by SPV

$$= \begin{cases} 0 & ; \text{when cloudy or night} \\ P_{G,SPV} & ; \text{otherwise} \end{cases}$$

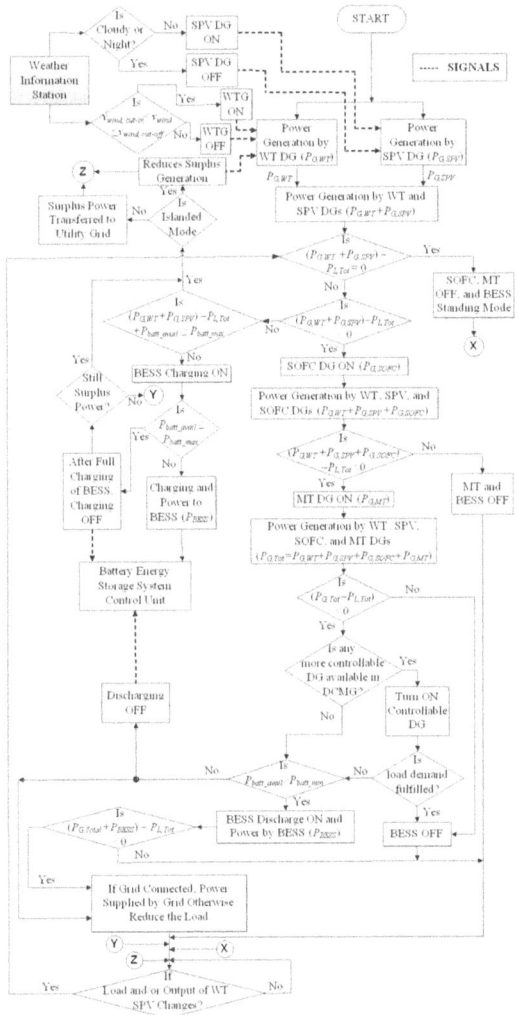

FIGURE 6.2   Flow chart of the proposed power control technique for power management.

Step 2. Check the output power of the SPV and WT DGs, and total load ($P_{L,Tot}$). In this work, it is assumed that the power loss in the DCMG is negligible, and the generations as well as the load measurements are available:

$$If\ (P_{G,WT} + P_{G,SPV}) - P_{L,Tot}$$

$$\left\{ \begin{array}{ll} = 0 & ;\ \text{SOFC, MT, BESS OFF, and go to step 10} \\ < 0 & ;\ \text{then SOFC ON and go to step 6} \\ > 0 & ;\ \text{then go to step 3} \end{array} \right\}$$

Step 3. Check the power available in the battery as:

$$\left(P_{G,WT} + P_{G,SPV} + P_{batt\_avail} \right) - P_{L,Tot} \geq P_{batt\_max}$$

$$\text{If} \begin{cases} \text{No} & ; \text{then BESS charging ON and go to step 4} \\ \text{Yes} & ; \text{then go to step 5} \end{cases}$$

Step 4. Check the charging status of battery as:

$$\text{If } P_{batt\_avail} \geq P_{batt\_max}$$

$$\begin{cases} \text{Yes} & ; \text{then BESS charging OFF and go to step 5} \\ \text{No} & ; \text{then charging ON} \end{cases}$$

Step 5. If the BESS is fully charged and there is still surplus power generation, then check the operating mode of the microgrid:

If islanded mode

$$\begin{cases} \text{Yes} & ; \text{Reduce WT generation, and go to step 10} \\ \text{No} & ; \text{Surplus power transferred to the grid and go to step 10} \end{cases}$$

Step 6. Check the difference between the output power of the WT, SPV, and SOFC ($P_{G,SOFC}$) DGs, and the total load:

$$\text{If } (P_{G,WT} + P_{G,SPV} + P_{G,SOFC}) - P_{L,Tot}$$

$$\begin{cases} = 0 & ; \text{then MT, BESS OFF, and go to step 10} \\ < 0 & ; \text{then MT ON and go to step 7} \end{cases}$$

Step 7. Check the difference between the total generations of all DGs ($P_{G,Tot}$) and the total load:

$$\text{If } (P_{G,Tot} - P_{L,Tot}) \begin{cases} = 0 & ; \text{then BESS OFF and go to step 10} \\ < 0 & ; \text{then BESS discharges and go to step 8} \end{cases}$$

where, $P_{G,Tot} = P_{G,WT} + P_{G,SPV} + P_{G,SOFC} + P_{G,MT}$

Step 8. Check the discharging status of the battery as:

$$\text{If } P_{batt\_avail} < P_{batt\_min}$$

$$\begin{cases} \text{Yes} & ; \text{BESS discharge OFF, power from the grid, and go to step 10} \\ \text{No} & ; \text{then discharging ON and go to step 9} \end{cases}$$

Step 9. Check the power difference between the total power and the total load:

If $(P_{G,Tot} + P_{batt\_avail} - P_{L,Tot})$

$$\begin{cases} < 0 & ; \text{then power taken from the grid and go to step 10} \\ = 0 & \qquad\qquad ; \text{then go to step 10} \end{cases}$$

Step 10. If the generation and/or the load changes then go to step 1, and repeat the steps (1–9).

6.1.3 DYNAMIC MODELING OF SINGLE-PHASE VSI WITH LC FILTER

Figure 6.3(a) shows a one-phase full-bridge VSI with a second-order LC filter, which is used to integrate the one-phase AC loads to the DC microgrid [1].

The voltage and current equations of the one-phase VSI are expressed by Equation (6.1), and its state space model is given by Equation (6.2):

$$v_{inv} = R_{L_f} i_L + L_f \frac{di_L}{dt} + v_a \quad \text{and} \quad 0 = i_L - i_o - C_f \frac{dv_a}{dt} \right\} \qquad (6.1)$$

$$\frac{d}{dt}\begin{bmatrix} i_L \\ v_a \end{bmatrix} = \begin{bmatrix} -R_{L_f}/L_f & -1/L_f \\ 1/C_f & 0 \end{bmatrix}\begin{bmatrix} i_L \\ v_a \end{bmatrix} + \begin{bmatrix} 1/L_f \\ 0 \end{bmatrix} v_{inv} + \begin{bmatrix} 0 \\ -1/C_f \end{bmatrix} i_o \qquad (6.2)$$

where v_{inv} is the inverter output voltage, v_a is load voltage, i_L is the filter inductor (L_f) current, i_o is load current, R_{L_f} is the resistance of filter inductor, C_f is the filter

(a)

(b)

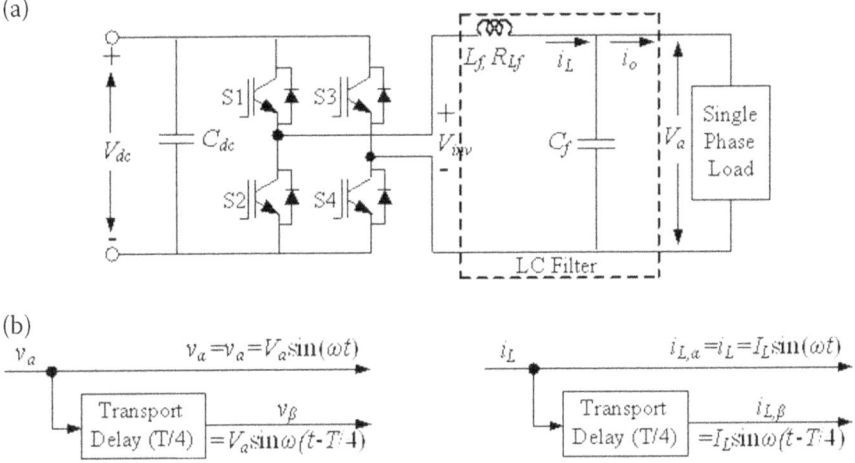

FIGURE 6.3 (a) A topology of one-phase full-bridge VSI. (b) Orthogonal vector transformations.

capacitance, $V_{inv}(=\delta V_{dc})$ is the inverter output voltage, and δ (=1, −1, or 0) is the control variables that depend on the switch states of one-phase VSI.

The one-phase VSI output voltage and current have been transformed into α-β stationary reference frame, as shown in Figure 6.3(b).

The voltage and current equations of the one-phase VSI, into two-phase (α-β) stationary reference frame as well as into rotating d-q SRF, are derived by Equations (6.3) and (6.4), respectively.

$$
\left.
\begin{aligned}
v_{inv,\alpha} &= R_{L_f} i_{L,\alpha} + L_f \frac{di_{L,\alpha}}{dt} + v_{a,\alpha}, \quad v_{inv,\beta} = R_{L_f} i_{L,\beta} + L_f \frac{di_{L,\beta}}{dt} + v_{a,\beta} \\
0 &= i_{L,\alpha} - i_{o,\alpha} - C_f \frac{dv_{a,\alpha}}{dt}, \qquad\qquad 0 = i_{L,\beta} - i_{o,\beta} - C_f \frac{dv_{a,\beta}}{dt}
\end{aligned}
\right\}
\tag{6.3}
$$

$$
\left.
\begin{aligned}
v_{inv,d} &= R_{L_f} i_{L,d} + L_f \frac{di_{L,d}}{dt} - \omega L_f i_{L,q} + v_{a,d} \\
v_{inv,q} &= R_{L_f} i_{L,q} + L_f \frac{di_{L,q}}{dt} + \omega L_f i_{L,d} + v_{a,q} \\
0 &= i_{L,d} - C_f \frac{dv_{a,d}}{dt} - i_{o,d} - \omega C_f v_{a,q}, \quad \text{and } 0 = i_{L,q} - C_f \frac{dv_{a,q}}{dt} - i_{o,q} + \omega C_f v_{a,d}
\end{aligned}
\right\}
\tag{6.4}
$$

where $v_{inv,\alpha}$ and $v_{inv,\beta}$ are α-axis and β-axis inverter voltage components; $v_{a,\alpha}$ and $v_{a,\beta}$ are α-axis and β-axis load voltage components; $i_{L,\alpha}$ and $i_{L,\beta}$ are α-axis and β-axis inductor current components; $i_{o,\alpha}$ and $i_{o,\beta}$ are α-axis and β-axis load current components; $v_{inv,d}$ and $v_{inv,q}$ are d-axis and q-axis VSI voltage components; $v_{a,d}$ and $v_{a,q}$ are d-axis and q-axis load voltage components; $i_{L,d}$ and $i_{L,q}$ are d-axis and q-axis inductor current components; $i_{o,d}$ and $i_{o,q}$ are d-axis and q-axis load current components; i_o (=V_a/Z); and Z are load impedance.

The state space model in d-axis and q-axis components has been expressed as:

$$
\left.
\begin{aligned}
\frac{d}{dt}\begin{bmatrix} i_{L,d} \\ v_{a,d} \end{bmatrix} &= \begin{bmatrix} -\frac{R_{L_f}}{L_f} & -\frac{1}{L_f} \\ 1/C_f & 0 \end{bmatrix} \begin{bmatrix} i_{L,d} \\ v_{a,d} \end{bmatrix} + \begin{bmatrix} \frac{1}{L_f} \\ 0 \end{bmatrix} v_{inv,d} + \begin{bmatrix} \omega \\ 0 \end{bmatrix} i_{L,q} + \begin{bmatrix} 0 \\ -\frac{1}{C_f} \end{bmatrix} i_{o,d} + \begin{bmatrix} 0 \\ -\omega \end{bmatrix} v_{a,q} \\
\frac{d}{dt}\begin{bmatrix} i_{L,q} \\ v_{a,q} \end{bmatrix} &= \begin{bmatrix} -\frac{R_{L_f}}{L_f} & -\frac{1}{L_f} \\ 1/C_f & 0 \end{bmatrix} \begin{bmatrix} i_{L,q} \\ v_{a,q} \end{bmatrix} + \begin{bmatrix} \frac{1}{L_f} \\ 0 \end{bmatrix} v_{inv,q} + \begin{bmatrix} -\omega \\ 0 \end{bmatrix} i_{L,d} + \begin{bmatrix} 0 \\ -\frac{1}{C_f} \end{bmatrix} i_{o,q} + \begin{bmatrix} 0 \\ \omega \end{bmatrix} v_{a,d}
\end{aligned}
\right\}
\tag{6.5}
$$

Consider small perturbations in the various quantities as $i_{o,dq} = I_{o,dq} + \tilde{i}_{o,dq}$, $v_{a,dq} = V_{a,dq} + \tilde{v}_{a,dq}$, and $v_{inv,dq} = V_{inv,dq} + \tilde{v}_{inv,dq}$. The quantities in small letters, $i_{L,dq}$, i_{odq}, $v_{a,dq}$, and $v_{inv,dq}$, are instantaneous values of the respective quantities. The quantities in capital letters, $I_{L,dq}$, I_{odq}, $V_{a,dq}$, and $V_{inv,dq}$, are average values. The quantities with tilde, $\tilde{i}_{L,dq}$, $\tilde{i}_{o,dq}$, $\tilde{v}_{a,dq}$, and $\tilde{v}_{inv,dq}$, are the perturbations in the various quantities.

By putting all these values in Equation (6.5) and solving these equations, then we get small signal state space models as follows:

$$\frac{d}{dt}\begin{bmatrix} \tilde{i}_{L,d} \\ \tilde{v}_{a,d} \end{bmatrix} = \begin{bmatrix} -\frac{R_{Lf}}{L_f} & -\frac{1}{L_f} \\ 1/C_f & 0 \end{bmatrix}\begin{bmatrix} \tilde{i}_{L,d} \\ \tilde{v}_{a,d} \end{bmatrix} + \begin{bmatrix} \frac{1}{L_f} \\ 0 \end{bmatrix}\tilde{v}_{inv,d} + \begin{bmatrix} \omega \\ 0 \end{bmatrix}\tilde{i}_{L,q} + \begin{bmatrix} 0 \\ -\frac{1}{C_f} \end{bmatrix}\tilde{i}_{o,d} + \begin{bmatrix} 0 \\ -\omega \end{bmatrix}\tilde{v}_{a,q}$$

$$\text{and,} \quad \text{output}(\mathbf{Y_1}) = \begin{bmatrix} \tilde{i}_{L,d} \\ \tilde{v}_{a,d} \end{bmatrix} = \begin{bmatrix} 1 & 0 \\ 0 & 1 \end{bmatrix}\begin{bmatrix} \tilde{i}_{L,d} \\ \tilde{v}_{a,d} \end{bmatrix}$$

$$(6.6)$$

$$\frac{d}{dt}\begin{bmatrix} \tilde{i}_{L,q} \\ \tilde{v}_{a,q} \end{bmatrix} = \begin{bmatrix} -\frac{R_{Lf}}{L_f} & -\frac{1}{L_f} \\ 1/C_f & 0 \end{bmatrix}\begin{bmatrix} \tilde{i}_{L,q} \\ \tilde{v}_{a,q} \end{bmatrix} + \begin{bmatrix} \frac{1}{L_f} \\ 0 \end{bmatrix}\tilde{v}_{inv,q} + \begin{bmatrix} -\omega \\ 0 \end{bmatrix}\tilde{i}_{L,d} + \begin{bmatrix} 0 \\ -\frac{1}{C_f} \end{bmatrix}\tilde{i}_{o,q} + \begin{bmatrix} 0 \\ \omega \end{bmatrix}\tilde{v}_{a,d}$$

$$\text{and,} \quad \text{output}(\mathbf{Y_2}) = \begin{bmatrix} \tilde{i}_{L,q} \\ \tilde{v}_{a,q} \end{bmatrix} = \begin{bmatrix} 1 & 0 \\ 0 & 1 \end{bmatrix}\begin{bmatrix} \tilde{i}_{L,q} \\ \tilde{v}_{a,q} \end{bmatrix}$$

$$(6.7)$$

6.1.4 DETERMINATION OF TRANSFER FUNCTIONS OF SINGLE-PHASE VSI

The standard state space equations and their transfer function have been expressed as [1]:

$$\dot{\mathbf{X}} = \mathbf{A}\mathbf{X} + \mathbf{B_1}U_1 + \mathbf{B_2}U_2 + \mathbf{B_3}U_3 + \mathbf{B_4}U_4, \quad \text{and} \quad \text{output,} \ \mathbf{Y} = \mathbf{C}\mathbf{X} \quad (6.8)$$

$$\mathbf{Y}(s) = \mathbf{C}[s\mathbf{I} - \mathbf{A}]^{-1}[\mathbf{B_1}U_1(s) + \mathbf{B_2}U_2(s) + \mathbf{B_3}U_3(s) + \mathbf{B_4}U_4(s)] \quad (6.9)$$

Comparing Equations (6.6) and (6.7) with Equation (6.8), one gets d-axis and q-axis components, and are expressed by Equations (6.10) and (6.11), respectively:

$$\left.\begin{aligned} \mathbf{X} &= \begin{bmatrix} \tilde{i}_{L,d} \\ \tilde{v}_{a,d} \end{bmatrix}, \quad \mathbf{A} = \begin{bmatrix} -\frac{R_{Lf}}{L_f} & -\frac{1}{L_f} \\ 1/C_f & 0 \end{bmatrix}, \quad \mathbf{C} = \begin{bmatrix} 1 & 0 \\ 0 & 1 \end{bmatrix} \\[2mm] \mathbf{B_1} &= \begin{bmatrix} \frac{1}{L_f} \\ 0 \end{bmatrix}, \quad \mathbf{B_2} = \begin{bmatrix} \omega \\ 0 \end{bmatrix}, \quad \mathbf{B_3} = \begin{bmatrix} 0 \\ -\frac{1}{C_f} \end{bmatrix}, \quad \mathbf{B_4} = \begin{bmatrix} 0 \\ -\omega \end{bmatrix} \\[2mm] U_1 &= \tilde{v}_{inv,d}, \quad U_2 = \tilde{i}_{L,q}, \quad U_3 = \tilde{i}_{o,d}, \quad U_4 = \tilde{v}_{a,q} \end{aligned}\right\} \quad (6.10)$$

$$\left.\begin{aligned} \mathbf{X} &= \begin{bmatrix} \tilde{i}_{L,q} \\ \tilde{v}_{a,q} \end{bmatrix}, \quad \mathbf{A} = \begin{bmatrix} -R_{Lf}/L_f & -1/L_f \\ 1/C_f & 0 \end{bmatrix}, \quad \mathbf{C} = \begin{bmatrix} 1 & 0 \\ 0 & 1 \end{bmatrix} \\[2mm] \mathbf{B_1} &= \begin{bmatrix} 1/L_f \\ 0 \end{bmatrix}, \quad \mathbf{B_2} = \begin{bmatrix} -\omega \\ 0 \end{bmatrix}, \quad \mathbf{B_3} = \begin{bmatrix} 0 \\ -1/C_f \end{bmatrix}, \quad \mathbf{B_4} = \begin{bmatrix} 0 \\ \omega \end{bmatrix} \\[2mm] U_1 &= \tilde{v}_{inv,q}, \quad U_2 = \tilde{i}_{L,d}, \quad U_3 = \tilde{i}_{o,q}, \quad U_4 = \tilde{v}_{a,d} \end{aligned}\right\} \quad (6.11)$$

The controllable transfer functions of d-axis and q-axis components are derived from Equations (6.6), (6.9), (6.10), and (6.7), (6.9), and (6.11), respectively, as:

$$
\left.
\begin{aligned}
G_{C,d}(s) &= \frac{Y_1(s)}{U_1(s)} = \frac{Y_1(s)}{\tilde{v}_{inv,d}(s)}\bigg|_{\tilde{i}_{L,q}=0,\tilde{i}_{o,d}=0,\tilde{v}_{a,q}=0} = C[sI-A]^{-1}B_1, \text{ where } Y_1(s) = \begin{bmatrix} \tilde{i}_{L,d}(s) \\ \tilde{v}_{a,d}(s) \end{bmatrix} \\
G_{C,q}(s) &= \frac{Y_2(s)}{U_1(s)} = \frac{Y_2(s)}{\tilde{v}_{inv,q}(s)}\bigg|_{\tilde{i}_{L,d}=0,\tilde{i}_{o,q}=0,\tilde{v}_{a,d}=0} = C[sI-A]^{-1}B_1, \text{ where } Y_2(s) = \begin{bmatrix} \tilde{i}_{L,q}(s) \\ \tilde{v}_{a,q}(s) \end{bmatrix}
\end{aligned}
\right\}
$$

(6.12)

and

$$
\left.
\begin{aligned}
[sI-A]^{-1} &= \left[\begin{array}{cc} \left(s+\dfrac{R_{Lf}}{L_f}\right) & \dfrac{1}{L_f} \\ -1/C_f & s \end{array} \right]^{-1} = \frac{1}{\Delta_1} \left[\begin{array}{cc} s & -1/L_f \\ 1/C_f & \left(s+\dfrac{R_{Lf}}{L_f}\right) \end{array} \right] \\
\text{where, } \Delta_1 &= \left(s^2 + \dfrac{R_{Lf}}{L_f}s + \dfrac{1}{C_f L_f}\right)
\end{aligned}
\right\}
$$

(6.13)

The controllable transfer functions of d-axis and q-axis components are expressed as follows, by solving Equations (6.6), (6.10), (6.12), (6.13), and (6.7) and (6.11–6.13):

$$
\left.
\begin{aligned}
G_{C1}(s) &= \frac{\tilde{v}_{a,d}(s)}{\tilde{v}_{inv,d}(s)}\bigg|_{\tilde{i}_{L,q}=0,\tilde{i}_{o,d}=0,\tilde{v}_{a,q}=0} = \frac{1}{C_f L_f s^2 + C_f R_{Lf} s + 1} = \frac{\tilde{v}_{a,q}(s)}{\tilde{v}_{inv,q}(s)}\bigg|_{\tilde{i}_{L,d}=0,\tilde{i}_{o,q}=0,\tilde{v}_{a,d}=0} \\
G_{C2}(s) &= \frac{\tilde{i}_{L,d}(s)}{\tilde{v}_{inv,d}(s)}\bigg|_{\tilde{i}_{L,q}=0,\tilde{i}_{o,d}=0,\tilde{v}_{a,q}=0} = \frac{C_f s}{C_f L_f s^2 + C_f R_{Lf} s + 1} = \frac{\tilde{i}_{L,q}(s)}{\tilde{v}_{inv,q}(s)}\bigg|_{\tilde{i}_{L,d}=0,\tilde{i}_{o,q}=0,\tilde{v}_{a,d}=0}
\end{aligned}
\right\}
$$

(6.14)

By solving Equations (6.6), (6.7), (6.9)–(6.11), and (6.13), the disturbance transfer functions for d-axis and q-axis components are expressed as follows:

$$
\left.
\begin{aligned}
G_{Di1}(s) &= \frac{Y_1(s)}{\tilde{i}_{o,q}(s)}\bigg|_{\tilde{v}_{inv,d}=\tilde{i}_{o,d}=\tilde{v}_{a,q}=0} = C[sI-A]^{-1}B_2 = \frac{Y_2(s)}{\tilde{i}_{o,d}(s)}\bigg|_{\tilde{v}_{inv,q}=\tilde{i}_{o,q}=\tilde{v}_{a,d}=0} \\
G_{Di1,1}(s) &= \frac{\tilde{v}_{a,d}(s)}{\tilde{i}_{o,q}(s)}\bigg|_{\tilde{v}_{inv,d}=\tilde{i}_{o,d}=\tilde{v}_{a,q}=0} = \frac{\omega L_f}{C_f L_f s^2 + C_f R_{Lf} s + 1} = -\frac{\tilde{v}_{a,q}(s)}{\tilde{i}_{o,d}(s)}\bigg|_{\tilde{v}_{inv,q}=\tilde{i}_{o,q}=\tilde{v}_{a,d}=0} \\
G_{Di1,2}(s) &= \frac{\tilde{i}_{L,d}(s)}{\tilde{i}_{o,q}(s)}\bigg|_{\tilde{v}_{inv,d}=\tilde{i}_{o,d}=\tilde{v}_{a,q}=0} = \frac{L_f C_f \omega s}{C_f L_f s^2 + C_f R_{Lf} s + 1} = -\frac{\tilde{i}_{L,q}(s)}{\tilde{i}_{o,d}(s)}\bigg|_{\tilde{v}_{inv,q}=\tilde{i}_{o,q}=\tilde{v}_{a,d}=0}
\end{aligned}
\right\}
$$

(6.15)

$$G_{Di2}(s) = \left.\frac{Y_1(s)}{\tilde{\imath}_{o,d}(s)}\right|_{\tilde{v}_{inv,d}=\tilde{\imath}_{L,q}=\tilde{v}_{a,q}=0} = C[sI-A]^{-1}B_3 = \left.\frac{Y_2(s)}{\tilde{\imath}_{o,q}(s)}\right|_{\tilde{v}_{inv,q}=\tilde{\imath}_{L,d}=\tilde{v}_{a,d}=0}$$

$$G_{Di2,1}(s) = \left.\frac{\tilde{v}_{a,d}(s)}{\tilde{\imath}_{o,d}(s)}\right|_{\tilde{v}_{inv,d}=\tilde{\imath}_{L,q}=\tilde{v}_{a,q}=0} = -\frac{L_f s + R_{Lf}}{C_f L_f s^2 + C_f R_{Lf} s + 1} = \left.\frac{\tilde{v}_{a,q}(s)}{\tilde{\imath}_{o,q}(s)}\right|_{\tilde{v}_{inv,q}=\tilde{\imath}_{L,d}=\tilde{v}_{a,d}=0}$$

$$G_{Di2,2}(s) = \left.\frac{\tilde{\imath}_{L,d}(s)}{\tilde{\imath}_{o,d}(s)}\right|_{\tilde{v}_{inv,d}=\tilde{\imath}_{L,q}=\tilde{v}_{a,q}=0} = \frac{1}{C_f L_f s^2 + C_f R_{Lf} s + 1} = \left.\frac{\tilde{\imath}_{L,q}(s)}{\tilde{\imath}_{o,q}(s)}\right|_{\tilde{v}_{inv,q}=\tilde{\imath}_{L,d}=,\tilde{v}_{a,d}=0}$$

$$(6.16)$$

$$G_{Di3}(s) = \left.\frac{Y_1(s)}{\tilde{v}_{a,q}(s)}\right|_{\tilde{v}_{inv,d}=\tilde{\imath}_{L,q}=\tilde{\imath}_{o,d}=0} = C[sI-A]^{-1}B_4 = \left.\frac{Y_2(s)}{\tilde{v}_{a,d}(s)}\right|_{\tilde{v}_{inv,q}=\tilde{\imath}_{L,d}=\tilde{\imath}_{o,q}=0}$$

$$G_{Di3,1}(s) = \left.\frac{\tilde{v}_{a,d}(s)}{\tilde{v}_{a,q}(s)}\right|_{\tilde{v}_{inv,d}=\tilde{\imath}_{L,q}=\tilde{\imath}_{o,d}=0} = -\frac{C_f \omega (L_f s + R_{Lf})}{C_f L_f s^2 + C_f R_{Lf} s + 1} = -\left.\frac{\tilde{v}_{a,q}(s)}{\tilde{v}_{a,d}(s)}\right|_{\tilde{v}_{inv,q}=\tilde{\imath}_{L,d}=\tilde{\imath}_{o,q}=0}$$

$$(6.17)$$

$$G_{Di3,2}(s) = \left.\frac{\tilde{\imath}_{L,d}(s)}{\tilde{v}_{a,q}(s)}\right|_{\tilde{v}_{inv,d}=\tilde{\imath}_{L,q}=\tilde{\imath}_{o,d}=0} = \frac{C_f \omega}{C_f L_f s^2 + C_f R_{Lf} s + 1} = -\left.\frac{\tilde{\imath}_{L,q}(s)}{\tilde{v}_{a,d}(s)}\right|_{\tilde{v}_{inv,q}=\tilde{\imath}_{L,d}=\tilde{\imath}_{o,q}=0}$$

6.2 DEVELOPED CONTROL TECHNIQUE OF SINGLE-PHASE VSI

A control technique of the one-phase full-bridge VSI for connecting the one-phase AC loads to the DCMG, is proposed, as in Figure 6.4(a). The proposed control technique, with the combination of the feedbacks and feedforward control loops, is implemented in a single rotating d-q SRF [1], [45]. The main highlights of the developed control strategy are as follows:

1. In general, only a single feedback control loop is used in control schemes available in the literature. However, the proposed control strategy uses two feedback control loops for controlling the d-q components of both the voltage and the current, independently.
2. DC voltage controller (as feedforward) is implemented in the developed control technique to provide the feedforward controlled reference current signal.

In the developed control technique, using orthogonal vector transformation, the AC voltage/current of the one-phase VSI have been transformed into two-phase (α-β) stationary reference frame, as in Figure 6.3(b). The α component of the voltage and current is equal to the measured voltage and current with no phase difference, and the β component of the voltage and current are determined by creating a phase delay of $\pi/2$ rad (i.e., transport delay time of $T/4$) on both voltage and current with the same magnitude. Using Park's transformation, the α-β components of voltage and current have been transformed into rotating d-q SRF. The phase-locked loop (PLL) has been used to find the phase angle information of the output AC voltage, and to track the frequency. The α-β voltages' components as the input signals are sent to the PLL, as shown in Figure 6.4(a).

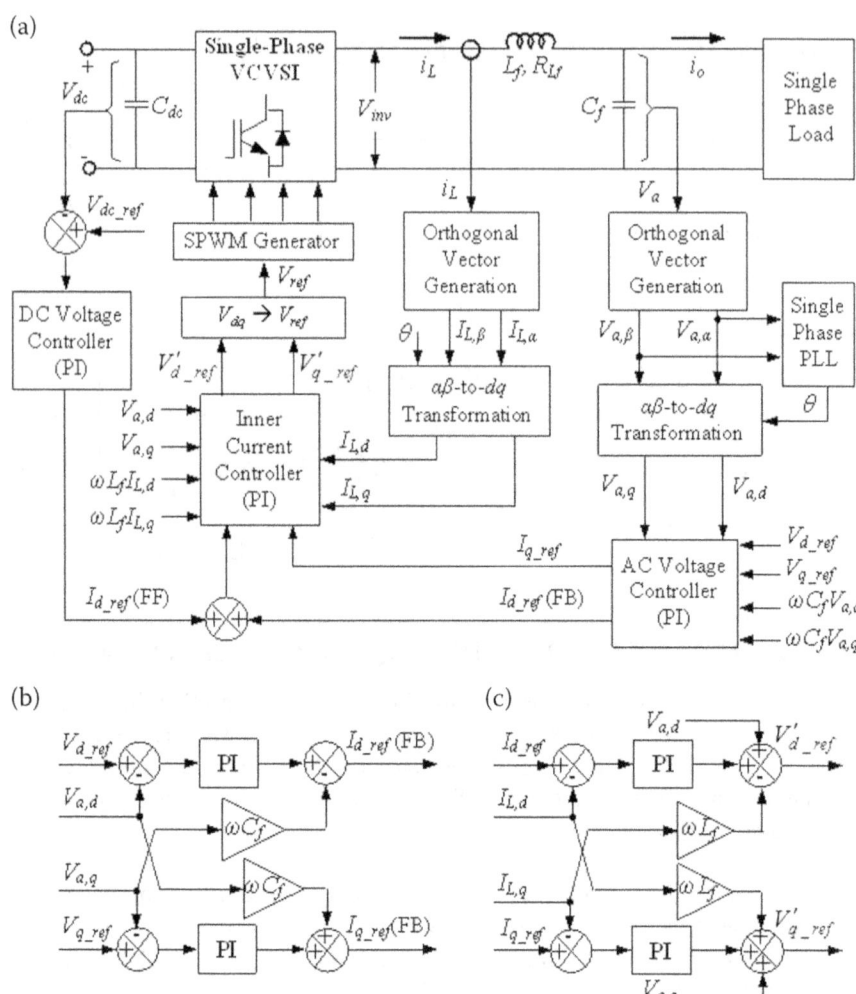

FIGURE 6.4 (a) Developed control technique of single-phase full-bridge VSI. (b) AC voltage controller. (c) Inner current controller.

The AC voltage regulator, as in Figure 6.4(b), is used to control the AC voltage. The d-q voltage components have been compared with the reference values ($V_{q\_ref} = 0$ and $V_{d\_ref} = 1.0$ pu), and the errors have been used as the input to the PI controllers, which provide the feedback controlled reference current signals (I_{dq_ref}(FB)). The DC voltage controller provides the feedforward controlled reference current signals (I_{dq_ref}(FF)), as given in Figure 6.4(a).

The inner current controller, as in Figure 6.4(c), is used to control the single-phase VSI output current. The d-q current components have been compared with the reference currents (I_{dq_ref}), which are obtained by the combination of the feedforward reference currents and feedback reference currents, as shown in Figure 6.4(a). These errors have been used as the input to the PI controllers, which provide the controlled voltage (V'_{dq_ref}) as reference signals to generate the controlled gate pulses for the one-phase VSI, using the PWM generator.

The feedforward control signal allows stabilizing DC voltage and to improve the stability of the system. The feedback control signals maintain better voltage regulation and improved power quality.

The voltage equations for the inner current controller in a single rotating d-q SRF are derived as:

$$\left.\begin{array}{l} V'_{d\_ref} = V_{a,d} + (I_{d\_ref} - I_{L,d})\left(K_{P_C} + \dfrac{K_{IC}}{s}\right) - \omega L_f I_{L,q} \\[2mm] V'_{q\_ref} = V_{a,q} + (I_{q\_ref} - I_{L,q})\left(K_{P_C} + \dfrac{K_{IC}}{s}\right) + \omega L_f I_{L,d} \end{array}\right\}$$ (6.18)

The PI controllers' parameters using controllable transfer functions, as given in Equation (6.14), have been determined for the single-phase VSI with 240 V and 50 Hz, which are given in Table 6.1. The LC filter parameters are determined as $L_f = 1.068$ mH, $R_f = 5.76$ mΩ, and $C_f = 30$ μF.

6.3   OPERATIONAL ANALYSIS WITH SIMULATION RESULTS

The performance analysis of the developed control techniques for the proposed DCMG have been carried out for different operational conditions, on the real-time digital simulator environment.

6.3.1   PERFORMANCE ANALYSIS OF SMART DCMG FOR POWER CONTROL TECHNIQUE

The smart DCMG, as shown in Figure 6.1, has been considered for the performance analysis of the developed control algorithm. The ratings of various DGs such as SOFC, MT, SPV, and WT have been taken as 100 kW (with the peak power capacity of 180 kW), 150 kW, 100 kW, and 600 kW, respectively, [1], [3]. The rated loads connected to the DCMG have been considered as follows: 100 kW three-phase load, 200 kW one-phase load, 100 kW DC load @ 220 V DC voltage, 100 kW EV load @ 500 V DC voltage, and 100 kW three-phase load in the local area of WT DG. The total rated load has been taken as 600 kW in all the cases. The variation of load [46] is

---

TABLE 6.1
PI Controllers' Parameters for One-Phase VSI

PI Controllers	Proportional Gain (K_P)	Integral Gain (K_I)
AC Voltage Controller	$K_{P\_Vac} = 6.5$	$K_{I\_Vac} = 500$ s^{-1}
Inner Current Controller	$K_{P_C} = 4$	$K_{IC} = 200$ s^{-1}
DC Voltage Controller	$K_{P\_Vdc} = 0.004$	$K_{I\_Vdc} = 1.2$ s^{-1}

TABLE 6.2

Variation of the Load

Time (Hrs.)	7:00 PM	9:00 PM	11:00 PM	1:00 AM	3:00 AM	5:00 AM
Load (pu)	1.0	0.85	0.7	0.55	0.47	0.4
Time (Hrs.)	7:00 AM	9:00 AM	11:00 AM	1:00 PM	3:00 PM	5:00 PM
Load (pu)	0.6	0.7	0.8	0.85	1.05	0.99

given Table 6.2. Two cases for the performance analysis have been carried to show the robustness of the developed power control technique, under different operational conditions.

In this case, under islanded mode, variable power generations and variable loads have been considered. Figure 6.5(a) shows the output power of the WT DG at variable wind speeds, and the actual output of WT DG under islanding mode. The output of the SPV DG is also variable, as shown in Figure 6.5(b). Figure 6.5(c) shows the variable output power of the MT as well as SOFC DGs. The nature of the three-phase load in the area of the WT DG and DC load are variable, as in Figure 6.5(d). The total load, one-phase AC load, and another three-phase load have been varied with assuming constant for the next two hours, as given in Table 6.2. Figure 6.6(a) shows the total output generation, total load, and battery output.

During 5:00–9:00 AM and 5:00–7:00 PM, there is more surplus power generation by the WT DG (600 kW) and BESS is already full charged. Therefore, it is necessary to reduce the output power generation of the WT DG for power balancing, as shown in Figure 6.5(a). The deficit as well as surplus power in the DCMG is fulfilled by the battery, as in Figure 6.6(a).

During some intervals, the available wind speeds is more than cut-in speed, and also higher than rated wind speeds, still the output power of the WT DG is zero because of some technical issues, as shown in Figure 6.5(a). This scenario is known as a loss of WT DG. In such cases, load demand is met by other DGs and BESS available in the DCMG, as shown in Figure 6.6(a). Similarly, if the output power of SPV DG is also zero along with the WT DG, then the load has been fulfilled by the remaining DGs and BESS. The simulation results show that the performance of the proposed smart DCMG with the proposed power control algorithm under islanded mode has been found satisfactorily by using the proper coordination of the DGs, for all cases. The DCMG voltage has established almost constant. The spikes in the DCMG voltage are due to the transients, which occur by changing the generation and/or the load, as shown in Figure 6.6(b).

6.3.1.1 Performance Analysis for Developed Control Technique of Single-Phase VSI

For the performance analysis of the developed control technique of one-phase VSI, the smart DCMG consists of three DGs such as WT, SPV, SOFC, battery storage system, and loads, as shown in Figure 6.1. The ratings of various DGs such as SOFC, SPV, and WT have been taken as 100 kW (with peak power capacity of 180 kW), 100 kW, and 300 kW, respectively [1]. The rated loads in the DCMG are taken as: 200 kW one-

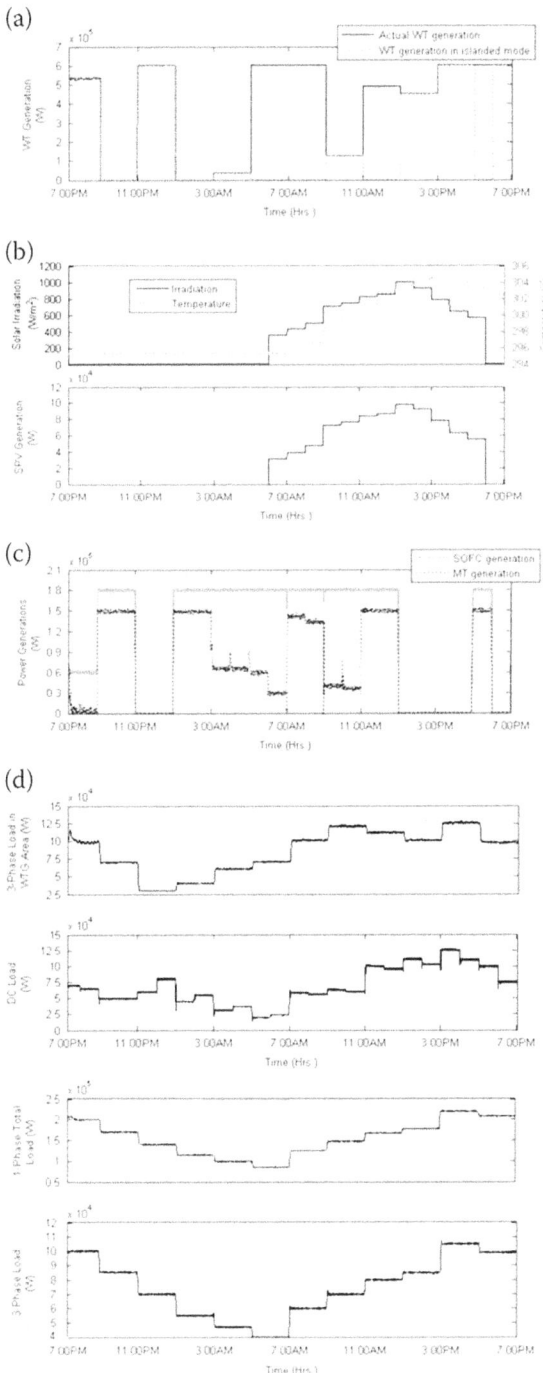

FIGURE 6.5 (a) Output power of WT DG with varying wind, and actual WT DG output in islanded mode. (b) Output power of SPV with varying temperature and SI. (c) Output power of MT and SOFC DGs. (d) Variation of the individual loads.

(a)

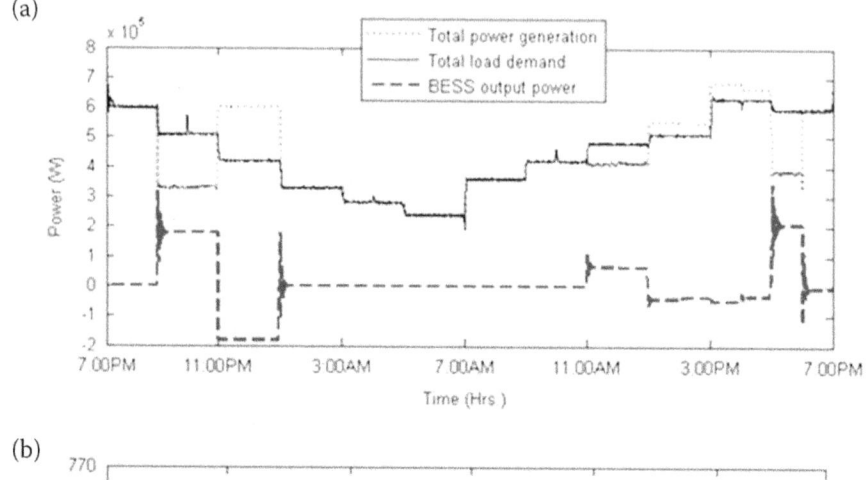

(b)

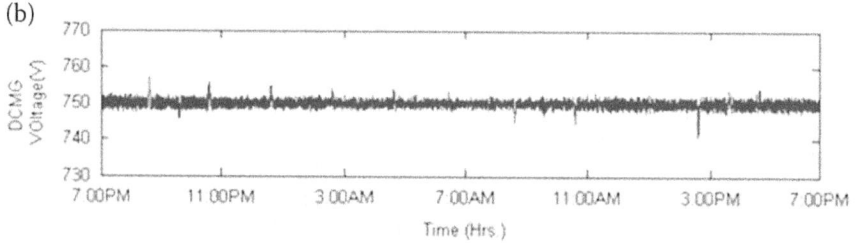

FIGURE 6.6   (a) Total load demand, power generation, and battery output. (b) Voltage of the DC microgrid.

phase load, 50 kW DC load @ 220 V DC voltage, 40 kW EV load @500 V DC voltage, and 10 kW telecommunication load @ 48 V. Two case studies under islanded mode are tested for showing the robustness of the developed control technique of one-phase VSI along with the proposed DCMG.

CASE A   *L-G FAULT ON AC TERMINALS*

The output powers of the WT, SPV, and SOFC DGs have been considered variable, as shown in Figure 6.7(a). The loads' variations are shown in Table 6.3. An L-G fault, with fault resistance ($R_f = 50$ mΩ), is occurred on one-phase VSI output AC terminals, at $t = 1.4$ s. This fault has decreased the AC voltage from 240 V to 65 V, as in Figure 6.7(b). The power supplied to the one-phase AC load, and power through one-phase VSI is also reduced, and a small amount of power (84.5 kW) dissipates in fault resistance, as shown in Figures 6.7(b) and 6.7(c). Hence, the SOFC output power has also decreased to zero, as in Figure 6.7(a), i.e., reduction in total power consumption, as shown in Figure 6.7(d). The difference between the generation and load demand is met by the battery, for all cases. The DC voltage of the microgrid is established constant, as given in Figure 6.7(e).

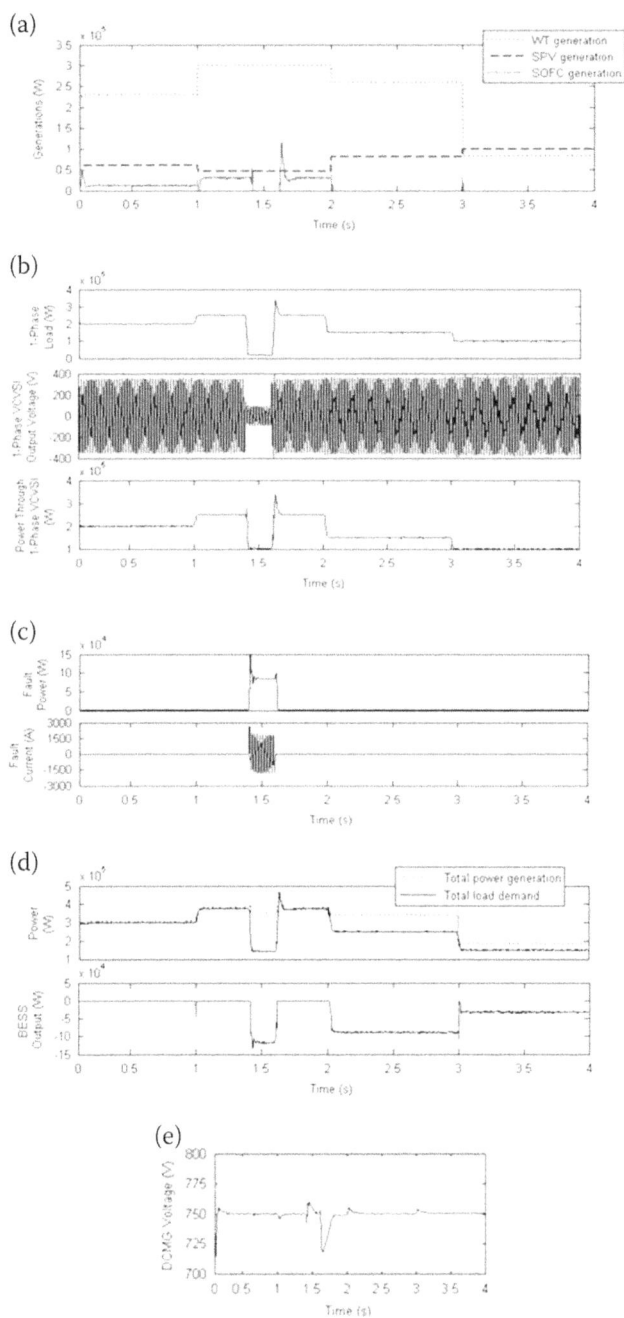

FIGURE 6.7 (a) Output Power of SOFC, SPV, and WT DGs (Case A). (b) One-phase load power, one-phase voltage, and power by one-phase VSI (Case A). (c) Power dissipated and fault current (Case A). (d) Total load demand, generation, and battery output (Case A). (e) Voltage of the DC microgrid (Case A).

TABLE 6.3

Individual Load Variations

Time (s)	One-Phase Load	DC Load	EV Load	Telecom Load	Total Load
0–1	200	50	40	10	300
1–2	250	75	45	5	375
2–3	150	35	60	5	250
3–4	100	15	20	15	150

Thus, during fault, the developed control technique of the one-phase VSI is not permitting any adverse impact on the remaining DCMG, except the faulted area. When the fault is cleared at t = 1.6s, the DCMG operates normally and recollects pre-fault condition, as given in Figures 6.7(b)–6.7(e). The proposed power control algorithm also manages the power balance in the DCMG with the proper coordination of the DGs for all cases, as in Figures 6.7(a) and 6.7(d).

CASE B   DC FAULT ON THE DCMG IN ISLANDED MODE

The output power generated by the SPV and WT DGs are constant, and output of a SOFC system is variable, as shown in Figure 6.8(a). The individual loads are varied, as given in Figure 6.8(b).

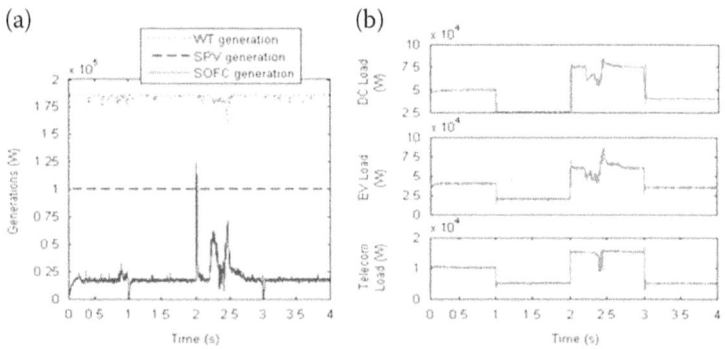

FIGURE 6.8   (a) Output power of SOFC, SPV, and WT DGs. (b) Variations of individual loads (Case B).

At t = 2.2 s, a DC fault with fault resistance (R_f = 200 mΩ) has occurred on the DCMG. This fault has reduced DCMG voltage, as in Figure 6.9(a) and the proposed control technique of the one-phase VSI has maintained the single-phase load power supply, and rated sinusoidal output voltage, as shown in Figure 6.9(b). The total power supplied to the load is reduced.

However, the total generation as well as SOFC output has increased, as shown in Figures 6.10 and 6.8(a). A few part of the power dissipates through the fault resistance, as shown in Figure 6.9(a). At $t = 2.4$ s, as the fault has been cleared, the DCMG has regained the pre-fault conditions and operates normally, as shown in Figures 6.8–6.10. The proposed power control technique maintains the power balance in the microgrid for all cases.

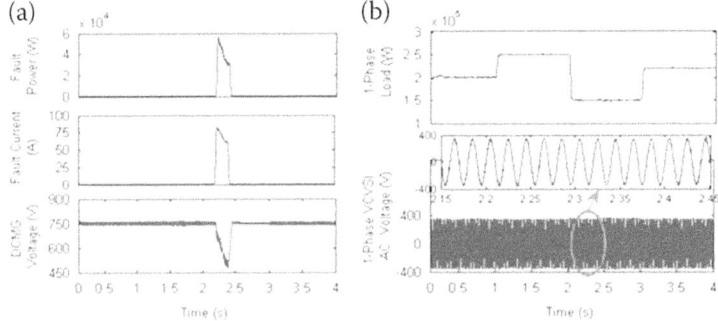

FIGURE 6.9   (a) Power dissipated, fault current, and DCMG voltage. (b) One-phase load power and one-phase AC voltage (Case B).

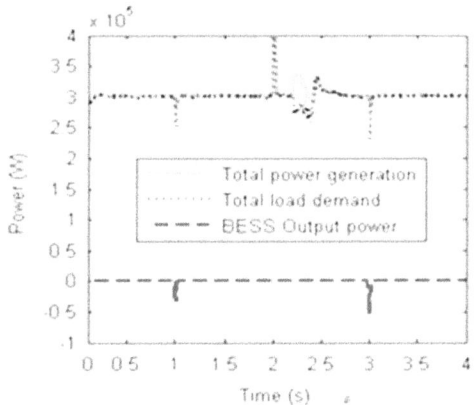

FIGURE 6.10   Total generation, total load, and battery output (Case B).

6.4   CONCLUSIONS

A power control technique, for proper generation scheduling and coordinated control of the DGs integrated to the microgrid, has been presented for balancing the power at the smart DCMG, in this chapter. Various simulation results have been carried out for showing the robustness of the developed power algorithm along with the proposed DCMG under different operating conditions. Thus, the proposed power control technique is able to maintain the energy balance in the DCMG in all cases.

In this chapter, the dynamic modeling of the one-phase full-bridge VSI, with LC filter, has been described into α-β stationary reference frame using orthogonal transformation, and into a single rotating d-q SRF. A control strategy of one-phase VSI into a single d-q SRF, using the combination of the feedforward and feedback loops, has also been presented for integrating the one-phase loads to the DC microgrid. The main contributions of this scheme are as follows:

1. It uses two feedback control loops for controlling both AC voltage and current, independently. The feedback signals maintain good voltage regulation.
2. It uses feedforward loop to provide the feedforward reference current signal for the inner current controller. This allows to stabilize the DC voltage effectively and to improve the stability of the system.

Various simulation results are conducted to show the effectiveness of the proposed control strategy of one-phase VSI, in islanded mode under various operating conditions. The proposed technique maintains rated output sinusoidal voltage, and needed power flow from the DCMG to the one-phase loads, even during the DC fault. In addition, the IEEE-1547 standard has been compiled by the developed control techniques.

The DCMG voltage has also been maintained constant at its rated value in all cases.

REFERENCES

[1] M. Kumar, "Development of control schemes for power management and operation of dc microgrids," Ph.D. Thesis, Electrical Engineering Department, Indian Institute of Technology Kanpur, India, Jan. 2015.
[2] B. Kroposki, C. Pink, R. DeBlasio, H. Thomas, M. Simoes, and P. K. Sen, "Benefits of power electronic interfaces for distributed energy systems," *IEEE Trans. Energy Convers.*, vol. 25, no. 3, pp. 901–908, Sep. 2010.
[3] M. Kumar, S. N. Singh, and S. C. Srivastava, "Design and control of smart DC microgrid for integration of renewable energy sources," in Proc. IEEE Power Energy Soc. Gen. Meeting, San Diego, CA, USA, 2012, pp. 1–7.
[4] A. Khorsandi, M. Ashourloo, and H. Mokhtari, "A decentralized control method for a low-voltage dc microgrid," *IEEE Trans. Energy Convers.*, vol. 29, no. 4, pp. 793–801, Dec. 2014.
[5] B. Khan and P. Singh, "Optimal power flow techniques under characterization of conventional and renewable energy sources: A comprehensive analysis," *J. Eng.*, vol. 2017, Article ID 9539506, 16 pages, 2017.
[6] P. Singh and B. Khan, "Smart microgrid energy management using a novel artificial shark optimization," *Complexity*, vol. 2017, Article ID 2158926, 22 pages, 2017.
[7] T. Molla, B. Khan, B. Moges, H. H. Alhelou, R. Zamani and P. Siano, "Integrated optimization of smart home appliances with cost-effective energy management system," *CSEE J. Power Energy Syst.*, vol. 5, no. 2, pp. 249–258, June 2019.

[8] Z. Tang, Y. Lin, M. Vosoogh, N. Parsa, A. Baziar and B. Khan, "Securing microgrid optimal energy management using deep generative model," *IEEE Access*, vol. 9, pp. 63377–63387, 2021.

[9] S. P. Bihari et al., "A comprehensive review of microgrid control mechanism and impact assessment for hybrid renewable energy integration," *IEEE Access*, vol. 9, pp. 88942–88958, 2021, doi: 10.1109/ACCESS.2021.3090266.

[10] B. Khan and P. Singh, "Selecting a meta-heuristic technique for smart micro-grid optimization problem: A comprehensive analysis," *IEEE Access*, vol. 5, pp. 13951–13977, 2017.

[11] O. P. Mahela et al., "Comprehensive overview of multi-agent systems for controlling smart grids," *CSEE J. Power Energy Syst.*, vol. 8, no. 1, pp. 115–131, Jan. 2022, doi: 10.17775/CSEEJPES.2020.03390.

[12] Z. Wang, B. Zhang, M. Mobtahej, A. Baziar and B. Khan, "Advanced reactive power compensation of wind power plant using PMU data," *IEEE Access*, vol. 9, pp. 67006–67014, 2021.

[13] S. Padmanaban, C. Dhanamjayulu and B. Khan, "Artificial Neural Network and Newton Raphson (ANN-NR) algorithm based selective harmonic elimination in cascaded multilevel inverter for PV applications," in *IEEE Access*, vol. 9, pp. 75058–75070, 2021.

[14] M. S. Fakhar et al., "Implementation of APSO and improved APSO on non-cascaded and cascaded short term hydrothermal scheduling," *IEEE Access*, vol. 9, pp. 77784–77797, 2021.

[15] P. Karlsson and J. Svensson, "DC bus voltage control for a distributed power system," *IEEE Trans. Power Electron.*, vol. 18, no. 6, pp. 1405–1412, Nov. 2003.

[16] D. Salomonsson and A. Sannino, "Low-voltage DC distribution system for commercial power systems with sensitive electronic loads," *IEEE Trans. Power Del.*, vol. 22, no. 3, pp. 1620–1627, Jul. 2007.

[17] D. D. C. Lu and V. G. Agelidis, "Photovoltaic-battery-powered dc bus system for common portable electronic devices," *IEEE Trans. Power Electron.*, vol. 24, no. 3, pp. 849–855, Mar. 2009.

[18] H. Kakigano, M. Nomura, and T. Ise, "Low-voltage bipolar-type dc microgrid for super high quality distribution," *IEEE Trans. Power Electron.*, vol. 25, no. 12, pp. 3066–3075, Dec. 2010.

[19] L. Xu and D. Chen, "Control and operation of a dc microgrid with variable generation and energy storage," *IEEE Trans. Power Del.*, vol. 26, no. 4, pp. 2513–2522, Oct. 2011.

[20] A. A. A. Radwan and Y. A. R. I. Mohamed, "Linear active stabilization of converter-dominated dc microgrids," *IEEE Trans. Smart Grid*, vol. 3, no. 1, pp. 203–216, Mar. 2012.

[21] D. Chen and L. Xu, "Autonomous dc voltage control of a dc microgrid with multiple slack terminals," *IEEE Trans. Power Syst.*, vol. 27, no. 4, pp. 1897–1905, Nov. 2012.

[22] B. Wang, M. Sechilariu, and F. Locment, "Intelligent dc microgrid with smart grid communications: Control strategy consideration and design," *IEEE Trans. Smart Grid*, vol. 3, no. 4, pp. 2148–2156, Dec. 2012.

[23] D. Chen, L. Xu, and L. Yao, "DC voltage variation based autonomous control of dc microgrids," *IEEE Trans. Power Del.*, vol. 28, no. 2, pp. 637–648, Apr. 2013.

[24] N. Eghtedarpour and E. Farjah, "Distributed charge/discharge control of energy storages in a renewable-energy-based dc micro-grid," *IET Renew. Power Gener.*, nol. 8, no. 1, pp. 45–57, Jan. 2014.

[25] C. Jin, P. Wang, J. Xiao, Y. Tang, and F. H. Choo, "Implementation of hierarchical control in dc microgrids," *IEEE Trans. Ind. Electron.*, vol. 61, no. 8, pp. 4032–4042, Aug. 2014.

[26] Y. Gu, W. Li, and X. He, "Frequency-coordinating virtual impedance for autonomous power management of dc microgrid," *IEEE Trans. Power Electron.*, vol. 30, no. 4, pp. 2328–2337, Apr. 2015.

[27] M. J. Ryan, W. E. Brumsickle, and R. D. Lorenz, "Control topology options for single-phase UPS inverters," *IEEE Trans. Ind. Appl.*, vol. 33, no. 2, pp. 493–501, Apr. 1997.

[28] M. C. Trigg and C. V. Nayar, "DC bus compensation for a sinusoidal voltage-source inverter with wave-shaping control," *IEEE Trans. Ind. Electron.*, vol. 55, no. 10, pp. 3661–3669, Oct. 2008.

[29] W. Tsai-Fu, S. Kun-Han, K. Chia-Ling, and C. Chih-Hao, "Predictive current controlled 5-kW single-phase bidirectional inverter with wide inductance variation for dc-microgrid applications," *IEEE Trans. Power Electron.*, vol. 25, no. 12, pp. 3076–3084, Dec. 2010.

[30] S. Dasgupta, S. K. Sahoo, S. K. Panda, and G. A. J. Amaratunga, "Single-phase inverter-control techniques for interfacing renewable energy sources with microgrid-part II: series-connected inverter topology to mitigate voltage-related problems along with active power flow control," *IEEE Trans. Power Electron.*, vol. 26, no. 3, pp. 732–746, Mar. 2011.

[31] B. Bahrani, A. Rufer, S. Kenzelmann, and L. A. C. Lopes, "Vector control of single-phase voltage-source converters based on fictive-axis emulation," *IEEE Trans. Ind. Appl.*, vol. 47, no. 2, pp. 831–839, Apr. 2011.

[32] N. A. Rahim, K. Chaniago, and J. Selvaraj, "Single-phase seven-level grid-connected inverter for photovoltaic system," *IEEE Trans. Ind. Electron.*, vol. 58, no. 6, pp. 2435–2443, Jun. 2011.

[33] L. Malesani, P. Mattavelli, and P. Tomasin, "Improved constant-frequency hysteresis current control of VSI inverters with simple feedforward bandwidth prediction," *IEEE Trans. Ind. Appl.*, vol. 33, no. 5, pp. 1194–1202, Oct. 1997.

[34] S. Buso, S. Fasolo, L. Malesani, and P. Mattavelli, "A dead-beat adaptive hysteresis current control," *IEEE Trans. Ind. Appl.*, vol. 36, no. 4, pp. 1174–1180, Aug. 2000.

[35] H. Mao, X. Yang, Z. Chen, and Z. Wang, "A hysteresis current controller for single-phase three-level voltage source inverters," *IEEE Trans. Power Electron.*, vol. 27, no. 7, pp. 3330–3339, Jul. 2012.

[36] Y. A. R. I. Mohamed and E. F. El-Saadany, "An improved deadbeat current control scheme with a novel adaptive self-tuning load model for a three-phase PWM voltage-source inverter," *IEEE Trans. Ind. Electron.*, vol. 54, no. 2, pp. 747–759, Apr. 2007.

[37] R. Teodorescu, F. Blaabjerg, M. Liserre, and P. C. Loh, "Proportional-resonant controllers and filters for grid-connected voltage-source converters," *Proc. IEE Electric Power Appllications*, vol. 153, no. 5, Sep. 2006, pp. 750–762.

[38] A. Vidal, F. D. Freijedo, A. G. Yepes, P. F. Comesaña, J. Malvar, Ó. López, and J. D. Gandoy, "Assessment and optimization of the transient response of proportional-resonant current controllers for distributed power generation systems," *IEEE Trans. Ind. Electron.*, vol. 60, no. 4, pp. 1367–1383, Apr. 2013.

[39] J. Rodríguez, J. Pontt, C. A. Silva, P. Correa, P. Lezana, P. Cortés, and U. Ammann, "Predictive current control of a voltage source inverter," *IEEE Trans. Ind. Electron.*, vol. 54, no. 1, pp. 495–503, Feb. 2007.

[40] P. M. Sanchez, O. Machado, E. J. B. Peña, F. J. Rodríguez, and F. J. Meca, "FPGA-based implementation of a predictive current controller for power converters," *IEEE Trans. Ind. Inf.*, vol. 9, no. 3, pp. 1312–1321, Aug. 2013.

[41] R. O. Ramírez, J. R. Espinoza, P. E. Melın, M. E. Reyes, E. E. Espinosa, C. Silva, and E. Maurelia, "Predictive controller for a three-phase/single-phase voltage source converter cell," *IEEE Trans. Ind. Inf.*, vol. 10, no. 3, pp. 1878–1889, Aug. 2014.

[42] B. Bahrani, S. Kenzelmann, and A. Rufer, "Multivariable-PI-based dq current control of voltage source converters with superior axis decoupling capability," *IEEE Trans. Ind. Electron.*, vol. 58, no. 7, pp. 3016–3026, Jul. 2011.

[43] M. C. Chandorkar, D. M. Divan, and R. Adapa, "Control of parallel connected inverters in standalone ac supply systems," *IEEE Trans. Ind. Appl.*, vol. 29, no. 1, pp. 136–143, Feb. 1993.

[44] S. Golestan, M. Joorabian, H. Rastegar, A. Roshan, and J. M. Guerrero, "Droop based control of parallel-connected single-phase inverters in d-q rotating frame," in *Proc. IEEE International Conference on Industrial Technology*, Gippsland, VIC, 10–13 Feb. 2009, pp. 1–6.

[45] M. Kumar, S. C. Srivastava, and S. N. Singh, "Dynamic performance analysis of dc microgrid with a proposed control strategy for single-phase VCVSI," *IEEE PES Transmission & Distribution Conference & Exposition*, Chicago, IL, USA, 14–17 Apr., 2014.

[46] T. A. Short, *Electric Power Distribution Handbook*. CRC Press, Boca Raton, FL, USA, 2004, pp. 47.

7 Analysis and Optimization of a PV-Integrated Rural Distribution Network

P. K. Bhatt
Department of Electrical Engineering, Amity University, Jaipur, Rajasthan, India

Om Prakash Mahela
Power System Study Division, Rajasthan Rajya Vidyut Prasarn Nigam Ltd., Jaipur, India

Baseem Khan
Department of Electrical Engineering, Hawassa University, Ethiopia

CONTENTS

7.1 INTRODUCTION

Electrical distribution systems are undergoing continuous changes with respect to increased penetration of renewable energy sources. The integration of PV poses various challenges in the power system [1]. Some major challenges are voltage rise, voltage unbalance, and reverse power flow [2,3]. Moreover, modern one-phase and

DOI: 10.1201/b22884-7

three-phase loads connected at the distribution networks are variable in nature. Therefore, each phase has different loading levels. Further, the PV integration affects the bus voltage level and branch current loading significant [4–13]. Due to unequal current in all phases, considerable neutral current flows through the neutral wire and poses a potential threat to the operation-sensitive equipment [14,15]. It is apparent that large-scale integration of a PV into the distribution grid affects its performance at various operating conditions [16]. Various studies have been performed earlier to improve the power system performance [17,18]. It is of utmost importance for the utility to provide better quality and reliability of the power supply to the customers. Therefore, the investigation of network optimal operational behavior under PV penetration is essential for understanding and mitigating these impacts.

The research in the field of optimal power flow (OPF) analysis has tremendous impetus in last decades [19–22]. Various stochastic optimization techniques have been suggested in the literature that includes harmony search (HS) algorithm [23], genetic algorithm [24,25], gravitational search algorithm (GSA) [26], particle swarm optimization (PSO), artificial bee colony algorithm [22,27,28], biogeography-based optimization BBO) [29,30], and cuckoo search [31]. These algorithms promise the required accuracy and do not trap in the local optima but suffer from the disadvantage of overburdened computational time and requirement of algorithmic specific controlling parameters.

The novel population-based JAYA algorithm [32] can effectively address the above issues. The remarkable feature of the JAYA algorithm is that it has the tendency to elude from local minima and provides the optimal solution in comparison to other population-based algorithm such as a teaching-learning—based optimization algorithm TLBO [33]. The novelty of this chapter is that it presents the real-time optimal power management strategy that is particularly useful for smart grid planners and engineers. In this chapter, a rural distribution network is modeled on electrical transient analyzer (ETAP) software and then JAYA algorithm is implemented to obtain the optimal performance of the network. The contribution ofthis chapter is threefold. Firstly, it examines the performance of an un-optimized real-time PV distribution network. Secondly, it proposes the JAYA multi-objective optimal power management algorithms for investigating the performance of the network. Thirdly, this chapter presents a performance comparison of this algorithm with the un-optimized base case.

The rest of this chapter is organized as follows. In Section 7.2, the modeling of a rural PV distribution network is presented. Section 7.3 presents problem formulation that includes the objective functions and various constraints. In Section 7.4, the solution methodology for implementation of a JAYA algorithm is presented. Section 7.5 presents the result and discussion of the work carried out. Finally, this chapter is concluded in Section 7.6 with an observations and results discussion.

7.2 MODELING OF A RURAL DISTRIBUTION NETWORK

A typical layout of a rural PV distribution network that supplies power in a residential area is shown in Figure 7.1. The network is modeled in electrical transient analyzer

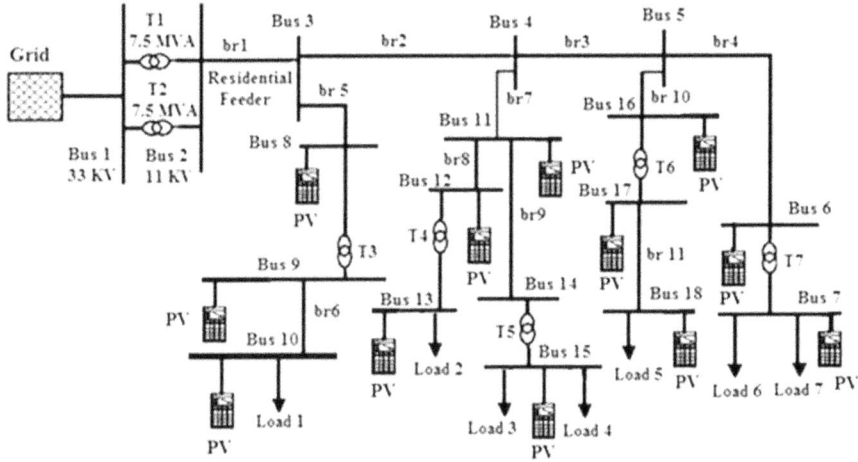

FIGURE 7.1 Single-line diagram of residential distribution network.

program (ETAP) software. Two transformers each of a capacity 33/11 kV, 7.5 MVA are installed to supply the entire locality, while five other transformers 11/0.433 kV are installed to meet the load demand of individual sections. Typical 11/0.433 kV transformer have different power ratings. The parameters of these transformers are presented in Table 7.1. These transformers are supplying seven load clusters. The capacity of a connected load at each bus is given in Table 7.2. The PV specification is given in Table 7.3 [33].

7.3   FORMULATION OF OPTIMAL POWER FLOW PROBLEM

The proposed multi-objective problem is formulated to minimize the total distribution power loss and improving the voltage profile of the remote consumers. The constraints of the problem include voltage, power, and physical limitations such as transformer tapping ratio, variation of grid impedance, and thermal loading of conductor. The necessary steps are presented in the given subsections.

TABLE 7.1
Transformer Specification

S. No.	Tr. No.	Capacity (MVA)	Voltage (kV)	Current (Amp.)	Z (%)	Vector Group	Grounding
1	T1-T2	7.5	33/11	150/525	7	Dy11	–
2	T3-T7	2	11/0.4	105/2887	5.70	Dyn11	Solid

TABLE 7.2

Connected Load on Distribution Network

Load Id.	Connected at Bus	Rating MW
1	10	1.2
2	13	1.2
3	15	0.8
4	15	0.4
5	18	0.8
6	7	0.8
7	7	0.4

TABLE 7.3

Solar PV Rating/Specification

System Component	Parameter	Rating/Specification
Solar PV Module	PV panel (watt/panel)	280 W
	PV panels in series	11 Nos.
	PV panels in parallel	161 Nos.
	Total no. of panels	1,771 Nos.
	Irradiance	1,000 (W/m^2)
Inverter	AC output power	600 kVA
	AC output current	800 A
	AC output voltage	433 V

7.3.1 FORMULATION OF OBJECTIVE FUNCTIONS

The objectives are to minimize the total distribution power loss and improving the voltage profile of the remote consumers. The constraints are related to voltage, power, and physical limitations such as transformer tapping ratio and thermal loading of conductor. This is framed as an optimization problem with the objective functions and constraints as given below.

Objective function 1: To minimize the power losses in the jth branch as given in equation (7.1):

$$\text{Minimize } J_1 = \left(\sum_{j=1}^{M} \left| I_j \right|^2 R_j \right) \tag{7.1}$$

Objective function 2: To improve the voltage profile of the ith bus of the consumer as given in equation (7.2):

$$\textbf{Minimize } J_2 = \left(\sum_{i=1}^{N} \frac{v_i - 1}{v_i^{max} - v_i^{min}} \right) p. \ u \qquad (7.2)$$

7.3.2 FORMULATION CONSTRAINTS FUNCTIONS

7.3.2.1 Equality Constraints

The equality constraints are taken as the system power flow equations. At any bus i the real and reactive power flow equations can be stated by equations (7.3) and (7.4), respectively:

$$P_{G,i} - P_{L,i} = \sum_{i=1}^{N_{bus}} |v_i^{(1)}||v_j^{(1)}||y_{i,j}^{(1)}| \cos\left(\theta_{i,j}^{(1)} - \delta_i^{(1)} + \delta_j^{(1)} \right) \qquad (7.3)$$

$$Q_{G,i} - Q_{L,i} = \sum_{i=1}^{N_{bus}} |v_i^{(1)}||v_j^{(1)}||y_{i,j}^{(1)}| \sin\left(\theta_{i,j}^{(1)} - \delta_i^{(1)} + \delta_j^{(1)} \right) \qquad (7.4)$$

7.3.2.2 Inequality Constraints

Inequality constraints that represent the control variable constraints and dependent variable constraints are presented here.

- **Bus voltage constraints:** The bus voltage magnitudes are bound within acceptable operating limits as given below:

$$V^{min} \leq V_i^{rms} \leq V^{max}$$

The deviation in voltage at the ith node is expressed by equation as given below:

$$V_{deviation} = \sum_{i=1}^{N} \left(\frac{|V_{specified} - V_{i,bus}|}{V_{specified}} \right) \quad \text{Where, } i = 2 \dots \dots (\text{No. of buses})$$

- **Branch current constraints:** The current flow through each distribution line must not surpass the thermal capacity (Ampacity) of the line:

$$I_{nj} \leq I_n^{max} \quad \forall j \in M$$

- **Active and reactive power constraints:** The limits for real and reactive power injected by solar PV at a node i are given as below:

$$P_{i,\,min}{}^{Solar} \leq P_i{}^{Solar} \leq P_{i,max}{}^{Solar} \ \forall\, i \in N$$

$$Q_{i,\,min}{}^{Solar} \leq Q_i{}^{Solar} \leq Q_{i,max}{}^{Solar} \ \forall\, i \in N$$

- **Transformer taps:** constraintsIn order to compensate the grid impedance variation the transformer taps must be within limits:

$$T_{i,\,min} \leq T_i \leq T_{i,max} \ \forall\, i \in NT$$

7.4 SOLUTION METHODOLOGY

This section presents the steps for implementation of JAYA algorithms to achieve the optimal solution. The JAYA involves only two controlling parameters: population size and number of generations. It shifts the solution towards the optimal values and discards the mediocre solution. The steps to implement the JAYA algorithm are given below:

Step 1. Read the network line and bus parameters.
Step 2. Implement the N-R power flow algorithm and compute an initial value of bus voltage magnitudes and phase angles.
Step 3. Initialize the number of control variables, population size, iterations count, and bounds of design variables. Generate an initial random population of data within the stated bounds of design variables. Evaluate objective functions considering the defined controlling parameters. This population is formulated as equation (7.5):

$$\text{Population} = \begin{vmatrix} Y_{1,1} & Y_{1,2} & \cdots\cdots & Y_{1,n} \\ Y_{2,1} & Y_{2,2} & \cdots\cdots & Y_{2,n} \\ \vdots & \vdots & \vdots & \vdots \\ Y_{m,1} & Y_{m,2} & \cdots\cdots & Y_{m,n} \end{vmatrix} \quad\quad (7.5)$$

With $k = 1, 2, 3.... \, m$ and $j = 1, 2, 3... \, n$

where n - control variables and m - candidate solutions. The value of jth control variable $Y_{k,j}$ in the kth candidate solution is expressed and written as equation (7.6):

$$Y_{k,j} = Y_j{}^{\min} + ran(.)\left[Y_j{}^{\max} - Y_j{}^{\min} \right] \tag{7.6}$$

Step 4. For each candidate solution, run the power flow program and compute the objective function as equations (7.1) and (7.2) corresponding to each solution.

Step 5. Classify the best and worst solutions from the obtained candidate solutions.

Step 6. Based on the best and worst solutions, modify all candidate solutions. The proposed modification is expressed as given in equation (7.7):

$$Y^M_{j,k,i} = Y_{j,k,i} + r_{1,j,i}\left[\left(Y_{j,best,i} \right) - \left(\left| Y_{j,k,i} \right| \right) \right] - r_{2,j,i}\left[\left(Y_{j,worst,i} \right) - \left(\left| Y_{j,k,i} \right| \right) \right] \tag{7.7}$$

where $Y^M_{j,k,i}$ is the modified value of jth designed variable, $r_{1,j,i}$ and $r_{2,j,i}$ are random numbers for the jth variable during the ith iteration of the term $\left[\left(Y_{j,best,i} \right) - \left(\left| Y_{j,k,i} \right| \right) \right]$ has the tendency to bring solution closer to solution while the term $r_{2,j,i}\left[\left(Y_{j,worst,i} \right) - \left(\left| Y_{j,k,i} \right| \right) \right]$ has tendency to avoid solution to move towards the worst solution. $Y_{j,k,i}$ is accepted if it gives better function values. All better function values become the input for next iteration.

Step 7. Check the control variable lower/upper limit violation for all updated solutions.

Step 8. Check the objective function values for the previous and updated solution. Retain the superior solution.

Step 9. Check the termination criterion for number of iterations and report the optimal solution.

7.5   RESULTS AND DISCUSSION

The efficacy of the proposed JAYA optimization algorithm is examined on a rural distribution network. The rated substation voltage is considered as 100%. The active and reactive power losses with base case and JAYA are given in Figures 7.2 and 7.3. The results show that both power losses are considerably reduced with the optimal parameters obtained from JAYA. Figure 7.4 shows that the power factor improves to almost 100% (1 pu) level, which is the best power factor, as a low power factor would increase the copper losses in the system.

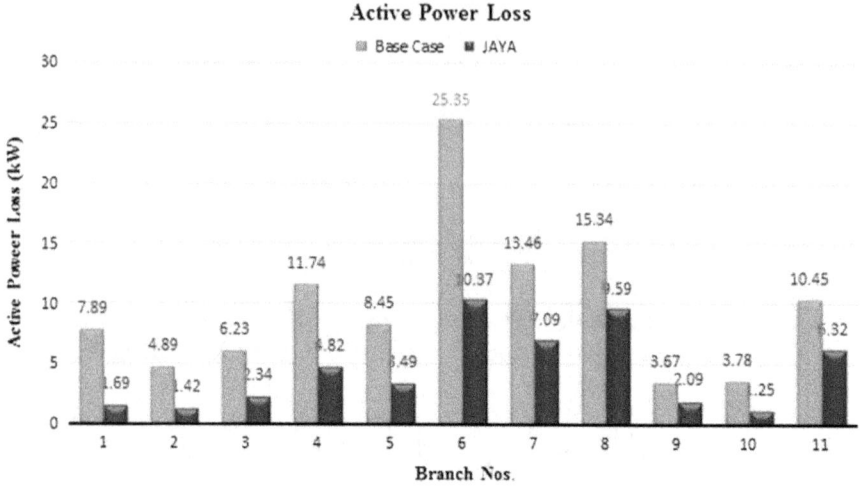

FIGURE 7.2 Active power loss in various branches with base case and JAYA.

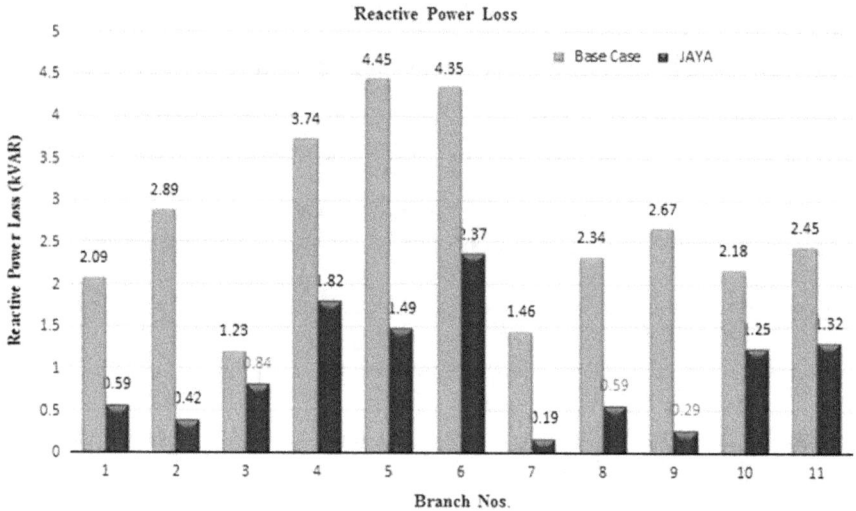

FIGURE 7.3 Reactive power loss in various branches with base case and JAYA.

The voltage profile of system for the base case and JAYA optimization algorithms are presented in Figure 7.5. Comparison results show that the proposed algorithm provides an almost flat voltage profile than the base case. Considerable improvement in the bus voltage profile has been achieved by a JAYA algorithm, which is very close to a linear profile and in the permissible specified range. It is notable that the voltage is improved to 100% by a JAYA algorithm. This shows the competency of algorithm and justifies the fact that it provides the optimal setting of

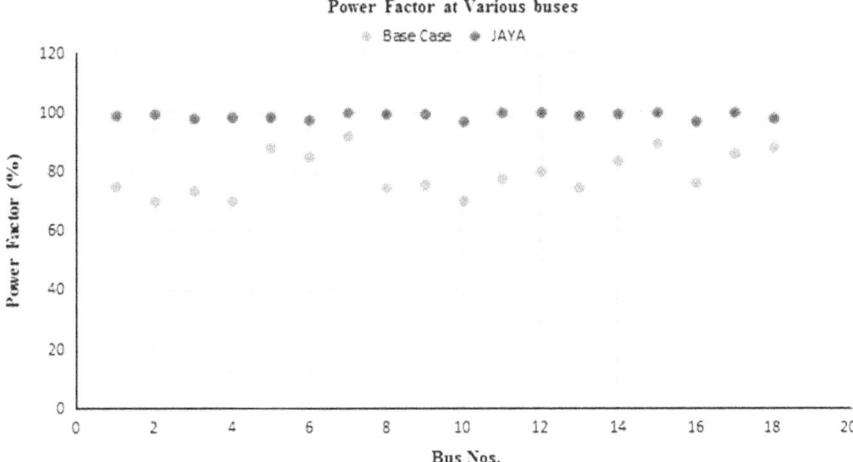

FIGURE 7.4 Power factor at various buses with base case and JAYA.

the controller and associated devices in order to give the best voltage profile for load variation to any limit. Figure 7.6 depicts the branch ampacity (thermal loading) comparison between the base case and JAYA for a given load level. It can be observed that a considerable reduction in thermal loading occurs in the branches. From Table 7.4 it is clear that the power losses reduce considerably by about 68.49% in branch 7.

The decrease in power loss in other branches can also be observed in Table 7.4. This in turn saves substantial money due to reduction in annual energy losses in the system.

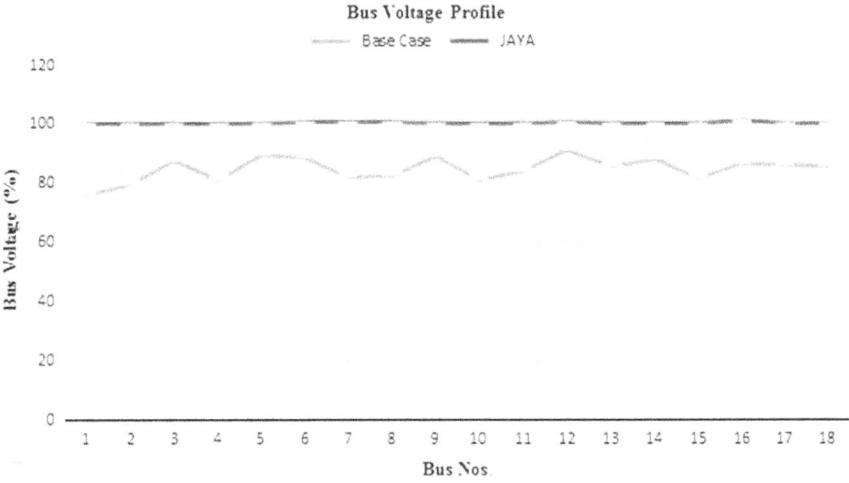

FIGURE 7.5 Bus voltage profile with base case and JAYA.

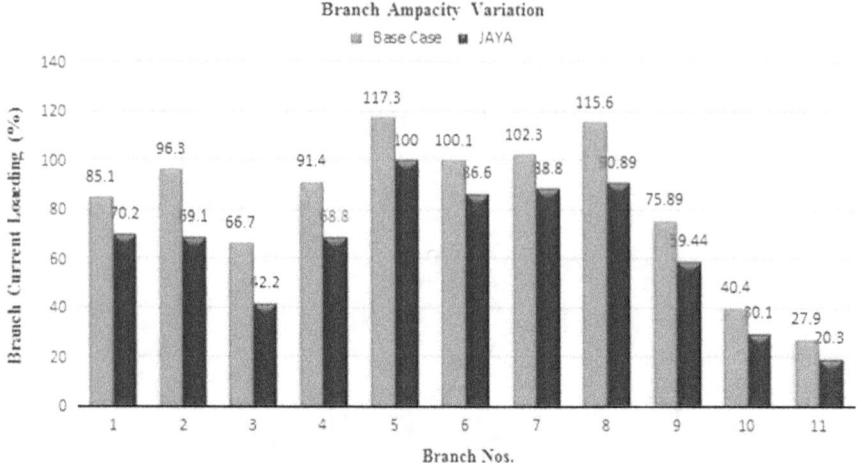

FIGURE 7.6 Comparison of branch ampacity variation with base case and JAYA.

TABLE 7.4
Power Loss Reduction with JAYA in Various Branches

Branch Id.	Base Case Power Loss (kW)	JAYA Power Loss (kW)	JAYA (%) Reduction in Power Loss
1	5.89	2.69	54.32
2	5.89	2.42	58.91
3	7.23	4.34	39.97
4	12.74	7.82	38.61
5	7.45	2.49	66.57
6	15.35	5.37	65.01
7	3.46	1.09	**68.49**
8	5.34	2.59	51.49
9	3.67	2.09	43.05
10	3.78	1.25	66.93
11	9.45	6.32	33.12

7.6 CONCLUSION

The existing networks are accommodating more and more PV energy sources due to their cost-effective and eco-friendly advantages. However, the optimal operation of distribution network is very difficult to achieve due to the varying nature of sources and loads. As the grids are entering in smart grid era, the operation optimization is a crucial need to achieve the grid objectives. This chapter has presented a comparative performance study between an existing un-optimized and optimized distribution network. The main factors used to compare the algorithms are the voltage profile at various buses, branch ampacity, and power losses. The JAYA algorithms are compared with base case in view of the above factors. It has been observed that the proposed algorithm is capable of providing optimal energy management with a cost-effective solution and has better convergence characteristics. The above algorithm can be used to obtain the online optimal parameters adjustments for any number of buses and helps the operator to plan and manage the power demand effectively.

REFERENCES

[1] R. A. Walling, R. Saint, R. C. Dugan, J. Burke, and L. A. Kojovic. Summary of distributed resources impact on power delivery systems, *IEEE Transactions on Power Delivery*, Vol. 23, No. 3, pp. 1636–1644, July 2008.

[2] A. Canova, L. Giaccone, F. Spertino, and M. Tartaglia, Electrical impact of photovoltaic plant in distributed network, *IEEE Transaction on Industry Applications*, Vol. 45, No. 1, pp. 341–347, February 2009.

[3] M. N. Kabir, Y. Mishra, G. Ledwich, Z. Xu, and R. C. Bansal, Improving voltage profile of residential distribution systems using rooftop PVs and battery energy storage systems, *Applied Energy*, Vol. 134, pp. 290–300, December 2014.

[4] T. F. Agajie, B. Khan, H. H. Alhelou, and O. P. Mahela, Optimal expansion planning of distribution system using grid-based multi-objective harmony search algorithm, *Computers & Electrical Engineering*, Vol. 87, pp. 106823, 2020.

[5] B. Khan, H. H. Alhelou, and F. Mebrahtu. A holistic analysis of distribution system reliability assessment methods with conventional and renewable energy sources. *AIMS Energy*, Vol. 7, No. 4, pp. 413–429, 2019.

[6] D. Anteneh and B. Khan, Reliability enhancement of distribution substation by using network reconfiguration a case study at Debre Berhan Distribution Substation, *International Journal of Economy, Energy and Environment*, Vol. 4, No. 2, pp. 33–40, 2019.

[7] R. K. Pachauri et al., Impact of partial shading on various PV array configurations and different modeling approaches: A comprehensive review, *IEEE Access*, Vol. 8, pp. 181375–181403, 2020.

[8] R. K. Pachauri, O. P. Mahela, B. Khan, A. Kumar, S. Agarwal, H. H. Alhelou, and J. Bai, Development of arduino assisted data acquisition system for solar photovoltaic array characterization under partial shading conditions, *Computers & Electrical Engineering*, Vol. 92, pp. 107175, 2021.

[9] R. K. Pachauri et al., Shade dispersion methodologies for performance improvement of classical total cross-tied photovoltaic array configuration under partial shading conditions, *IET Renewable Power Generation*, Vol. 15, pp. 1796–1811, 2021.

[10] M. T. Yeshalem and B. Khan, Design of an off-grid hybrid PV/wind power system for remote mobile base station: A case study. *AIMS Energy*, Vol. 5, No. 1, pp. 96–112, 2017.

[11] Y. Kifle, B. Khan, and P. Singh, Assessment and enhancement of distribution system reliability by renewable energy sources and energy storage, *Journal of Green Engineering*, Vol. 8, Issue 3, No. 2, pp. 219–262, July 2018.

[12] S. Kiros, B. Khan, S. Padmanaban, H. Haes Alhelou, Z. Leonowicz, O. P. Mahela, and J. B. Holm-Nielsen, Development of stand-alone green hybrid system for rural areas. *Sustainability*, Vol. 12, pp. 3808, 2020.

[13] M. Jariso, B. Khan, S. Tanwar, S. Tyagi, and V. Rishiwal, Hybrid energy system for upgrading the rural environment, 2018 IEEE Globecom Workshops (GC Wkshps), Abu Dhabi, United Arab Emirates, 2018, pp. 1–6.

[14] R. M. Ciric, A. P. Feltrin, and L. F. Ochoa, Power flow in four-wire distribution networks-general approach, *IEEE Transaction in Power Systems*, Vol. 18, No. 4, pp. 1283–1290, Nov. 2003.

[15] T. M. Gruzs, A survey of neutral currents in three-phase computer power system, *IEEE Transaction on Industry Applications*, Vol. 26, No. 4, pp. 719–725, July/Aug. 1990.

[16] M. Thomson and D. G. Infield, Impact of wide spread photovoltaic generation on distribution systems, *IET Renewable Power Generation*, Vol. 1, No. 1, pp. 33–40, March 2007.

[17] P. K. Bhatt and S. Y. Kumar, Investigations on operational characteristics of a PV integrated unbalance distribution system for energy management studies, *Current Alternative Energy*, Vol. 2, No. 1, pp. 72–80, 2018.

[18] D. Q. Hung, N. Mithulananthan, and R.C. Bansal, Integration of PV and BES units in commercial distribution systems considering energy loss and voltage stability, *Applied Energy*, Vol. 113, pp. 1163–1170, Jan. 2014.

[19] G. Carpinelli, G. Celli, S. Mocci, F. Mottola, F. Pilo, and D. Proto, Optimal integration of distributed energy storage devices in smart grids, *IEEE Transaction on Smart Grid*, Vol. 4, No. 2, pp. 985–995, June 2013.

[20] N. Jayasekara, M. A. S. Masoum, and P. J. Wolfs, Optimal operation of distributed energy storage systems to improve distribution network load and generation hosting capability, *IEEE Transaction on Sustainable Energy*, Vol. 7, No. 1, pp. 250–261, Jan 2016.

[21] H. Ahmadi and J. R. Martí, Distribution system optimization based on a linear power-flow formulation, *IEEE Transaction on Power Delivery*, Vol. 30, No. 1, pp. 25–33, Feb. 2015.

[22] M. R. Adaryani and A. Karami, Artificial bee colony algorithm for solving multi-objective optimal power flow problem, *International Journal of Electrical Power and Energy System*, Vol. 53, pp. 219–230, Dec. 2013.

[23] S. Sivasubramani and K. S. Swarup, Multi-objective harmony search algorithm for optimal power flow problem, *International Journal of Electrical Power and Energy System*, Vol. 33, pp. 745–752, March 2011.

[24] A. G. Bakirtzis, P. N. Biskas, C. E. Zoumas, and V. Petridis, Optimal power flow by enhanced genetic algorithm, *IEEE Transaction on Power Systems*, Vol. 17, No. 2, pp. 229–236, May 2002.

[25] M. Sailaja Kumari and S. Maheswarapu, Enhanced genetic algorithm based computation technique for multi-objective optimal power flow solution, *International Journal of Electrical Power and Energy System*, Vol. 32, pp. 736–742, July 2010.

[26] S. Duman, U. Guvenc, U. Guvenc, Y. Sönmez, and N. Yörükeren, Optimal power flow using gravitational search algorithm, *Energy Conversion and Management*, Vol. 59, pp. 86–95, July 2012.

[27] X. He, W. Wang, J. Jiang, and L. Xu, An improved artificial bee colony algorithm and its application to multi-objective optimal power flow, *Energies*, Vol. 8, No. 4, pp. 2412–2437, March 2015.

[28] N. Kanwar, N. Gupta, K. R. Niazi, A. Swarnkar, and R. C. Bansal, Simultaneous allocation of distributed energy resource using improved particle swarm optimization, *Applied Energy*, Vol. 185, No. 2, pp. 1684–1693, 2017.

[29] A. Bhattacharya and P. K. Chattopadhyay, Application of biogeography-based optimization to solve different optimal power flow problems, *IET Generation Transmission and Distribution*, Vol. 5, No. 1, pp. 70–80, Jan. 2011.

[30] A. Ananthi Christy, P. Ajay, and D. Vimal Raj, Adaptive biogeography based predator–prey optimization technique for optimal power flow, *International Journal of Electrical Power and Energy System*, Vol. 62, pp. 344–352, Nov. 2014.

[31] K. V. Kumar Kavuturu and P. V. R. L. Narasimham, Multi-objective economic operation of modern power system considering weather variability using adaptive cuckoo search algorithm, *Journal of Electrical Systems and Information Technology*, Vol. 7, No. 11, 2020.

[32] R. V. Rao, V. J. Savsani and D. P. Vakharia, Teaching-learning-based optimization: A novel method for constrained mechanical design optimization problems, *Computer-Aided Design*, Vol. 43, No. 3, pp. 303–315, March 2011.

[33] R. V. Rao Jaya, A simple and new optimization algorithm for solving constrained and unconstrained optimization problems, *International Journal of Industrial Engineering Computations*, Vol. 7, No. 1, pp. 19–34, 2016.

NOMENCLATURE

NT	Number of transformer taps
$P_{G,i}, Q_{G,i}$	Real and reactive power generation at bus i
$P_{L,i}\ Q_{L,i}$	Active and reactive power of load at bus i
$P_i^{solar}\ Q_i^{solar}$	Real and reactive power generated by solar PV at bus i
$rand\,(\bullet)$	Randomly generated numbers between [0 1]
V_i	Voltage of ith bus
V^{min}	Lower bound of voltage (95%)
V^{max}	Upper bound of voltage (105%)
$|y_{i,j}^{(1)}|$	Magnitude of admittance matrix for $(i, j)^{th}$ element
Y_j^{min}	Minimum limits of the jth control variable
Y_j^{max}	Maximum limits of the jth control variable
$\delta_i^{(1)}$	Voltage angle at bus i
$\theta_{i,j}^{(1)}$	Angle of the admittance matrix for $(i, j)^{th}$ element

8 Fuzzy C-Means Clustering and K-NN Regression-Based Protection Scheme for Transmission Lines

Amit Kumar Gangwar and Abdul Gafoor Shaik
Department of Electrical Engineering Indian Institute of
Technology Jodhpur, India

CONTENTS

8.1 INTRODUCTION

The transmission line is the most vital component of the power system, because of exposure to the environment, the chance of fault occurrence is high. To provide uninterrupted power supply to customers, fast and reliable relay algorithms are required. Due to the development of fast computational hardware, there are a large number of computational intelligence techniques. A lot of research has been reported in the literature regarding protection of transmission lines. There are mainly four techniques available in the computational intelligence for fault diagnosis in the transmission lines. These are knowledge-based, systems based on data, optimization-based systems, and hybrid systems. Knowledge-based systems also are known as the expert-based system; there are uncertainties in the data present during solving the problem. These uncertainties are solved by statistical models and probability theory. Fuzzy logic is a good alternative dealing with imprecise information. In [1], Zadeh introduced the concept of fuzzy set theory; a variety of problems in the literature on transmission line fault diagnosis are solved by fuzzy logic. In [2], a fuzzy logic–based approach is used to classify the type of faults.

Data-driven systems are widely used in the fault detection and classification of the transmission lines. These systems mainly depend on signal processing techniques, statistical methods, and machine learning techniques. Signal processing techniques consist of mathematical rules used for the analysis of signals of a different nature. A lot of research has been reported on the application of signal processing techniques for fault detection, classification, and location in the transmission lines. Most of the literature uses fast Fourier transform (FFT) and wavelet transform (WT) for the analysis of oscillography data. In [3–5], uses wavelet transform to detect the faults in transmission lines. [6] uses adaptive wavelet transform along with the Bayesian classifier to classify the faults in the transmission lines. Traveling wave theory along with wavelet transform is used to locate the fault [7,8]. Statistical methods use the concept of statistics (mean, median) to diagnosis the fault in the transmission line. In [9], the median is calculated using a sliding window over voltage and current. The Kalman filter is a statistical method that is also used in the fault diagnosis of the transmission lines. In [10], uses the Kalman filter for fault diagnosis in the transmission line. The machine learning technique is widely used for fault detection and classification. In the literature reported, there are several types of artificial neural networks for fault detection and classification such as multilayer perceptron (MLP) and radial basis function (RBF) networks. In [11], the author uses RBFs for fault classification using current and voltage signals as input. Similarly, in [12], uses MLPs for fault classification of the transmission line. A support vector machine (SVM) is another machine learning tool widely used in fault classification. A SVM uses a hyperplane for classification and maximizes the margin of separation among the different classes. In [13], the author uses a SVM for fault classification, considering the input samples of three phase currents. In [14], the author uses hyperbolic S-transform to extract features from current and voltage, which is input to RBFs for fault classification. A hybrid systems–based technique is widely reported in the literature; it combines more than one technique for identification of fault in the system such as fuzzy sets and artificial neural networks–based systems [15]. In [16], the author presents neural-fuzzy networks in the protection of transmission lines. In [17], the author proposed the hybrid system based on the combination of fuzzy logic, adaptive resonance theory, and neural networks for the protection scheme. In [18], the author combines artificial neural network, particle swarm optimization (PSO), and wavelet transform for fault detection, classification, and location in the transmission lines. The author used a time-frequency approach of current signal to detect and classify the transmission line fault [17]. Centroid difference and support vector regression–based protection scheme of a transmission line is done [18].

The main contribution of this chapter is presented as follows:

1. This chapter used fuzzy C-means clustering and K-NN regression-based protection scheme of transmission line.

2. Fuzzy C-means clustering is used for the fault detection and classification and K-NN regression for the fault location estimation.
3. The first contribution is to reduce the fault detection time to 1/4th cycle after the incidence of fault.
4. The algorithm performs well in the presence of high impedance fault and noise, respectively.
5. The error of the fault location is less than 3% for all types of faults, as shown in Table 8.2.

This chapter is organized in the following way: Section 8.2 mentions the machine leaning algorithm used for the proposed algorithm. Section 8.3 explains the proposed algorithm of fault detection and fault location estimation. Simulation results of fault detection and classification are given in Section 8.4. Finally, Section 8.5 concludes the research work.

8.2 MACHINE LEARNING ALGORITHM USED IN THE PROPOSED ALGORITHM

The proposed algorithm of fault detection and classification used is fuzzy C-means clustering and for fault location estimation K-nearest neighbor regression is used. The mathematics and algorithm steps of both machine learning algorithms is given below.

8.2.1 FUZZY C-MEANS CLUSTERING (FCM)

Let a finite set of data given by $X = x_1, x_2..........x_n$, the FCM algorithm gives a set of cluster centers $C = c_1, c_2.................c_c$.
   The steps of the FCM algorithm are given below:

1. Initially randomly select "c" cluster center.
2. Computes the fuzzy membership w_{ij} by the following formula:

$$w_{ij} = \frac{1}{\sum\limits_{k=1}^{c}\left(\frac{\| xi - cj \|}{\| xi - ck \|}\right)^{\frac{2}{m-1}}} \tag{8.1}$$

where "m" is the fuzziness coefficient.
3. Computes the fuzzy centers c_j:

$$c_j = \frac{\sum\limits_{i=1}^{n} w_{ij}^m x_i}{\sum\limits_{i=1}^{n} w_{ij}^m} \tag{8.2}$$

for $j = 1, 2.........c$
4. Minimizes the following objective function:

$$J(W, C) = \sum_{i=1}^{n} \sum_{j=1}^{c} w_{ij}^{m} \|x_i - c_j\|^2 \tag{8.3}$$

5. Repeat the objective function (2),(3), until the objective function J is minimized.

8.2.2   K-NEAREST NEIGHBOR (K-NN)

K-nearest neighbor is the machine leaning algorithm works on the basis of distance of a test data point from the K-nearest neighbor. The calculated distance from the K-nearest neighbor is arranged in ascending order. The target class of the K-nearest neighbor is computed. To get the output value, the average of the target class is computed.

Let x_i be the feature vector and y_i be the test vector whose label is to be predicted using K-nearest neighbors in the feature space for $i = 1, 2, 3...........n$.

Euclidean distance functions used in the K-NN regression are given below:

$$d(x, y) = \sum_{i=1}^{n}(x_i - y_i)^2 \tag{8.4}$$

8.3   PROPOSED ALGORITHM

The system considered for simulation is given in Figure 8.1. The current and voltage is sampled from both of the buses of the transmission line that is synchronized with a GPS clock. The fault detection and classification is done by fuzzy C-means clustering and K-NN regression is used to compute the fault location. The algorithm is explained in the following sections:

1. Fault detection and classification
2. Fault location estimation

8.3.1   DETECTION AND CLASSIFICATION METHOD

The flow chart of the proposed fault detection and classification is given in Figure 8.2. The three phase currents are obtained from MATLAB® Simulink software; wavelet transform is applied to decompose the signal to get third-level approximate coefficients. Using a moving window, the current quarter cycle approximate coefficients is subtracted from previous quarter cycle approximate coefficients to compute resultant approximate coefficients. Fuzzy C-means clustering is applied on the resultant approximate coefficients to compute two centroids. The difference of centroids is computed to obtain the centroid difference (C.D), as shown in equation (8.5), (8.6), and (8.7) for all three phases. Similarly, the centroid difference of the second terminal of the transmission line is computed. The fault index of phase A is computed by adding the centroid difference of both terminals of the transmission line, as shown in equation (8.8). Similarly, the fault index of phase B and C are calculated. The computed fault index is compared with the threshold value to detect the fault. As shown in equation (8.10), if the fault index of any phase exceeds the threshold value,

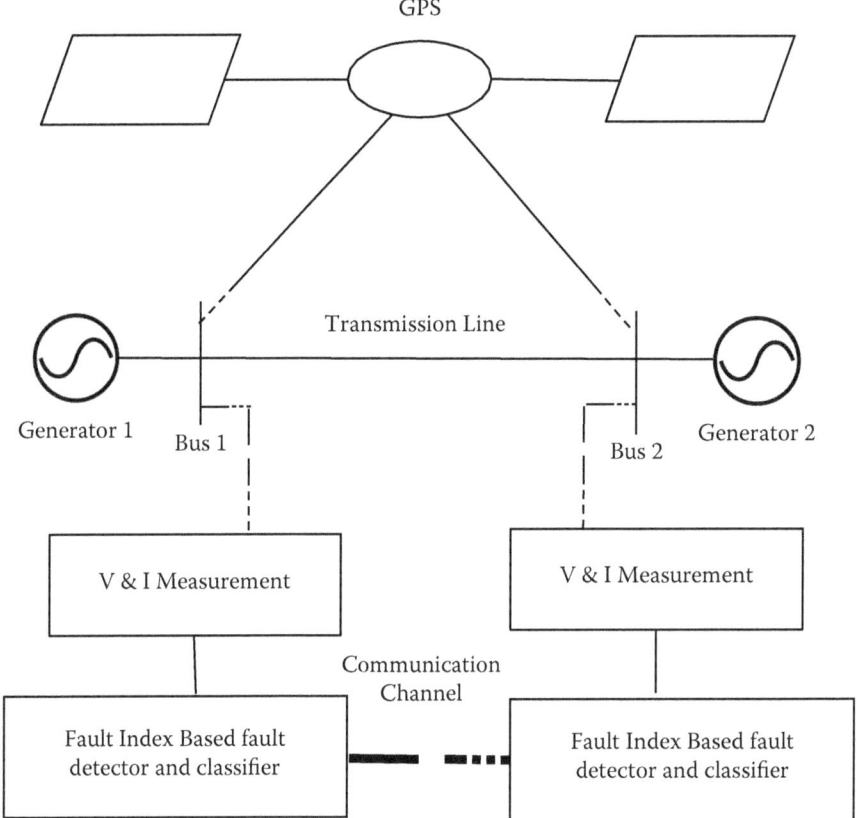

FIGURE 8.1 Two-bus test system incorporated with proposed algorithm.

the fault is detected in the transmission line. For healthy conditions, the the fault index of all the three phases must be below the threshold value, as given in equation (8.11).

For the fault classification, the fault index of the ground current is computed as given in equation (8.9). The ground current is the sum of all three phase currents. On the basis of the fault index, all three phases and fault index of the ground current of the fault is classified. The fault classification is done on the basis of the number of phases for which a fault index exceeds the threshold value, as shown in Figure 8.2.

The centroid difference of phases A, B, and C at bus 1 is calculated by:

$$CD_A^1 = abs\,(|C_1| - |C_2|) \tag{8.5}$$

$$CD_B^1 = abs\,(|C_1| - |C_2|) \tag{8.6}$$

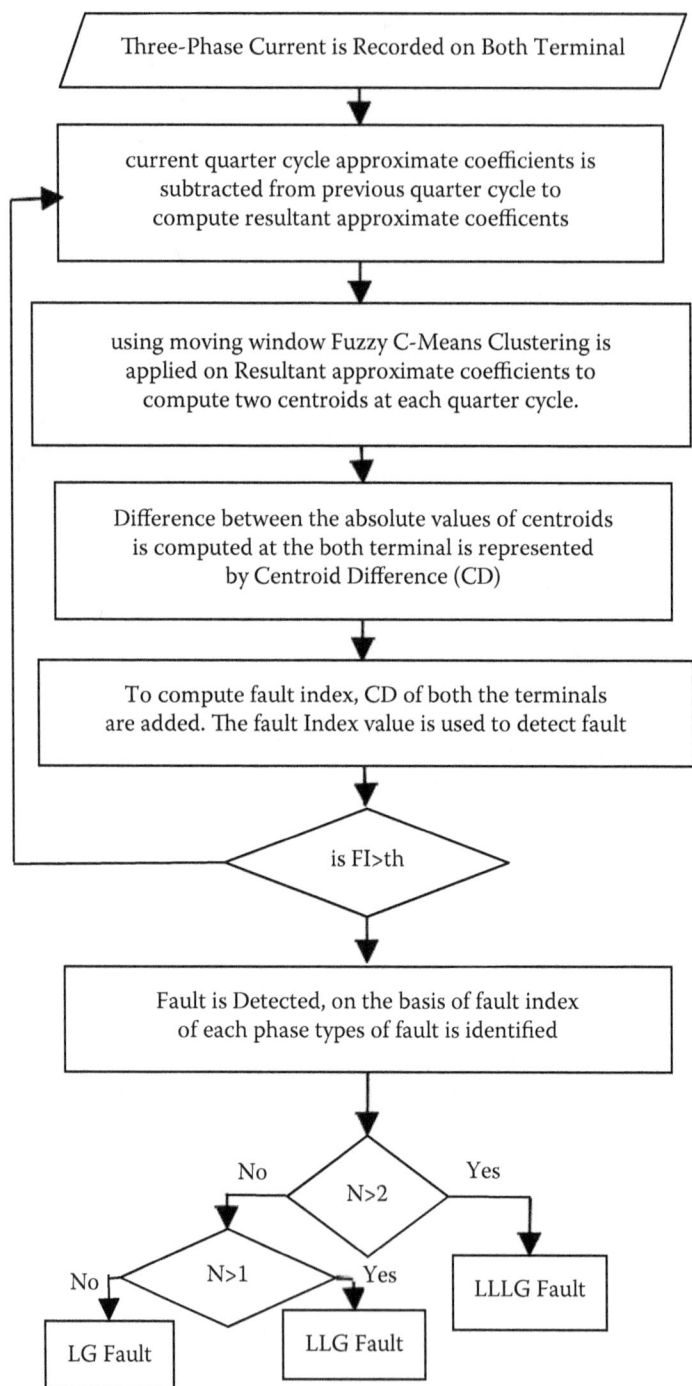

FIGURE 8.2 Flow chart of the proposed algorithm.

$$CD_C^1 = abs\,(|C_1| - |C_2|) \tag{8.7}$$

$$FI_A = CD_A^1 + CD_A^2 \tag{8.8}$$

$$FI_G = abs\,(|C_1| - |C_2|) \tag{8.9}$$

Whether a fault is present or not depends on the following two conditions:

$$Faulty = FI_A \| FI_B \| FI_C > th \tag{8.10}$$

$$Healthy = FI_A \&\& FI_B \&\& FI_C < th \tag{8.11}$$

8.3.2 FAULT LOCATION DETECTION

The fault location is estimated by the K-NN regression method, as shown in Figure 8.3. The ratio of approximate coefficients of voltage and current signals of quarter cycle along with the type of fault information (classified in the previous step) are used to train the K-NN regression model. The training data is obtained by simulating the fault at every 10-km distance. For each type of fault, one K-NN regression model is trained; therefore, 11 K-NN regression models are assigned for 11 types of faults. These are the following steps to estimate the fault location using K-NN regression:

1. Calculate the two nearest neighbors of the test data from the trained data.
2. Calculate the distances of test data from the two nearest neighbors' training data.
3. Assign the class of these two nearest neighbors' training data.
4. Compute the average value of two assigned classes to get the fault location.

$$Location = \frac{Cl_1 + Cl_2}{2}; \tag{8.12}$$

8.4 SIMULATION AND RESULTS

The simulation model parameters are given in Table 8.1. The simulation is done for the different fault parameters, such as various fault impedance, fault incidence angle, and fault location. The effect of noise is also studied.

8.4.1 VARIOUS CASE STUDIES OF FAULT IDENTIFICATION AND CLASSIFICATION

The fault index variation of AG fault is given in Figure 8.4. It is observed that the fault index of phase A exceeds the threshold, while the fault index of phases B and C are

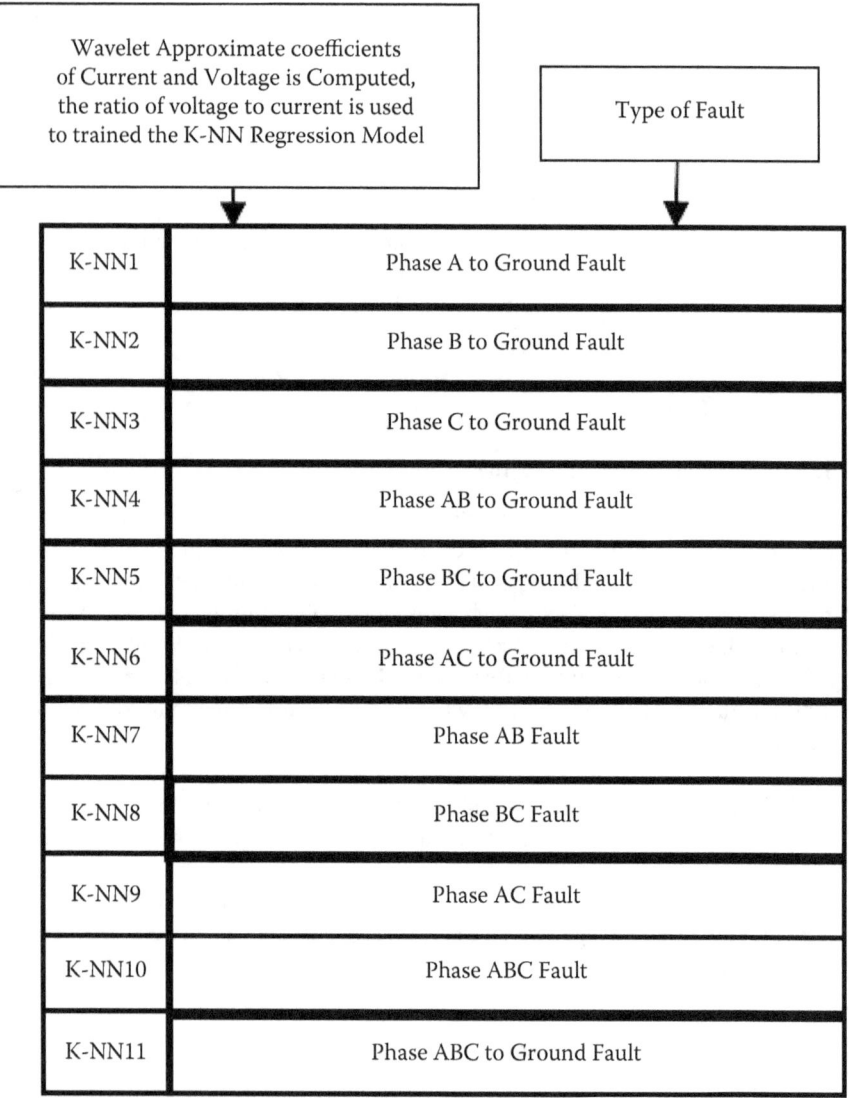

K-NN1	Phase A to Ground Fault
K-NN2	Phase B to Ground Fault
K-NN3	Phase C to Ground Fault
K-NN4	Phase AB to Ground Fault
K-NN5	Phase BC to Ground Fault
K-NN6	Phase AC to Ground Fault
K-NN7	Phase AB Fault
K-NN8	Phase BC Fault
K-NN9	Phase AC Fault
K-NN10	Phase ABC Fault
K-NN11	Phase ABC to Ground Fault

FIGURE 8.3 Block diagram of fault location.

below the threshold value. It indicates the presence of phase A to ground fault present in the line. Similarly, in Figure 8.5, the fault index of phase A, B exceeds the threshold, while the fault index of phase C is below the threshold. It shows the presence of the ABG fault in the line. Similarly, Figure 8.6 shows that all the phases are above the threshold value, which confirmed the presence of the ABCG fault in the line.

The case studies of variation of fault location are done. The fault varies with respect to fault location for the effect of fault location on the proposed algorithm. In Figure 8.7, the fault index variation with respect to fault location is shown for phase A to the ground fault. The fault index of phase A exceeds the threshold all along the

TABLE 8.1
Parameters of the Simulated Model

Parameter of Simulated Model	Value
Length of the Line (km)	230
AC Source 1 (*kV*)	500 ∠ 30
Source-1 equivalent impedance (Ω)	17.177 + j45.529
AC Source 2 (*kV*)	500 ∠ 0
Source-2 equivalent impedance (Ω)	15.31 + j45.925
Positive and Zero Sequence impedance of line (Ω)	4.983 + j117.83 and 12.682 + j364.196
Positive and Zero sequence admittance of line (*mMho*)	j1.468 and j1.099
Apparent Power	433.63 + j294.52 MVA

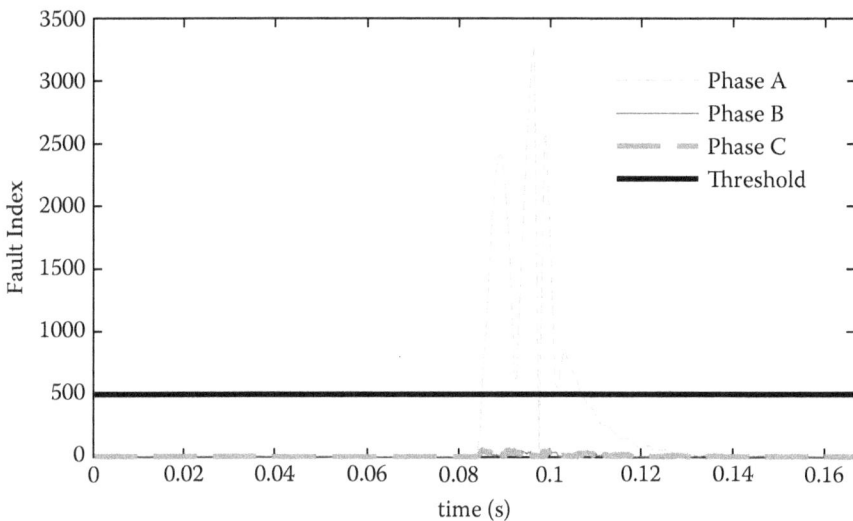

FIGURE 8.4   AG fault at 50% of the line.

line, while for phases B and C is below the threshold value. It shows that the algorithm is robust to variation of fault location. Similarly, Figures 8.8 and 8.9 show variation of the fault index with respect to fault location for ABG fault and ABCG fault, respectively. The case studies of variation of fault index with respect to fault incidence angle is done. Figure 8.10 shows the variation of fault index with respect to fault incidence angle. It is observed that the fault index of a faulty phase i.e., phase A remains above the threshold, for all the incidence angle. While the fault index of phase B and C remains below the threshold. Similarly, in Figure 8.11, the fault index of phases A and B is above the threshold for ABG fault. The fault index of all the phases exceeds the threshold, with respect to variation of fault incidence

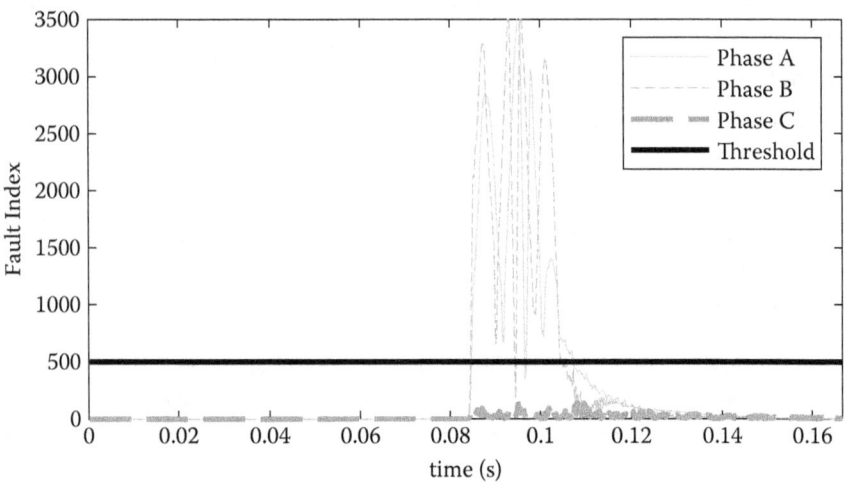

FIGURE 8.5 ABG fault at 50% of the line.

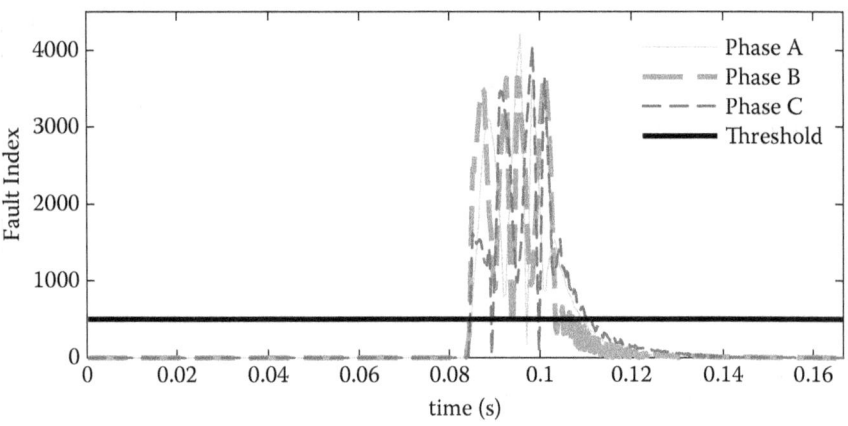

FIGURE 8.6 ABCG fault at 50% of the line.

angle for ABCG fault, as shown in Figure 8.12. Therefore, from the above case studies, it is validated that the proposed algorithm is robust to variation of fault incidence angle. The algorithm is able to detect the fault for any fault incidence angle. The effect of noise on the proposed algorithm of fault detection and classification is studied. The additive white Gaussian noise is added in the sampled current and voltage samples to verify the robustness of the algorithm with respect to noise. Figure 8.13 shows the variation of the fault index with respect to time for AG fault added with 30dB SNR. In this case, the performance of the proposed algorithm is not affected in the presence of noise. Similarly, Figure 8.14 shows the effect of noise on the performance of the algorithm for AG fault at 25dB SNR. In this case also, the algorithm detects and classifies the fault accurately.

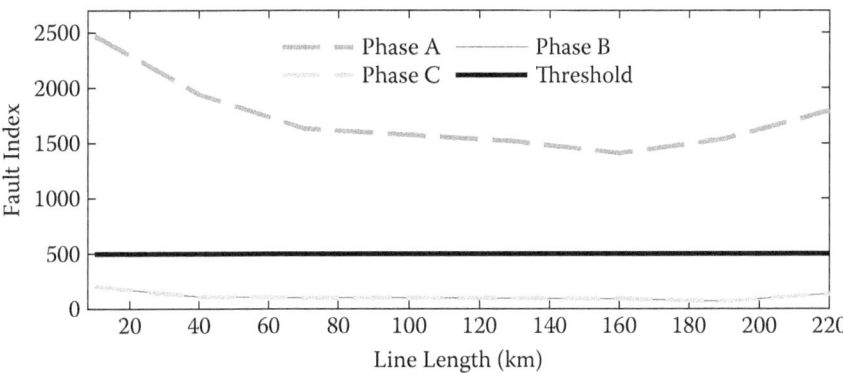

FIGURE 8.7   Variation of fault location for AG fault.

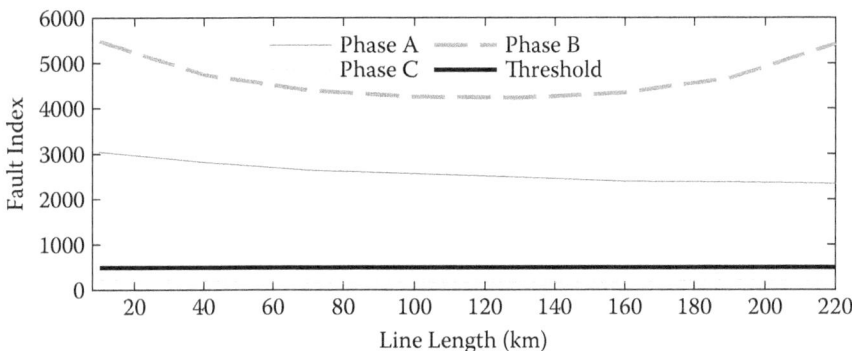

FIGURE 8.8   Variation of fault location for ABG fault.

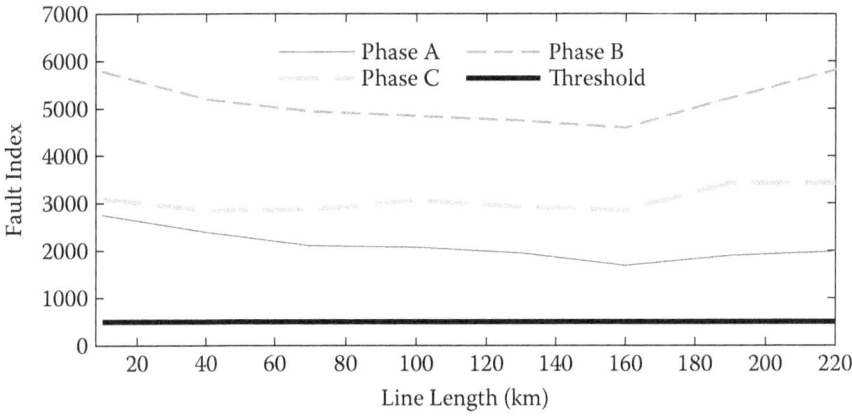

FIGURE 8.9   Variation of fault location for ABCG fault.

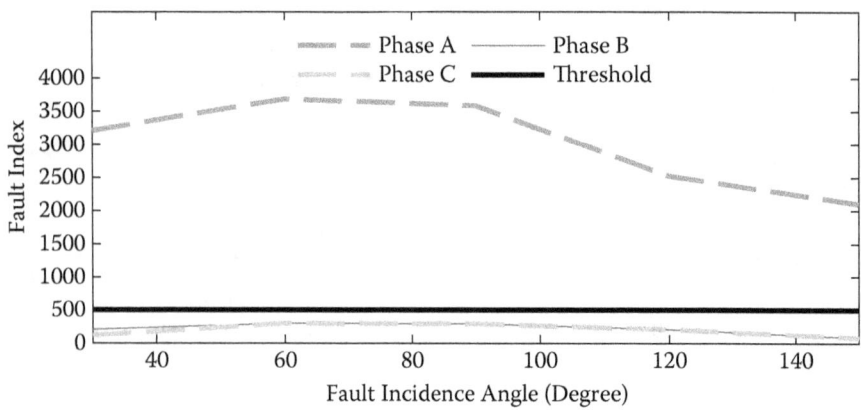

FIGURE 8.10   Variation of fault incidence angle for AG fault.

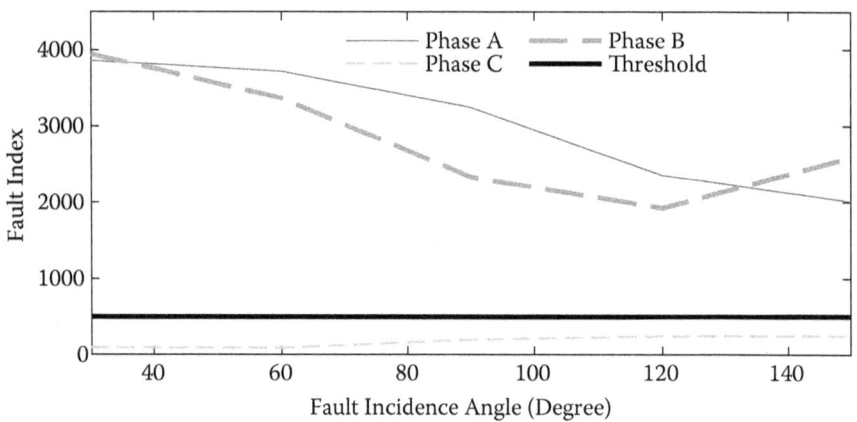

FIGURE 8.11   Variation of fault incidence angle for ABG fault.

8.4.2   SIMULATION OF FAULT LOCATION ESTIMATION

Fault location is estimated by the K-NN regression technique; 11 K-NN regression models are trained from the data of AP_V [n]/AP_I [n] obtained by simulating the faults at every 10 km along the transmission line. Each K-NN regression model is trained with the data of one type of fault. To compute the accuracy of each K-NN regression model, 100 test data are used to predict the fault location for each type of fault. Test data are used, which is not used in the training of K-NN regression. The obtained accuracy of the K-NN regression method is more than 97% for LG fault, as given in Table 8.2. The accuracy of fault location for other types of faults is greater than 98%, given in Table 8.2.

Average absolute error (A.A.E) of one weighted K-NN regression model is calculated by:

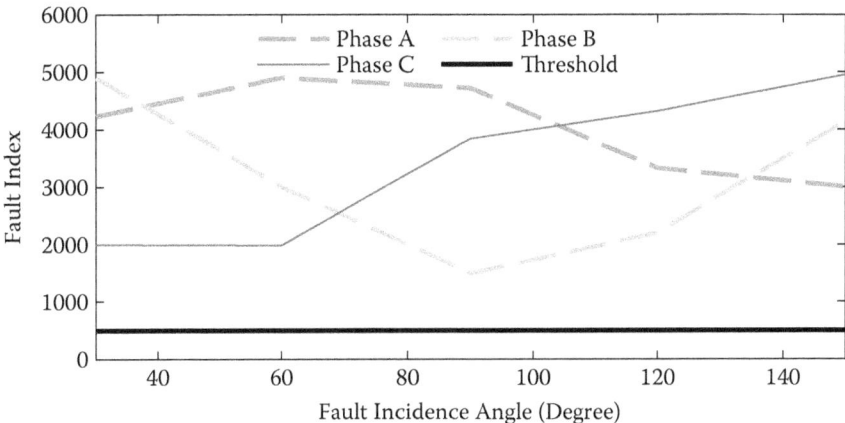

FIGURE 8.12   Variation of fault incidence angle for ABCG fault.

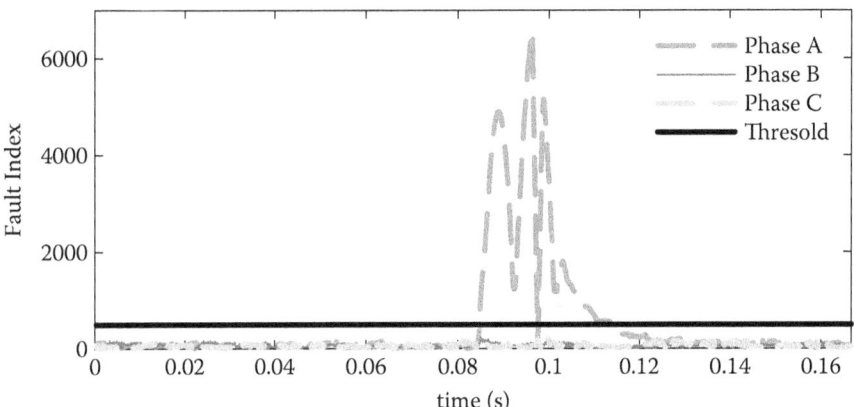

FIGURE 8.13   Effect of AWG noise at 30dB SNR for phase A to ground fault.

$$A.\,A.\,E = \frac{\sum_{i=1}^{n}|L_A - L_P|}{n} \qquad (8.13)$$

where n is the number of test samples, L_A represents the actual fault location, and L_P represents the predicted fault location. The accuracy percent ($AP\,\%$) of each K-NN predictor is calculated by:

$$A.\,P\% = \left(1 - \frac{A.\,A.\,E}{TL}\right) * 100 \qquad (8.14)$$

where TL represents the length of the transmission line.

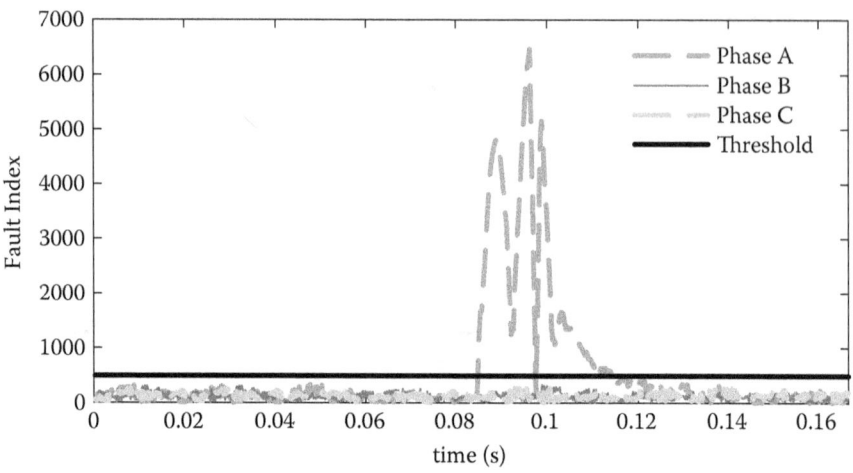

FIGURE 8.14   Effect of AWG noise at 25dB SNR for phase A to ground fault.

TABLE 8.2
Accuracy of the K-NN Regression Model

K-NN Model	Type of Fault	Obtained Accuracy (%)
K-NN1	Phase A to ground fault	97.14
K-NN2	Phase B to ground fault	97.25
K-NN3	Phase C to ground fault	97.27
K-NN4	Phase A, B to ground fault	98.10
K-NN5	Phase A, C to ground fault	98.22
K-NN6	Phase B, C to ground fault	98.11
K-NN7	Phase A, B fault	98.04
K-NN8	Phase B, C fault	98.12
K-NN9	Phase A, C fault	98.15
K-NN10	Phase A, B, C fault	98.40
K-NN11	Phase A, B, C to ground fault	98.55

8.5   CONCLUSION

Fuzzy C-means clustering and K-nearest neighbor—based protection scheme of transmission lines are successfully implemented. The fault detection and classification accuracy is 100% for all types of faults. Various case studies were performed, such as variation of the fault location and variation of the fault incidence angle to verify the robustness of the algorithm. The effect of noise was also studied on the performance of the algorithm. The proposed algorithm gives accurate results up to 25dB SNR. The proposed fault location K-NN regression-based algorithm gives satisfactory performance for all types of faults. The accuracy of the K-NN regression method is greater than 97% for the AG fault, while the obtained accuracy of ABG and ABCG faults is greater than 98%.

REFERENCES

[1] L. A. Zadeh, "Fuzzy sets," *Information and Control*, vol. 8, no. 3, pp. 338–353, 1965.

[2] A. Ferrero, S. Sangiovanni, and E. Zappitelli, "A fuzzy-set approach to fault-type identification in digital relaying," *IEEE Transactions on Power Delivery*, vol. 10, no. 1, pp. 169–175, 1995.

[3] A. Ukil and R. Zivanovic, "Abrupt change detection in power system fault analysis using adaptive whitening filter and wavelet transform," *Electric Power Systems Research*, vol. 76, no. 9, pp. 815–823, 2006.

[4] C. Aguilera, E. Orduna, and G. Ratta, "Fault detection, classification and faulted phase selection approach based on high-frequency voltage signals applied to a series-compensated line," *IEE Proceedings-Generation, Transmission and Distribution*, vol. 153, no. 4, pp. 469–475, 2006.

[5] D. Khalil Ibrahim, E. M. Aboul-Zahab, S. M. Saleh et al., "Real time evaluation of dwt-based high impedance fault detection in ehv transmission," *Electric Power Systems Research*, vol. 80, no. 8, pp. 907–914, 2010.

[6] F. Perez, R. Aguilar, E. Orduna, J. Jäger, and G. Guidi, "High-speed non-unit transmission line protection using single-phase measurements and an adaptive wavelet: zone detection and fault classification," *IET Generation, Transmission & Distribution*, vol. 6, no. 7, pp. 593–604, 2012.

[7] A. Abur and F. Magnago, "Use of time delays between modal components in wavelet based fault location," *International Journal of Electrical Power & Energy Systems*, vol. 22, no. 6, pp. 397–403, 2000.

[8] E. Ngu and K. Ramar, "A combined impedance and traveling wave based fault location method for multi-terminal transmission lines," *International Journal of Electrical Power & Energy Systems*, vol. 33, no. 10, pp. 1767–1775, 2011.

[9] D. Gilbert and I. Morrison, "A statistical method for the detection of power system faults," *International Journal of Electrical Power & Energy Systems*, vol. 19, no. 4, pp. 269–275, 1997.

[10] F. Chowdhury, J. P. Christensen, and J. L. Aravena, "Power system fault detection and state estimation using kalman filter with hypothesis testing," *IEEE Transactions on Power Delivery*, vol. 6, no. 3, pp. 1025–1030, 1991.

[11] W.-M. Lin, C.-D. Yang, J.-H. Lin, and M.-T. Tsay, "A fault classification method by RBF neural network with OLS learning procedure," *IEEE Transactions on Power Delivery*, vol. 16, no. 4, pp. 473–477, 2001.

[12] T. Dalstein and B. Kulicke, "Neural network approach to fault classification for high speed protective relaying," *IEEE Transactions on Power Delivery*, vol. 10, no. 2, pp. 1002–1011, 1995.

[13] U. B. Parikh, B. Das, and R. Maheshwari, "Fault classification technique for series compensated transmission line using support vector machine," *International Journal of Electrical Power & Energy Systems*, vol. 32, no. 6, pp. 629–636, 2010.

[14] S. Samantaray, P. Dash, and G. Panda, "Fault classification and location using hs-transform and radial basis function neural network," *Electric Power Systems Research*, vol. 76, no. 9, pp. 897–905, 2006.

[15] J. C. S. Souza , M. Rodrigues, M. T. Schilling, and M. B. Do Coutto Filho, "Fault location in electrical power systems using intelligent systems techniques," *IEEE Transactions on Power Delivery*, vol. 16, no. 1, pp. 59–67, 2001.

[16] P. Dash, A. Pradhan, and G. Panda, "A novel fuzzy neural network based distance relaying scheme," *IEEE Transactions on Power Delivery*, vol. 15, no. 3, pp. 902–907, 2000.

[17] N. Zhang and M. Kezunovic, "A real time fault analysis tool for monitoring operation of transmission line protective relay," *Electric Power Systems Research*, vol. 77, no. 3, pp. 361–370, 2007.

[18] J. Upendar, C. Gupta, G. Singh, and G. Ramakrishna, "PSO and ANN-based fault classification for protective relaying," *IET Generation, Transmission & Distribution*, vol. 4, no. 10, pp. 1197–1212, 2010.

9 Estimation of Solar Insolation Along with Worldwide Airports Situated on Different Latitude Locations: A Case Study of Rajasthan State, India

Gori Shanakr Sharma
Department of Electrical Engineering, Apex Institute of Engineering and Technology, Jaipur, India

Om Prakash Mahela
Power System Planning Division, Rajasthan Rajya Vidyut Prasarn Nigam Ltd., Jaipur, India

Baseem Khan
Department of Electrical and Computer Engineering, Hawassa University, Ethiopia

Akhil Ranjan Garg
Department of Electrical Engineering, Jai Narain Vyas University, Jodhpur, India

CONTENTS

DOI: 10.1201/b22884-9

9.1 INTRODUCTION

Solar energy is renewable energy that is clean in nature and available in abundance. The solar radiation can be used to provide heat energy, light, and electrical power for domestic as well as industrial applications. Sun radiation from the generation level reaches to the ground level and can be converted into electrical energy using different types of solar cells. Due to easy installation with all necessary accessories in a competitive market, photovoltaic power is the best and least expensive option today in the small power range in remote and rural areas [1–5].

Sun releases radiant energy toward Earth. This radiant energy from the sun, which travels to Earth, is in the form of waves known as electromagnetic radiation (EM). This EM is a stream of photons that carries energy. When this wave of energy strikes a photovoltaic cell, it is converted into electricity. Because of Earth's tilt on its axis, its elliptical orbit around the sun and associated location of the sun between the Tropic of Cancer and the Tropic of Capricorn during a year, the amount of radiant energy falling on Earth varies accordingly. Also, sunrise to sunset movement of sun's location in the sky results in the change of angle of incidence of the EM wave on the solar modules. The maximum energy absorption by solar modules takes place when it faces the sun directly/perpendicularly. To understand the term solar insolation, users need to be familiar with terms like solar energy, radiation, irradiance, solar constant, and effect of atmosphere on solar radiation, which are explained one by one in the following subsections [6–10].

9.1.1 SOLAR ENERGY

The sun is a composition of hydrogen and small amount of helium gas. Under the influence of gravity and electromagnetic fields, these gaseous clouds are found in the form of a heated core. By the nuclear fusion chain reaction, hydrogen gas

transforms into the heavier element helium, which when burned results in the re-
lease of an enormous amount of energy that radiates outward. The radiated energy,
in the form of waves or particles, travels the sun's visible surface (or photosphere)
and escapes into space in the form of heat radiation and light [11].

9.1.2 SOLAR RADIATION

In the core of the sun, by the fusion process, energy radiates from the surface of the
sun in all directions. Because of various atmospheric and environmental factors,
solar radiation travels towards Earth also. Sun energy is released in the form of
photons that strike the Earth. When these photons strike the PN junction, electrons
are released that travel to the junction and concentrate there [12–16].

9.1.3 SOLAR IRRADIANCE

Solar irradiance is the intensity of solar energy impacting an imaginary unit surface
area. It is expressed in watt per square meter (W/m^2) or kilowatt per square meter
(kW/m^2). Solar irradiance is an instantaneous value of energy and does not represent cumulated energy over a period of time. Solar irradiance is used to measure
peak power output performance of any solar power device or a PV module.

9.1.4 SOLAR CONSTANT

In the Earth's outside atmosphere, because of the absence of atmospheric pollutants,
clouds and water particles (which scatter and absorb solar energy) have a solar
irradiance value of approximately 1,367 W/m^2 that is relatively constant and varies
insignificantly over a long period. So 1,367 W/m^2 is called the solar constant [17].

9.1.5 EFFECT OF ATMOSPHERE ON SOLAR RADIATION

When solar radiation enters the Earth's atmosphere, it is partially absorbed and
scattered by ozone, water vapor, carbon dioxide, dust particles, and gases. Some
other factors that diminish the solar radiation strength are cloud, dust storm, pollution, and volcanic eruptions. Radiation impacting the Earth's surface is classified
into two categories: direct radiation and diffuse radiation. Yet another source of
diffusion radiation, called albedo radiation or albedo reflectance, results when the
sun's direct radiation is reflected back into the atmosphere. Direct radiation is
unobstructed solar radiation that travels directly from the sun and impacts the
Earth's surface without scattering. Diffuse solar radiation is dispersed and scattered
and is received at the Earth's surface from many directions. The combined form of
direct and diffuse radiation is called global radiation. At a particular location,
diffuse radiation is the same at all tilt angles of the PV module. To get maximum
output of PV module, it's demanded that PV module get the maximum direct radiation perpendicularly [18,19].

9.1.6 INSOLATION

The amount of solar energy that actually passes through the atmosphere and strikes a given area on the Earth over a specific time is known as the insolation (incident solar radiation). When the sun is directly overhead the insolation, that is the incident energy arriving on a surface on the ground perpendicular to the sun's rays, is typically 1,000 watts per square meter. This is due to

1. Radiation reflected off the atmosphere back into space.
2. Radiation reflected off clouds in the stratosphere.
3. The Earth's surface itself reflects sunlight.

This is due to the absorption of the sun's energy by the Earth's atmosphere, which dissipates about 25% to 30% of the radiant energy. Insolation varies with latitude and with the seasons as well as the weather [20–22].

9.1.7 THE SUN-EARTH RELATIONSHIP

To get maximum PV module electrical output, it is necessary to maximize direct radiation hit perpendicular to the PV module surface. But due to the spherical shape of Earth and also the polar axis tilted on 23.45°, it's a typical task to set the tilt angle of the PV module at different locations. Due to Earth's tilt on 23.45°, a particular location gets different insolation in different seasons. Earth's inclination and rotational path around the sun are explained in Figure 9.1.

- **Tilt angle:** The angle between the horizontal plane and the solar panel is called the tilt angle.
- **Azimuth angle:** The angle between the true south and the point on the horizon directly below the sun is the azimuth angle.

For estimation of solar insolation and the energy in kilowatt hour per square meter (kWh/m^2) per day following at a given locality on the Earth's surface, we need to understand the Earth's centric viewpoint and the parameters related to it. The sun's height above the horizon is its altitude and it changes based on time and season of the year, which is known as Earth's centric view and is explained in Figure 9.2. Based on the sun's altitude changes, the tilt angle of a solar module with respect to the sun must be carefully considered during module or array installations [23].

9.1.8 SOLAR ARRAY ORIENTATION

In Earth's centric viewpoint in Figure 9.2, it is easily seen that the sun's altitude changes according to the day of the year. The sun goes towards the north up to 23.45° (Tropic of Cancer) and toward the south up to 23.45° (Tropic of Capricorn) with reference to the equator. To get absorption of direct radiation at a fixed array, the tilt angle must be equal to the latitude. For better absorption of direct radiation, it is recommended that two seasonal tilt angles, an average tilt angle be adjusted to

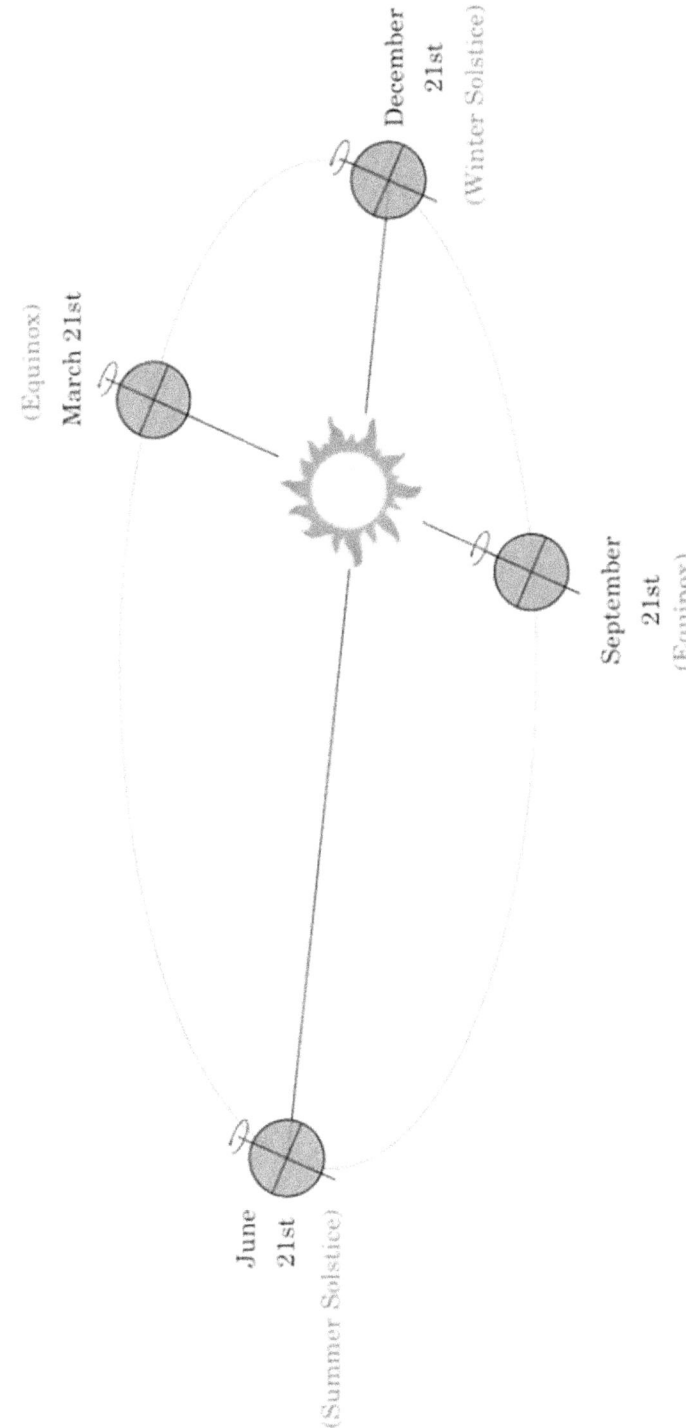

FIGURE 9.1 Sun-Earth geometry.

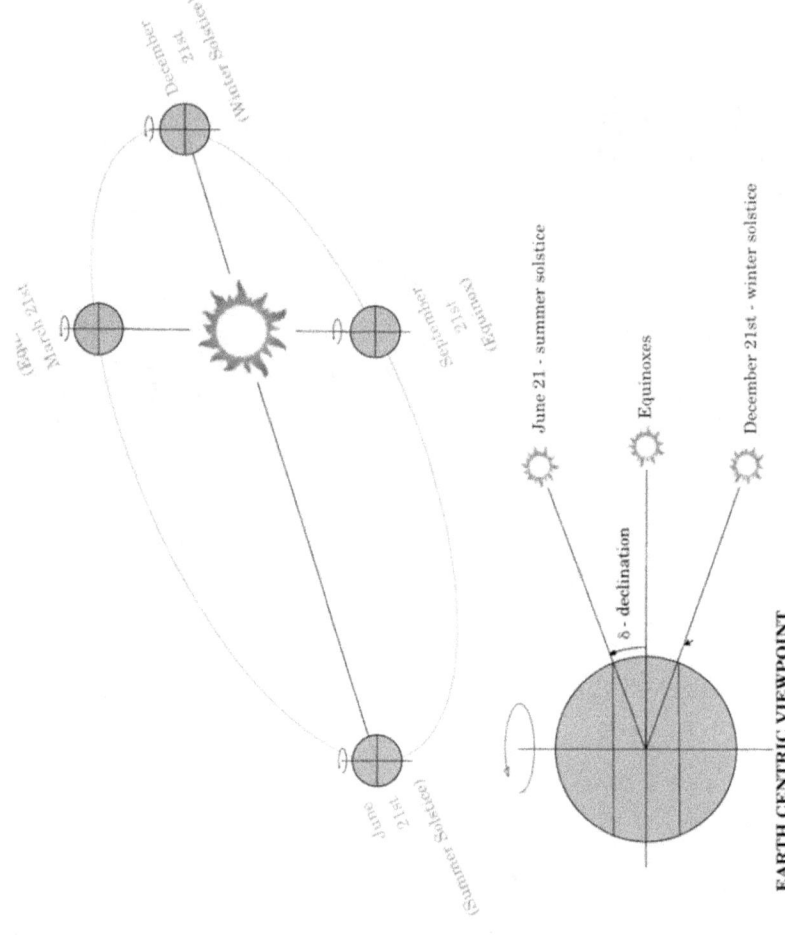

FIGURE 9.2   Earth's centric view of sun.

latitude +15° during winter and latitude −15° during summer, to get the average radiation during a season [24].

9.1.9 SOLAR RADIATION MEASUREMENT DEVICE

Solar irradiance varies moment to moment and is instantaneous. To calculate an accurate energy output performance, there is demand for a solar radiation measurement device.

A. Pyranometer

A pyranometer instrument is used to measure global solar irradiance within the solar aperture or field of view. The instrument is used to measure direct and diffuse radiation on the plain of the PV module. How it works is it converts the striking photon energy to millivolts and gives its value on the scale calibrated in watts/meters.

B. Pyrheliometer

A pyrheliometer is an instrument that measures only direct solar irradiance in the field of view. It does not measure any component of diffuse irradiance.

9.2 CALCULATION FOR CUMULATIVE INCLINED RADIATION BY USING INSTANT RADIATION WITH CLASSICAL APPROACH (AVERAGE METHOD)

For a specific day or hour or minute period, cumulative radiation is found by using a pyranometer provided instantaneous radiation (W/m²) data. By setting the time period of the pyranometer, we can get radiation in a time period according to demand (every 5 min or 10 min or 15 min). Cumulative inclined radiation is calculated by finding the area under the curve (between time and inclined radiation W/m²).

Generally, to calculate the area under the curve, an average method is used. For calculation of cumulative inclined radiation, pyranometer data of radiation is taken from the weather station from Rajasthan in Table 9.1. A pyranometer logger provides instantaneous data in each 10-minute variation; if any error occurs in the fetch data then the logger fetches the data in the next minute.

In Figure 9.3, a graph is shown with radiation (w/m²) and time (in minute). So by the average method, the area calculated is as shown by the graph.

Average inclined radiation = 476.43 W/m²

Time duration (from 6.58 to 17.49) = 10:51 hours

$$\text{So, cumulative inclined radiation} = \frac{476.43 \left(\frac{w}{m^2} \right)}{1000} * \left(10 + \frac{51}{60} \right) hrs$$

$$= 5.169347 \left(\frac{kWh}{m^2} \right)$$

TABLE 9.1

Inclined Radiation Details on 17.10.2018 at Bikaner

Entry Time (IST)	Inclined Radiation (W/m²)	Entry Time (IST)	Inclined Radiation (W/m²)	Entry Time (IST)	Inclined Radiation (W/m²)
17-10-2018 06:00	0	17-10-2018 10:25	663	17-10-2018 14:43	599
17-10-2018 06:10	0	17-10-2018 10:36	686	17-10-2018 14:54	570
17-10-2018 06:21	0	17-10-2018 10:46	700	17-10-2018 15:04	538
17-10-2018 06:31	0	17-10-2018 10:56	719	17-10-2018 15:14	503
17-10-2018 06:41	0	17-10-2018 11:07	731	17-10-2018 15:25	469
17-10-2018 06:58	5	17-10-2018 11:17	749	17-10-2018 15:35	439
17-10-2018 07:08	24	17-10-2018 11:27	765	17-10-2018 15:45	404
17-10-2018 07:18	52	17-10-2018 11:38	770	17-10-2018 15:56	365
17-10-2018 07:28	82	17-10-2018 11:49	768	17-10-2018 16:06	332
17-10-2018 07:39	117	17-10-2018 11:59	788	17-10-2018 16:16	293
17-10-2018 07:49	158	17-10-2018 12:09	788	17-10-2018 16:27	260
17-10-2018 08:00	209	17-10-2018 12:20	796	17-10-2018 16:37	219
17-10-2018 08:10	248	17-10-2018 12:30	791	17-10-2018 16:47	188
17-10-2018 08:20	283	17-10-2018 12:40	793	17-10-2018 16:57	149
17-10-2018 08:31	320	17-10-2018 12:51	788	17-10-2018 17:08	110
17-10-2018 08:41	355	17-10-2018 13:01	772	17-10-2018 17:18	70
17-10-2018 08:51	387	17-10-2018 13:11	768	17-10-2018 17:29	42
17-10-2018 09:02	427	17-10-2018 13:21	767	17-10-2018 17:39	21
17-10-2018 09:12	459	17-10-2018 13:32	749	17-10-2018 17:49	1
17-10-2018 09:23	492	17-10-2018 13:42	730	17-10-2018 18:00	0
17-10-2018 09:33	520	17-10-2018 13:52	705	17-10-2018 20:43	0
17-10-2018 09:44	554	17-10-2018 14:02	689	17-10-2018 20:53	0
17-10-2018 09:54	580	17-10-2018 14:13	668	17-10-2018 21:04	0
17-10-2018 10:04	608	17-10-2018 14:23	638	17-10-2018 21:15	0
17-10-2018 10:15	635	17-10-2018 14:33	624	17-10-2018 21:25	0

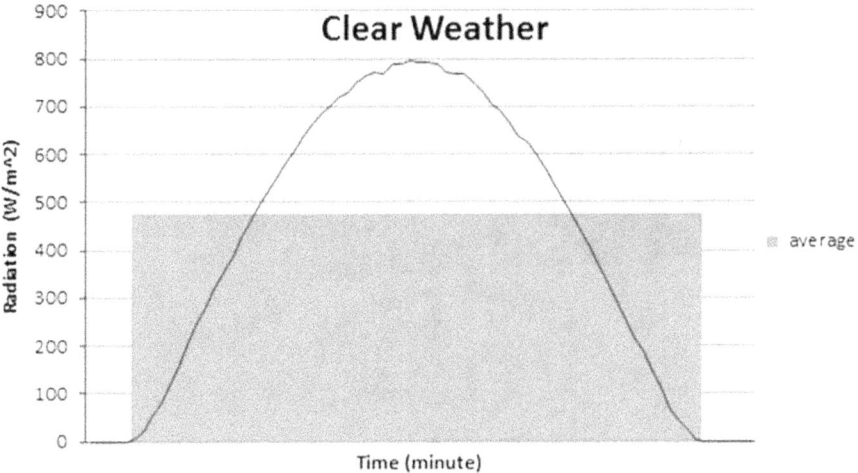

FIGURE 9.3 Graphical representation of cumulative radiation using the average method.

9.3 RELATIVE STUDY OF GLOBAL SOLAR INSOLATION IN BIKANER, RAJASTHAN DESERT

A solar PV plant completely depends upon solar insolation falling perpendicular on the module. As the sun moves from east to west during the day, it is difficult for the module to face the sun rays directly for the whole day. To achieve this, the tilt angle is decided at each location. Alternatively, we can move the module by tracking the sun and solar insolation depending on the day of the year, hour of day, and tilt angle of the PV array. To get maximum solar insolation, there are many possibilities in tilt angle and azimuth angle, as shown in Figure 9.4.

In dual axis tracking, tracking east to west (hours of day) with north to south (day of year according to sun movement from Tropic of Cancer to Capricorn or vice versa) direction. In single axis tracking, it can track only in the east to west direction. Due to very high installation and operation and maintenance cost of single and dual axis tracking plants, in Rajasthan (India), mainly prefer a fix tilt angle and two or four tilt seasonal angles. But due to unavailability of skilled labor to change the tilt angle in the desert area (careless handling of PV module mounting structure (MMS) have micro-cracks and life of MMS reduced and also PV module get damaged), mainly the fix tilt angle or maximum two seasonal (summer and winter) tilt angles for a large MW capacity solar PV plant is preferred. So for a maximum radiation and maximum electricity generation, it is necessary to study solar insolation on different tilt angles and azimuth angles.

9.3.1 SELECTION OF PV MODULE DIRECTION (AZIMUTH ANGLE) TO GET MAXIMUM SOLAR INSOLATION

Fix tilt angle solar PV plants have low installation and maintenance costs compared to seasonal solar PV plant costs (extra cost of seasonal tilt MMS structure and cost

FIGURE 9.4 Directions and tilt angle of the solar PV module.

in change tilt angle procedure). So for a fix tilt angle plant to study of sun-earth geometry, we can predict the angle equal to the latitude of location and direction of PV module in the south (if the location is in the Northern Hemisphere) or in the north (if the site location is in the Southern Hemisphere), but it is difficult to predict the unit electricity generation in digital format. So for prediction of solar radiation in the KWh/m^2 format, we use the PVSYST software. PVSYST can provide solar global radiation (direct + diffuse radiation) very easily according to the azimuth and tilt angle of a solar PV plant. PVSYST considers true south as zero degrees and true Nnorth is +180 degrees. Towards the east it's a considered positive angle and the west direction is considered negative angle. In detail, it is explained in Table 9.2.

By considering a particular location (LAT, LONG), we can find the cumulative inclined radiation for the location Bikaner (28N, 73.1E), Rajasthan (India), Figure 9.5 shows radiation values (KWh/m^2) on different azimuth and different tilt angles. In this figure, in the x-direction (left to right) is the azimuth angle and in the y-direction (top to bottom) is the tilt angle (degree) in increasing format. In practical considerations, due to use of land, there is also a limitation on tilt angle also for getting maximum direct radiation. Solar insolation in different directions and tilt angles is provided in Figure 9.5.

Figure 9.5 shows the global inclined radiation (kWh/m^2) azimuth angle vs. tilt angle of module for a pictorial representation according to their values. In this figure, the red color represents maximum value radiation, yellow color represents medium value radiation, and the green color represents low-value radiation.

For the Bikaner location (28N, 73.1E), we conclude results are explained below one by one:

TABLE 9.2

Relation between Compass Head and Azimuth Angle

Compass Head	Azimuth Angle
North	180
Northeast	135
East	90
Southeast	45
True South	0
Southwest	−45
West	−90
Northwest	−135
North	−180

1. There is zero-degree tilt angle that has the same values 1,903 KWh/m^2 in all azimuth angles because it represents global horizontal radiation and direct radiation are the same.
2. For maximum radiation, the best direction is south (zero-degree azimuth angle, shows red color).
3. To get maximum radiation, we increase the tilt angle and get a maximum solar radiation 2,117 KWh/m^2 on 30°S, nearly equal to latitude. So it is the best angle for maximum radiation at 28 North latitude with south orientation.
4. When we increase the tilt angle more than 30°, then there is a decrease in radiation values due to the sun's path.
5. When the shifting direction of the PV module towards the southeast or -west, then there is a decrease in radiation values.
6. If the PV module orientation is in the north direction, then we increase the tilt angle, and then lose more direct radiation.
7. To get maximum radiation in the north direction, the best suitable tilt angle is 0°.
8. In the north direction, we can get better efficiency using a thin film module (give good generation in diffuse radiation).
9. The poly and mono crystalline Si PV module give better efficiency on direct radiation, so for maximum electrical output, select the south orientation.

9.3.2 SELECTION OF TILT ANGLE OF A SOLAR PV MODULE AT A PARTICULAR LOCATION

By previous analysis, we can understand that the orientation for maximum radiation for a fixed tilt angle plants is south. For the seasonal tilt angle, only change the tilt angle

| Direction | North | | North-East | | East | | South-East | | True South | | south-West | | west | | North-West | | North |
Azimuth Angle \ Module Tilt	180	150	135	120	90	60	45	30	0	-30	-45	-60	-90	-120	-135	-150	-180
0	1903	1903	1903	1903	1903	1903	1903	1903	1903	1903	1903	1903	1903	1903	1903	1903	1903
5	1840	1849	1859	1873	1905	1937	1951	1961	1969	1960	1950	1936	1904	1871	1858	1848	1840
10	1766	1784	1806	1833	1898	1961	1987	2007	2024	2006	1985	1958	1895	1831	1804	1783	1766
15	1682	1712	1745	1787	1887	1978	2014	2042	2066	2040	2012	1975	1883	1784	1742	1710	1682
20	1590	1632	1679	1738	1867	1985	2032	2067	2095	2064	2028	1981	1863	1734	1676	1630	1590
25	1489	1548	1609	1684	1885	1985	2038	2079	2112	2076	2034	1980	1840	1679	1605	1545	1489
30	1384	1459	1537	1628	1815	1975	2035	2080	2117	2077	2031	1969	1809	1622	1532	1455	1384
35	1285	1376	1467	1572	1783	1950	2014	2062	2099	2058	2008	1945	1776	1566	1463	1372	1285
40	1187	1289	1394	1513	1742	1924	1990	2039	2077	2035	1984	1918	1735	1507	1388	1285	1187
45	1099	1202	1319	1451	1702	1887	1956	2005	2043	2000	1950	1880	1694	1444	1314	1198	1099
50	1011	1120	1249	1390	1650	1845	1911	1960	1997	1956	1905	1838	1641	1383	1243	1115	1011
55	931	1040	1178	1326	1601	1793	1859	1905	1938	1900	1852	1785	1593	1319	1171	1036	931

FIGURE 9.5   Solar insolation in different directions and tilt angles.

and freeze the azimuth angle in the south direction (in the Northern Hemisphere). So there is a need to calculate the tilt angle for different seasons according to month. To find a suitable seasonal tilt angle, use the METEONORM software. METEONORM software is a data source for engineering simulation programs in the passive, active, and photovoltaic application of solar energy with comprehensive data interfaces. It is used for climatological calculations. METEONORM has 8,350 weather stations all over the world. The METEONORM software contains climatological data for solar engineering applications at every location. It represents an average year of the selected climatological time period based on the user's settings. The METEONORM radiation database is based on 20-year measurement periods; the other meteorological parameters are mainly 1991–2010. METEONORM is a standardization tool that permits developers and users of engineering design programs access to a comprehensive, uniform meteorological database. METEONORM is a meteorological reference for environmental research, agriculture, and forestry. METEONORM software can provide a database according to the user. It has feasibility to edit the location (latitude, longitude), tilt angle, and azimuth angle and it provides data in graph and table formats.

By selecting a particular location, like (28N, 73.1E) Bikaner, Rajasthan, India, and by changing the value of the tilt angle, we get the value of radiation of all 12 months (January to December). Figure 9.6 considers the tilt angle on the x-axis (left to right) and the month on the y-axis (top to bottom, September to December to January to August for simplicity of the seasons summer and winter).

In Figure 9.6, the global inclined radiation (kWh/m^2), each month V/s tilt angle of module, is presented in a pictorial form according to their values. In this figure, the red color represents the maximum value radiation, the yellow color represents the mean value radiation, and the green color represents low-value radiation.

For the Bikaner location (28N, 73.1E), final results are in Table 9.3 and explained one by one:

1. For a fixed tilt angle, we had a maximum radiation on angle 30° is 2,120 KWh/m^2.
2. For two seasonal tilts, there are three choices:
   a. By using tilt angle (5° and 30°), we can increase radiation 2 to 2.5%.
   b. By using tilt angle (13° and 43°), we can increase radiation 3.2%.
   c. It also follows the rule for two seasonal tilt angles equal to (latitude +15°) in the winter season and (latitude −15°) for the summer season.
3. By using three seasonal tilt angles (5°, 28°, 45°), we can increase 4.6% radiation. It's a great increment in radiation.
4. By using four seasonal tilt angles (5°, 23°, 38°, 45°), we can increase 4.7% radiation. It's also good increment as compare to fix and two seasonal tilt angles, but compared to three seasonal tilt angles, there is only 0.1% increment, so it's not preferable.
5. For multiple seasonal tilt angles at a solar PV plant, only selection of tilt angle is not important; along the tilt angle month selection is more important.

BIKANER (28, 73.1) SOLAR INSOLATION ANALYSIS ON DIFFERENT ANGLE

TILT ANGLE	1	3	5	8	10	13	15	18	20	23	25	28	30	32	35	38	40	43	45	48	50	53	55	58	60
Sep	171	173	175	178	179	181	182	183	184	185	185	185	185	184	183	182	181	179	177	174	172	169	167	163	160
Oct	148	151	154	159	161	165	167	170	172	174	176	178	178	179	180	181	181	180	180	179	178	177	175	173	171
Nov	125	130	134	140	144	149	153	158	161	165	168	171	173	175	178	180	181	183	184	185	185	185	185	184	183
Dec	106	110	114	119	123	128	131	136	139	143	145	149	151	153	156	158	159	161	162	163	164	164	164	164	164
Jan	115	119	123	129	133	138	141	146	149	153	156	159	162	164	166	169	170	172	173	174	174	174	174	174	174
Feb	134	138	142	147	150	155	158	162	165	168	170	173	175	176	178	179	180	181	181	181	181	180	179	178	177
Mar	172	174	177	181	183	186	188	190	192	193	194	195	195	195	195	195	194	193	192	190	188	185	183	180	177
Apr	199	200	202	204	205	206	206	206	206	206	205	204	203	202	199	197	195	191	189	184	182	177	173	168	164
May	209	209	209	209	209	208	207	205	204	202	200	197	194	192	188	184	181	176	173	167	163	157	153	147	142
Jun	193	193	193	192	191	190	189	186	185	182	180	177	174	172	168	164	160	156	152	147	143	137	133	127	123
Jul	177	177	177	176	176	175	174	172	171	168	167	164	162	160	156	152	150	146	143	138	135	130	126	121	117
Aug	173	174	174	175	175	175	175	174	173	172	171	169	168	166	163	161	158	155	153	149	146	141	138	134	130
Total	1922	1949	1974	2008	2028	2055	2070	2089	2099	2111	2116	2120	2120	2119	2112	2101	2091	2072	2057	2031	2011	1977	1952	1912	1883

FIGURE 9.6   Global inclined radiation month wise with different tilt angles with true south.

TABLE 9.3

Summary of Global Inclined Radiation in Different Seasons

S. No.	Number of Tilt Angles	Tilt Angle	Tilt Angle Month	Radiation in Month at That Tilt Angle (kWh/m2)	Yearly Radiation (kWh/m2)	Radiation % Increment after Increasing Number of Tilt Angles
1	Fixed tilt angle	30°	All 12 months		2,120	
2	Two seasonal tilts					
	(i) A. Summer Tilt Angle	5°	(Apr, May, June, July, Aug, Sept)	1,128	2,162	2.0%
	B. Winter Tilt Angle	30°	(Oct, Nov, Dec, Jan, Feb, March)	1,034		
	(ii) A. Summer Tilt Angle	5°	(May, June, July, Aug)	752	2,174	2.5%
	B. Winter Tilt Angle	30°	(Sept, Oct, Nov, Dec, Jan, Feb, March, April)	1,422		
	(iii) A. Summer Tilt Angle	13°	(May, June, July, Aug)	748	2,188	3.2%
	B. Winter Tilt Angle	43°	(Sept, Oct, Nov, Dec, Jan, Feb, March, April)	1,440		
3	Three seasonal tilt angles	5°	(May, June, July, Aug)	753	2,217	4.6%
		28°	(Mar, Apr, Sept)	584		
		45°	(Oct, Nov, Dec, Jan, Feb)	880		
4	Four seasonal tilt angles	5°	(May, June, July, Aug)	753	2,220	4.7%
		23°	(Sept, Apr)	391		
		38°	(Mar, Oct)	376		
		45°	(Nov, Dec, Jan, Feb)	700		

9.4 RELATIVE STUDY OF CUMULATIVE INCLINED RADIATION ON DIFFERENT LATITUDES USING METEONORM SOFTWARE

Favorable data:

1. Selection of Tilt Angle
   A. For Fixed Tilt Angle – To get absorption of direct radiation at a fixed array, the tilt angle must be equal to the latitude
   B. For Two Seasonal Tilt Angles – To get better absorption of direct radiation, it is recommended that an average tilt angle be adjusted to latitude +15° during winter and latitude –15° during summer
2. Azimuth angle for Northern Hemisphere is 0° (true south) and for the Southern Hemisphere is 180°S (true north)

To study radiation on different locations of the world according to their latitude from the North Pole to South Pole along with the effect of two seasonal tilt angles, 14 places are selected.

These select locations are airports of different countries. In short, all 14 locations are shown in Table 9.4 with country name, city name, latitude, and longitude. By using METEONORM software, according to their tilt angle, global radiation data are collected in a simple format. In the table format from Table 9.4, all 14 locations showed global radiation in kWh/m². according to suitable tilt angle along with month-wise seasonal tilt angle. The global radiation also affects the altitude of location, so altitude is also taken along with location coordinate. Maximum considerable tilt angle is 90°. Beyond 90°S, it's not allowable due to practical consideration; the PV module cell will be facing toward the ground, not the sun along with pitch (distance between back to back modules).

For first 10 locations of the Northern Hemisphere tilt angle (LAT – 15 used from Apr–Sept and LAT + 15 used from Oct–March) and for last four locations tilt angle (LAT – 15 used from Oct–Mar and LAT + 15 used from Apr–Sept).

Table 9.4 was prepared according to latitude and location Northern Hemisphere to Southern Hemisphere (+90° to 0° to –90°) locations where human or land availability are a possibility. In Table 9.4, this has also been considered the ambient average temperature, taken by use of the METEONORM software.

Figure 9.7 presents the graphical representation of radiation on fixed and two seasonal tilt angles worldwide. The results are as follows:

1. By graph, it's easily understood by using two seasonal tilt angles that it can get more global radiation as compared to fix tilt angle.
2. By using two seasonal tilt angles, it can get 3.75% more radiation at any latitude as compared to fixed tilt.
3. Moving toward the Earth pole (both North Pole and South Pole) from the equator, there is decrease in radiation at a fixed tilt surface.
4. In the selected 14 places, maximum radiation reaches Earth's surface on fixed tilt surface at Cape Town, South Africa. Cape Town gets 5% more radiation as compared to Rajasthan.

TABLE 9.4

Global Radiation (kW/m²) Comparison of Different Latitude Locations

S. NO.	LOCATION NAME		LATITUDE	LONGITUDE	ALTIT-UDE	FIXED TILT ANGLE All 12 Months	TWO TILT ANGLES (Degree) Apr to Sep	TWO TILT ANGLES (Degree) Oct to Mar	AMBIENT TEMP (°C)	Global Solar Radiation (kWh/m²) Fixed Tilt Radiation	Global Solar Radiation (kWh/m²) Two Seasonal Tilt Radiation	% Radiation Increment with Two Seasonal Tilt Angles
	Country	City										
1	North Pole	Greenland	82.3872°N	35.7086°W	797 m	82°	67°	90°	-20.4	1,186	12,22	3.04%
2	Russia	Moscow	55.7494°N	37.3523°E	150 m	55°	45°	70°	6	1,204	1,266	5.15%
3	Germany	Berlin	52.5069°N	13.1445°E	40 m	52°	37°	67°	10.2	1,227	1,273	3.75%
4	Canada	Ottawa	45.4145°N	75.7120°W	53 m	45°	30°	60°	7	1,625	1,698	4.49%
5	Italy	Rome	41.9102°N	12.3959°E	52 m	41°	26°	56°	16.6	1,670	1,735	3.89%
6	China	Beijing	40.0799°N	116.6031°E	26 m	40°	25°	55°	12.9	1,560	1,617	3.65%
7	USA	Washington	38.9071°N	77.0369°W	29 m	38°	23°	53°	14.1	1,658	1,722	3.86%
8	India	Ladakh	34.1662°N	77.4966°E	3,960 m	34°	19°	49°	2.1	2,023	2,184	7.96%
9	India	Rajasthan	28.0440°N	73.0807°E	225 m	28°	13°	43°	27.5	2,122	2,205	3.91%
10	India	Maharashtra	19.0896°N	72.8656°E	45 m	19°	4°	34°	27.6	1,953	2,028	3.84%
11	South America	Brazil	23.4306°S	46.4730°W	743 m	23°	38°	8°	19.6	1,524	1,582	3.81%
12	South Africa	Cape Town	33.9715°S	18.6021°E	45 m	33°	49°	19°	16.9	2,233	2,331	4.39%
13	Australia	Canberra	35.3052°S	149.1934°E	570 m	35°	50°	20°	13.8	2,018	2,106	4.36%
14	South Pole	Antarctica	78.630°S	46.758°E	3,320 m	81°	90°	63°	-28.1	1,053	1,068	1.42%

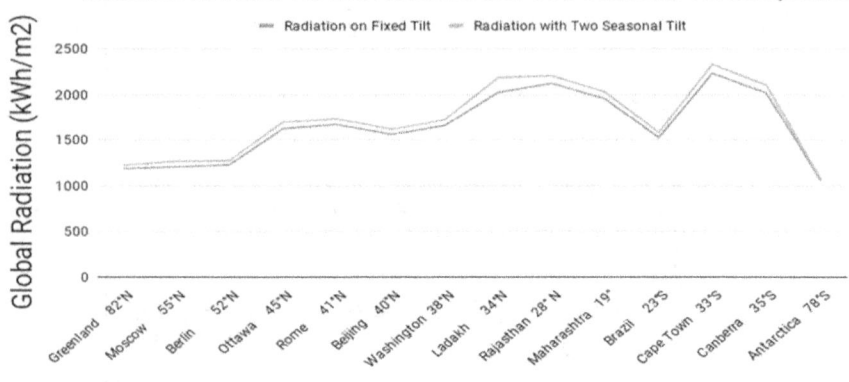

FIGURE 9.7 Graphical representation of radiation on fixed and two seasonal tilt angles worldwide.

5. Compared to Rajasthan (225 mtr altitude), India (9 hours' sunshine duration), and Ladakh (7.5 hours' sunshine duration), both places receive nearly the same amount of radiation.
   • Ladakh has less sunshine hours and also has a foggy season in January and February, but due to high altitude (3,960 meter), it receives more radiation.
   • Ladakh and Rajasthan are separated by 6° latitude, horizontal separate by 1,000 km and altitude difference of 3,735 m.
6. Maharashtra, India, is near the sea, but gets less radiation, due to the rainy season (humidity) in July, August, and September.
7. By using two seasonal tilt surfaces, we can get 3.91% more radiation in Rajasthan (India).
8. In Ladakh, by using two seasonal tilt surfaces, we get a maximum increment rate of 7.96% per year due to high altitude.
9. Due to high altitude and cold weather, a PV plant gives better performance at the Ladakh location.

9.5   CONCLUSION

This chapter presents the study of solar insolation using METEONORM and PVSYST software, for the Bikaner Rajasthan Desert location and helps to select the tilt angle for a fixed tilt solar PV plant and for two seasonal tilt angles suggest the tilt angle along with month. Also shown is the increment in solar insolation with the increment in the number of tilt angles along with their suitable month. In worldwide located airport locations, there is also an increment of 3% to 8% according to their latitude and altitude.

REFERENCES

[1] T. F. Agajie, B. Khan, H. H. Alhelou, O. P. Mahela, Optimal expansion planning of distribution system using grid-based multi-objective harmony search algorithm, *Computers & Electrical Engineering*, vol. 87, pp. 106823, 2020.

[2] B. Khan, H. H. Alhelou, and F. Mebrahtu. A holistic analysis of distribution system reliability assessment methods with conventional and renewable energy sources, *AIMS Energy*, vol. 7, issue 4, pp. 413–429, 2019.

[3] D. Anteneh, and B. Khan, Reliability enhancement of distribution substation by using network reconfiguration a case study at Debre Berhan distribution substation, *International Journal of Economy, Energy and Environment*, vol. 4, issue 2, pp. 33–40, 2019.

[4] R. K. Pachauri et al., Impact of partial shading on various PV array configurations and different modeling approaches: A comprehensive review, *IEEE Access*, vol. 8, pp. 181375–181403, 2020.

[5] R. K. Pachauri, O. P. Mahela, B. Khan, A. Kumar, S. Agarwal, H. H. Alhelou, and J. Bai, Development of arduino assisted data acquisition system for solar photovoltaic array characterization under partial shading conditions, *Computers & Electrical Engineering*, vol. 92, pp. 107175, 2021.

[6] R. K. Pachauri et al., Shade dispersion methodologies for performance improvement of classical total cross-tied photovoltaic array configuration under partial shading conditions, *IET Renewable Power Generation*. Vol. 15, pp. 1796–1811, 2021.

[7] M. T. Yeshalem, and B. Khan Design of an off-grid hybrid PV/wind power system for remote mobile base station: A case study, *AIMS Energy*, vol. 5, issue 1, pp. 96–112, 2017.

[8] Y. Kifle, B. Khan, and P. Singh, Assessment and enhancement of distribution system reliability by renewable energy sources and energy storage, *Journal of Green Engineering*, vol. 8, Issue 3, pp. 219–262, July 2018.

[9] S. Kiros, B. Khan, S. Padmanaban, H. Haes Alhelou, Z. Leonowicz, O. P. Mahela, and J. B. Holm-Nielsen. Development of stand-alone green hybrid system for rural areas, *Sustainability*, vol. 12, pp. 3808, 2020.

[10] M. Jariso, B. Khan, S. Tanwar, S. Tyagi and V. Rishiwal, Hybrid energy system for upgrading the rural environment, *2018 IEEE Globecom Workshops (GC Wkshps)*, Abu Dhabi, United Arab Emirates, pp. 1–6, 2018.

[11] H. D. Kambezidis, The solar radiation climate of Athens: Variations and tendencies in the period 1992–2017, the brightening era, *Solar Energy*, vol. 173, pp. 328–347, 2018.

[12] F. Scarpa, V. Bianco and L. A. Tagliafico, A clear sky physical based solar radiation decomposition model, *Thermal Science and Engineering Progress*, vol. 6, pp. 323–329, 2017.

[13] A. Choudhary, All about solar energy, *International Journal for Scientific Research & Development*, vol. 5, Issue 09, pp. 708–711, 2017.

[14] A. Z. Hafez, A. Soliman, K. A. El-Metwally and I. M. Ismail, Tilt and azimuth angles in solar energy applications – A review, *Renewable and Sustainable Energy Reviews*, vol. 77, pp. 147–168, 2017.

[15] H. Ibrahim, and N. Anani, Variations of PV module parameters with irradiance and temperature *Energy Procedia*, vol. 134, pp. 276–285, 2017.

[16] L. Mohanty, and S. K. Wittkopf, Effect of diffusion of light on thin-film photovoltaic laminates, *Results in Physics*, vol. 6, pp. 61–66, 2016.

[17] M. Alonso, F. Chenlo, F. Fabero, M. A. Ariza and E. Mejuto, "Measurement of irradiance sensors for PR calculation in PV plants," 29th European Photovoltaic Solar Energy Conference and Exhibition, 2015.

[18] K. Gairaa and Y. Bakelli. A comparative study of some regression models to estimate the global solar radiation on a horizontal surface from sunshine duration and meteorological parameters for Gharda Site, Algeri, *ISRN Renewable Energy*, Article ID 754956, 2013.

[19] M. Chegaar, P. Petit, A. Hamzaoui, M. Aillerie, A. Namoda and A. H'erguth, Effect of illumination intensity on solar cells parameters, *Energy Proceedia*, vol. 36, pp. 722–729, 2013.

[20] M. Blumthaler, Solar radiation of the high Alps, *Plants in Alpine Regions Cell Physiology of Adaption and Survival Strategies*, Springer-Verlag/Wien, 10.1007/978-3-7091-0136-0_2, 2012.

[21] W. H. Terjung and P. O'Rourke, A worldwide examination of solar beam-slope angle values, *Solar Energy*, vol. 31, Issue 2, pp. 217–221, 1983.

[22] L. Umanand, Design of photovoltaic systems, *NPTEL Online Course, IISC Bangalore*, 2018.

[23] P. Gevorkian, *Large-Scale Solar Power System Design*. The McGraw-Hill Companies, 2011.

[24] A. S. Kapur, *A Practical Guide for Total Engineering of MW Capacity Solar PV Power Project*. White Falcon Publishing, 2015.

10 An Algorithm for Identification of Multiple Power Quality Disturbances

Om Prakash Mahela
Power System Planning Division, Rajasthan Rajya Vidyut Prasarn Nigam Ltd., Jaipur, India

Shruti Rathore and Shoyab Ali
Department of Electrical Engineering, Vedant College of Engineering and Technology, Bundi, India

Baseem Khan
Department of Electrical and Computer Engineering, Hawassa University, Ethiopia

CONTENTS

DOI: 10.1201/b22884-10

10.1 INTRODUCTION

Power quality (PQ) is generated due to the intensive use of power electronics–supported equipment, microprocessor-supported devices, controllers used in the industry, and loads of non-linear nature and extensive use of computers [1–10]. Power quality issues may lead to damage or improper operation of equipment connected to the grid. Simultaneous incidence of two or PQ issues will result in the deterioration of quality of power to a great extent [11]. Signal processing methods are being employed for the identification and classification of the power quality issues [12–16]. In [17], authors detailed a rule-based method to classify PQ issues. In this study, a disturbed signal is first characterized using the multi-resolution based on the S-transform and features of the signal are extracted. Then, a simple but robust rule-based algorithm is used to identify disturbances. This algorithm uses parabolic and linear rules as pattern classifiers where decision boundaries are established by a heuristic search. The classification algorithm has a modular structure where each and every module works separately to detect particular disturbances. In [18], a method detects, localizes, and investigates the suitability for classifying different natures of PQ events. This scheme has been supported by analysis of signals using WT as well as dyadic-orthonormal WT (DOWT). The proposed method is effective for decomposition of signals with disturbances into different signals used to represent a fine version with detailed features of the original signal. In [19], the authors introduced a method for online detection of disturbances in voltage signals using the wavelet transform. This method is effective for the identification of disturbances in voltage signals, discrimination of different types of PQ disturbances associated with voltage signals, and the proposed approach is relatively faster and more precise for discriminating various natures of PQ events such as voltage-supported disturbance detection methods. In [20], a de-noising technique combined with WT for a PQ monitoring scheme is introduced. A developed scheme is effective for detecting and localizing PQ events even in a noisy environment. Hence, a scheme can be used for the storage of proper signals used for encompassing PQ events and subsequent analysis. In [21], performance of the wavelet-based scheme for online detection of a voltage-based PQ disturbance has been evaluated. A scheme is designed to achieve the following two objectives: (i) establish the effectiveness of the proposed method for identification of faults and PQ disturbances in the network of power system and (ii) to compare the performance of the method with the reported conventional methods. In [22], an approach for the data compression of PQ transients has been reported. The analyzed data can be utilized for analyzing the classification of PQ events. The original data have been reconstituted from the compressed data of PQ events and subsequently analyzed with the help of an advanced version of WT and ST. In [23], the authors introduced an approach aimed to recognize PQ events associated with the network of power. This approach is based on the application of wavelet transform and radial basis function neural network (RBFNN). In [24], the authors suggested a technique that is effective for detection and to classify PQ issues by the use of Stockwell's transform (ST) and fuzzy expert system (FES). In [25], the authors presented an automatic scheme for identification of PQ events supported by ST and extreme learning scheme (ELS). In [26], the authors introduced a method to detect and classify disturbances of power quality with the help of WT and wavelet networks (WN). This has been implemented for recognition of PQ events that have been generated in accordance with the standard IEEE-1159. In [27], the authors suggested a method based on the application of S-transform and decision rules to detect and classify the PQ issues that are simplistic in nature. In [28], the authors introduced a method for recognition of PQ issues by the application of fuzzy C-means clustering and S-transform. In [29], the authors introduced a technique for recognition of PQ issues by the application of an image processing–based technique. Various techniques for identification of PQ issues are reported in [30–33].

After detailed analysis of the research work discussed previously, it is observed that the feature selection is an important task of the recognition of complex nature PQ disturbances, which needs to be investigated in detail. Further, the complex nature of PQ disturbances have not been fully investigated and research is under process. Hence, this research work has considered a technique using Stockwell transform and decision rules for classification of the single stage and complex stage PQ issues with a minimum number of features.

Research work included in this chapter is aimed for identification and classification of complex PQ issues. The proposed approach is based on optimal use of the features extracted from the voltage signals by the use of Stockwell transform and decision rules. Research work included in this chapter has the following contributions:

- A technique supported by Stockwell transforms and decision rules for identification of complex PQ events have been proposed in this work.
- The Stockwell transform and decision rules supported approach is effective in identification of complex PQ issues.
- A power quality index (PQI) and a PQ time location index (PQTLI) are proposed for the identification of different types of complex PQ issues.
- Six statistical features are computed from the PQI and PQTLI, which are considered as the input to a rule-supported decision tree to classify the complex nature of PQ disturbances.
- The classification accuracy of the complex PQ issues has been achieved as high as 96.2%.
- A study is performed in the MATLAB®/Simulink environment.

The manuscript is organized as follows: Section 10.2 discusses the formulation of PQ issues and methodology utilized in this work. Section 10.3 presents the results with a detailed discussion. Classification results for complex PQ issues and features extraction are discussed in Section 10.4. Validation of the performance of the algorithm is detailed in Section 10.5. Section 10.6 presents the conclusion of this chapter.

10.2   FORMULATION OF PQ ISSUES AND METHODOLOGY

The proposed algorithm used for recognition and classification of the single stage and complex PQ events is detailed in this chapter. Generation of the single stage and complex PQ events by the use of standard mathematical formulations is also described in this chapter.

10.2.1   FORMULATION OF PQ DISTURBANCES

The mathematical relations reported in [34],[35] are utilized for the generation of complex PQ issues. These mathematical relations are formulated in MATLAB software. Complex PQ issues are formulated by different combinations of single-stage PQ issues. These complex PQ issues include voltage signal with sag and harmonics, voltage signal with swell and harmonics, voltage signal with momentary interruption (MI) and harmonics, voltage signal with oscillatory transient (OT) and voltage sag, voltage signal with oscillatory transient and voltage swell, voltage signal with impulsive transient (IT) and voltage sag, voltage signal with impulsive transient and voltage swell, voltage signal with simultaneous occurrence of voltage swell and OT and harmonics, voltage signal with simultaneous occurrence of voltage sag and OT and harmonics, voltage signal with simultaneous incidence of OT and IT and voltage sag and harmonics. These signals with associated complex PQ issues are processed using the proposed algorithm.

10.2.2   ALGORITHM ADOPTED TO IDENTIFY AND CATEGORIZE THE PQ DISTURBANCES

The algorithm adopted to identify and categorize the PQ issues is described in Figure 10.1. This algorithm is used for the recognition of the complex PQ events. This algorithm can be implemented with the following steps:

- Simulate the PQ events associated with the voltage signal using the mathematical formulation available in the [34], [35].

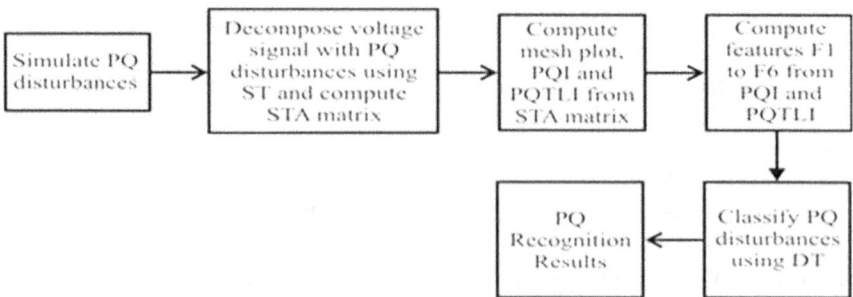

FIGURE 10.1   Proposed PQ recognition algorithm.

- Decompose voltage signal with PQ events using Stockwell transform with 1.6 kHz sampling frequency (32 samples in every cycle) and compute output matrix STM, which is complex in nature.
- Compute the absolute values of each element of the matrix STM to compute a matrix with absolute values and with the same dimensions and designed as STMA. The below mentioned command is used:

$$STMA = abs(STM)$$

- Compute the maximum magnitude of every column of the STMA matrix using the following command:

$$AP = max(STA)$$

- Compute the summation of each column of the STMA matrix using the following command:

$$SP = sum(STA)$$

- Compute the angle of the magnitude of every column of the STMA matrix using the following command:

$$PP = angle(max(SS))$$

- Compute the median of each column of the STMA matrix using the following command:

$$M = median(STA)$$

- Compute the weight factor using the following command:

$$C = max(SP)100;$$

- Compute the power quality index (PQI) using the following command:

$$PQI2 = (AP. * SP. * PP)/1.75;$$

- Compute the power quality time location index (PQTLI) using the following command:

$$PQI1 = C. * (AP. * SP. * M);$$

- Compute the mesh plot from the STMA matrix.

- Plot the mesh plot, PQI, and PQTLI. Compare these plots forthe complex nature PQ disturbance with the respective plots of the pure sine wave to recognize the associated PQ disturbances.
- Extract features from the PQI and PQTLI plots. The values of these features are considered as input to the rule-based decision tree for the classification of the PQ disturbances.

10.3 RESULTS AND DISCUSSION

The results of simulation for the identification of the complex nature PQ issues using the proposed approach are discussed in this section. Signals with the associated complex PQ event are processed using the Stockwell transform to compute the output matrix. Proposed power quality index (PQI), PQ time location index (PQTLI) and mesh plots are computed from this matrix to identify the PQ disturbances associated with the voltage signals. Features are computed from the PQI and PQTLI and considered as input to the rule-supported decision tree to categorize the different PQ events.

10.3.1 VOLTAGE SIGNAL WITH SAG AND HARMONICS

A voltage signal with sag between the time 0.06 s to 0.14 s and harmonics of the order 3rd, 5th, and 7th is simulated and processed using the Stockwell transform and absolute values output matrix STMA is computed. A mesh plot of this matrix is obtained that represents the time on the x-axis, normalized frequency on the y-axis, and amplitude on the z-axis, which is shown in Figure 10.2. From this figure, it is observed that for the normalized frequency corresponding to the fundamental

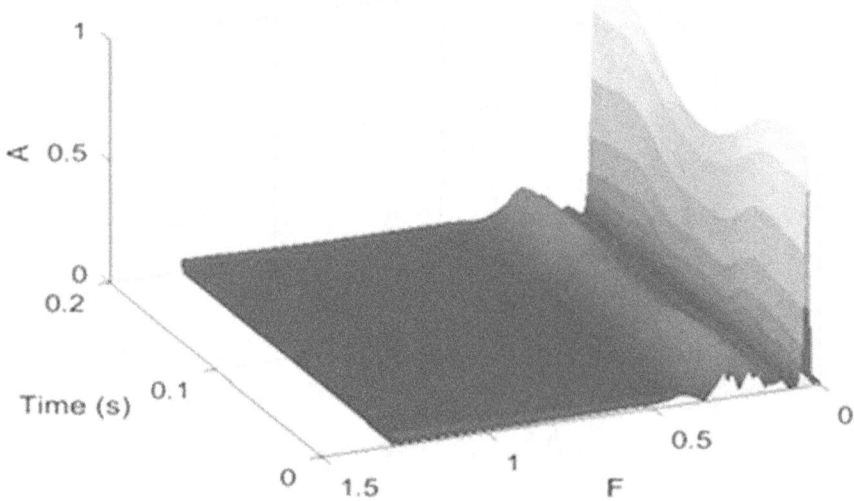

FIGURE 10.2   Mesh plot for the voltage waveform with sag and harmonics.

frequency of 50 Hz, the amplitude except between the time 0.06 s to 0.14 s is unity. During the time 0.06 s to 0.14 s, the magnitude corresponding to the fundamental frequency decreases, which indicates the presence of the voltage sag. Further, it is also observed that the amplitude for the frequency corresponding to the 3rd, 5th, and 7th harmonic components is also high for the entire time range. Hence, sag and harmonics associated with the voltage signal have been detected effectively. The mesh plot gives the visual inspection of the available voltage sag with harmonics.

The voltage signal with a PQ disturbance of sag and harmonics of the orders 3rd, 5th, and 7th is processed using the Stockwell transform and proposed power quality index (PQI) and proposed power quality time location index (PQTLI) are computed. Voltage signal, PQI, and PQTLI are shown in Figure 10.3(a), (b) and (c), respectively. From Figure 10.3(a), it is observed that the magnitude of the voltage waveform has decreased from 0.06 s to 0.14 s. It is also observed that the magnitude of the voltage waveform has continuous ripples over the entire time range, which indicates the presence of harmonics.

Figure 10.3(b) indicates that the magnitude of the PQI has decreased between 0.06 s to 0.14 s, which indicates the presence of the voltage sag. Figure 10.3(b) also indicates that magnitude of the PQI has continuous ripples of high magnitude and a fixed pattern over the entire time range, which indicates the presence of the harmonics. Figure 10.3(c) shows that the PQTLI has sharp magnitude peaks at the time instant of 0.06 s and 0.14 s, which indicates the start and end of the sag in voltage. Figure 10.3(c) also shows that the PQTLI has continuous ripples with sharp tips and a fixed pattern over the entire time range, which indicates presence of harmonics with the signal. Hence, it is established that the proposed approach effectively identified and localized the PQ disturbance of sag and harmonics associated with the voltage signal.

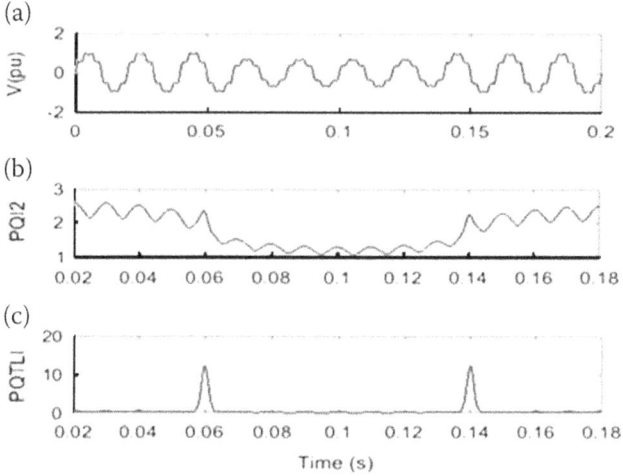

FIGURE 10.3   Voltage waveform with sag and harmonics: (a) voltage waveform, (b) PQ index, and (c) PQ time location index.

10.3.2 Voltage Signal with Swell and Harmonics

The voltage signal with sag between 0.06 s to 0.14 s and harmonics of the orders 3rd, 5th, and 7th is simulated and processed using the Stockwell transform and absolute values output matrix STMA is computed. A mesh plot of this matrix is obtained, which represents the time on the x-axis, normalized frequency on the y-axis, and amplitude on the z-axis, which is shown in Figure 10.4. From this figure, it is observed that for the normalized frequency corresponding to the fundamental frequency of 50 Hz, the amplitude except between 0.06 s to 0.14 s is unity. During the time 0.06 s to 0.14 s, the magnitude corresponding to the fundamental frequency increases, which indicates the presence of the voltage swell. Further, it is also observed that the amplitude for the frequency corresponding to the 3rd, 5th, and 7th harmonic components is also high for the entire time range. Hence, swell and harmonics associated with the voltage signal have been detected effectively. The mesh plot gives the visual inspection of the available voltage swell with harmonics.

The voltage signal with a PQ disturbance of swell and harmonics of the orders 3rd, 5th, and 7th is processed using the Stockwell transform and proposed power quality index (PQI) and proposed power quality time location index (PQTLI) are computed. Voltage signal, PQI, and PQTLI are shown in Figure 10.5(a), (b), and (c), respectively. From Figure 10.5(a), it is observed that magnitude of the voltage waveform has increased between 0.06 s to 0.14 s. It is also observed that the magnitude of the voltage waveform has continuous ripples over the entire time range, which indicates the presence of harmonics.

Figure 10.5(b) indicates that the magnitude of the PQI has increased between 0.06 s to 0.14 s, which indicates the presence of the voltage swell. Figure 10.5(b) also indicates that the magnitude of the PQI has continuous ripples of high magnitude and fixed pattern over the entire time range, which indicates the presence of the harmonics. Figure 10.5(c) shows that the PQTLI has sharp magnitude peaks at

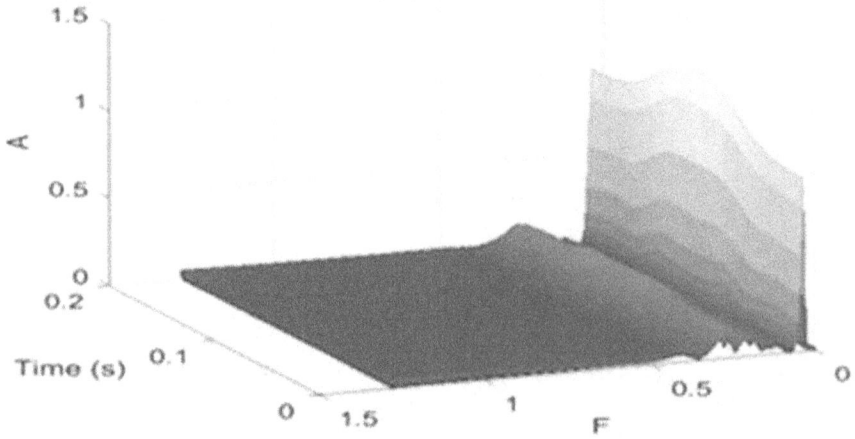

FIGURE 10.4 Mesh plot for the voltage waveform with swell and harmonics.

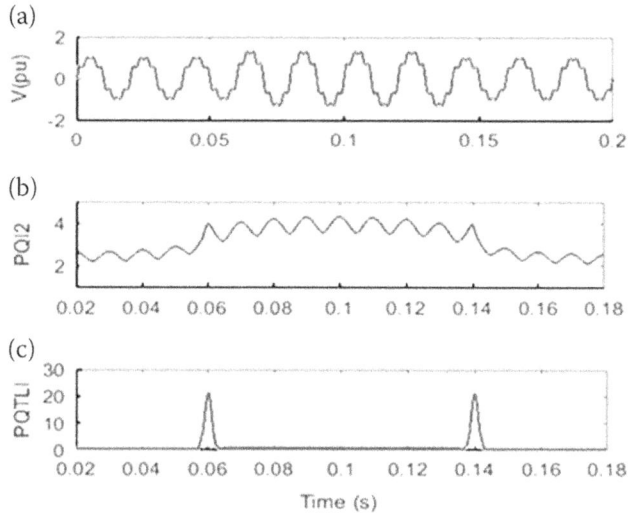

FIGURE 10.5 Voltage waveform with swell and harmonics: (a) voltage waveform, (b) PQ index, and (c) PQ time location index.

the time instant of 0.06 s and 0.14 s, which indicates the start and end of the swell in voltage. Figure 10.5(c) also shows that the PQTLI has continuous ripples with sharp tips and a fixed pattern over the entire time range, which indicates presence of harmonics with the signal. Hence, it is established that the proposed approach effectively identified and localized the PQ disturbance of swell and harmonics associated with the voltage signal.

10.3.3 VOLTAGE SIGNAL WITH MOMENTARY INTERRUPTION AND HARMONICS

The voltage signal with momentary interruption (MI) between 0.06 s to 0.14 s and harmonics of the orders 3rd, 5th, and 7th is simulated and processed using the Stockwell transform and absolute values output matrix STMA is computed. A mesh plot of this matrix is obtained, which represents the time on the x-axis, normalized frequency on the y-axis, and amplitude on the z-axis, which is shown in Figure 10.6. From this figure, it is observed that for the normalized frequency corresponding to the fundamental frequency of 50 Hz, the amplitude except between 0.06 s to 0.14 s is unity. During 0.06 s to 0.14 s, the magnitude corresponding to the fundamental frequency decreases below 10%, which indicates the presence of MI. Further, it is also observed that the amplitude for the frequency corresponding to the 3rd, 5th, and 7th harmonic components is also high for the entire time range. Hence, MI and harmonics associated with the voltage signal have been detected effectively. The mesh plot gives the visual inspection of the available MI with harmonics.

Voltage signal with a PQ disturbance of MI and harmonics of the orders 3rd, 5th, and 7th is processed using the Stockwell transform and proposed power quality index (PQI) and proposed power quality time location index (PQTLI) are computed. Voltage signal, PQI, and PQTLI are shown in Figure 10.7(a), (b), and (c), respectively. From

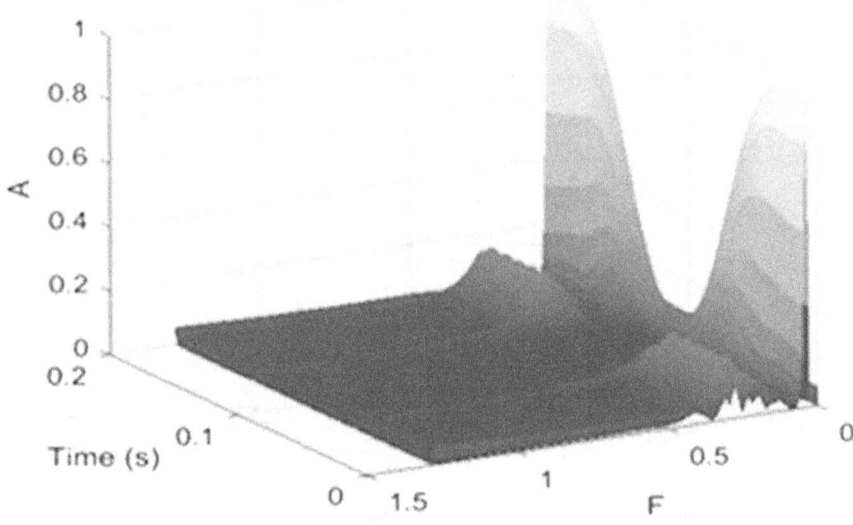

FIGURE 10.6   Mesh plot for the voltage waveform with momentary interruption and harmonics.

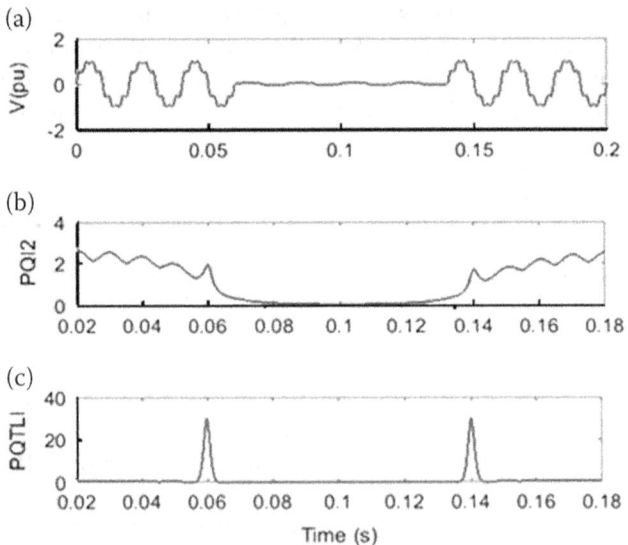

FIGURE 10.7   Voltage waveform with momentary interruption and harmonics: (a) voltage waveform, (b) PQ index, and (c) PQ time location index.

Figure 10.7(a), it is observed that the magnitude of the voltage waveform has decreased below 10% between the time 0.06 s to 0.14 s. It is also observed that the magnitude of the voltage waveform has continuous ripples over the entire time range, which indicates the presence of harmonics.

Figure 10.7(b) indicates that the magnitude of the PQI has decreased between 0.06 s to 0.14 s to a value nearly zero, which indicates the presence of the MI. Figure 10.7(b) also indicates that the magnitude of the PQI has continuous ripples of high magnitude and a fixed pattern over the entire time range, which indicates the presence of the harmonics. Figure 10.7(c) shows that the PQTLI has sharp magnitude peaks at the time instant of 0.06 s and 0.14 s, which indicates the start and end of the MI in voltage. Figure 10.7(c) also shows that the PQTLI has continuous ripples with sharp tips and a fixed pattern over the entire time range, which indicates the presence of harmonics with the signal. Hence, it is established that the proposed approach effectively identified and localized the PQ disturbance of MI and harmonics associated with the voltage signal.

10.3.4 VOLTAGE SIGNAL WITH OSCILLATORY TRANSIENT AND VOLTAGE SAG

The voltage signal with sag between 0.01 s to 0.16 s and oscillatory transient between the time duration 0.06 s to 0.08 s is simulated and processed using the Stockwell transform and absolute value output matrix STMA is computed. A mesh plot of this matrix is obtained, which represents the time on the x-axis, normalized frequency on the y-axis and amplitude on the z-axis, which is shown in Figure 10.8. From this figure, it is observed that for the normalized frequency corresponding to the fundamental frequency of 50 Hz, the amplitude except between the time 0.01 s to 0.16 s is unity. During the time 0.01 s to 0.16 s, the magnitude corresponding to the fundamental frequency decreases, which indicates the presence of the voltage sag. Further, it is also observed that a sharp magnitude peak is observed between the

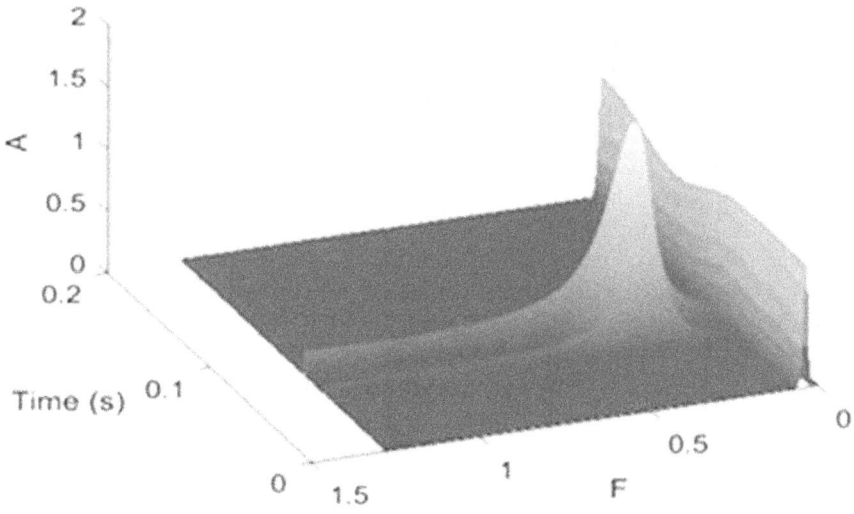

FIGURE 10.8 Mesh plot for the voltage waveform with oscillatory transient and voltage sag.

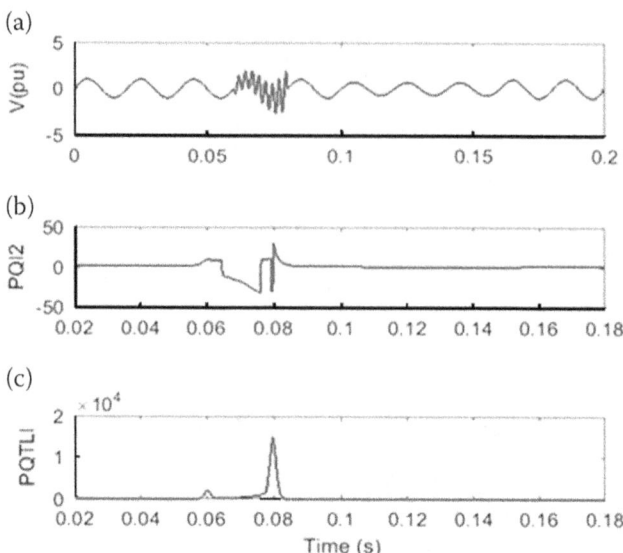

FIGURE 10.9 Voltage waveform with oscillatory transient and voltage sag: (a) voltage waveform, (b) PQ index, and (c) PQ time location index.

time duration 0.08 s to 0.10 s, indicating the presence of the OT. It is also observed that due to the presence of OT, low magnitude finite values are present throughout the time range, which indicates that all the frequencies are associated with the OT. Hence, sag and OT associated with the voltage signal have been detected effectively. The mesh plot gives the visual inspection of the available voltage sag with OT.

The voltage signal with a PQ disturbance of sag between 0.01 s to 0.16 s and oscillatory transient between the time duration 0.06 s to 0.08 s is processed using the Stockwell transform and proposed power quality index (PQI) and the proposed power quality time location index (PQTLI) are computed. Voltage signal, PQI, and PQTLI are shown in Figure 10.9(a), (b), and (c), respectively. From Figure 10.9(a), it is observed that the magnitude of the voltage waveform has decreased between 0.01 s to 0.16 s.

It is also observed from Figure 10.9(a) that the magnitude of the waveform oscillates for the duration of 0.08 s to 0.10 s, indicating that OT is associated with the voltage waveform. Figure 10.9(b) indicates that the magnitude of the PQI has decreased between 0.01 s to 0.16 s, which indicates the presence of the voltage sag. Figure 10.9(b) also indicates that the magnitude of the PQI changes abruptly during the time of 0.08 s to 0.10 s, which indicates the presence of the OT associated with the voltage signal. Figure 10.9(c) shows that the PQTLI has sharp magnitude peaks at the time instant of 0.10 s and 0.16 s, which indicates the start and end of the sag in voltage. Figure 10.9(c) shows that the PQTLI has high magnitude peaks at the moments 0.08 s and 0.10 s, which localizes the OT by identifying the start of the OT and end of the OT. Hence, it is established that the proposed approach effectively identified and localized the PQ disturbance of sag and OT associated with the voltage signal.

10.3.5 VOLTAGE SIGNAL WITH OSCILLATORY TRANSIENT AND VOLTAGE SWELL

The voltage signal with a swell between 0.01 s to 0.16 s and oscillatory transient between the time duration 0.06 s to 0.08 s is simulated and processed using the Stockwell transform and absolute values output matrix STMA is computed. A mesh plot of this matrix is obtained that represents the time on the x-axis, normalized frequency on the y-axis, and amplitude on the z-axis, which is shown in Figure 10.10. From this figure, it is observed that for the normalized frequency corresponding to the fundamental frequency of 50 Hz, the amplitude except between 0.01 s to 0.16 s is unity. During the time 0.01 s to 0.16 s, the magnitude corresponding to the fundamental frequency increases, which indicates the presence of the voltage swell. Further, it is also observed that a sharp magnitude peak is observed between the time duration 0.08 s to 0.10 s, indicating the presence of the OT. It is also observed that due to the presence of OT, low magnitude finite values are present throughout the time range, which indicates that all the frequencies are associated with the OT. Hence, swell and OT associated with the voltage signal have been detected effectively. The mesh plot gives the visual inspection of the available voltage swell with OT.

Voltage signal with a PQ disturbance of swell between time 0.01 s to 0.16 s and OT between the time duration 0.06 s to 0.08 s is processed using the Stockwell transform and proposed PQI and proposed PQTLI are computed. Voltage signal, PQI, and PQTLI are shown in Figure 10.11(a), (b), and (c), respectively. From

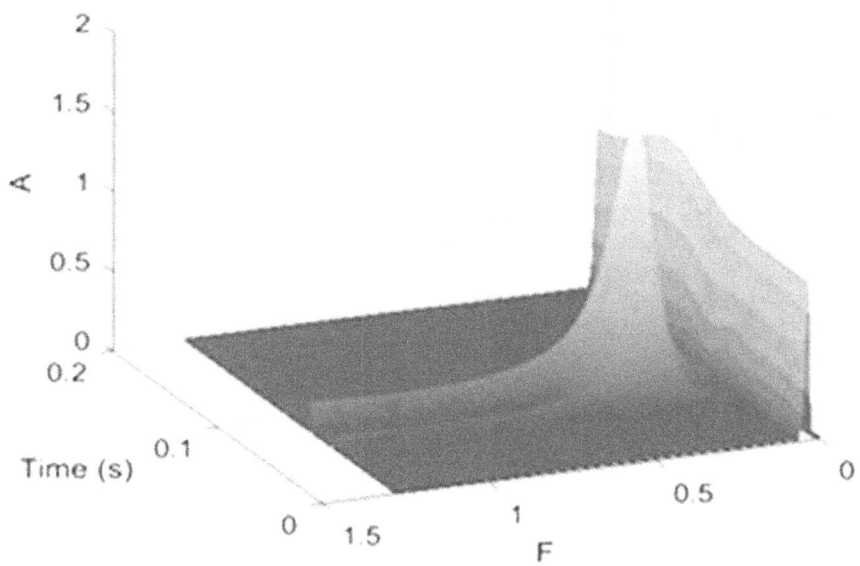

FIGURE 10.10 Mesh plot for the voltage waveform with oscillatory transient and voltage swell.

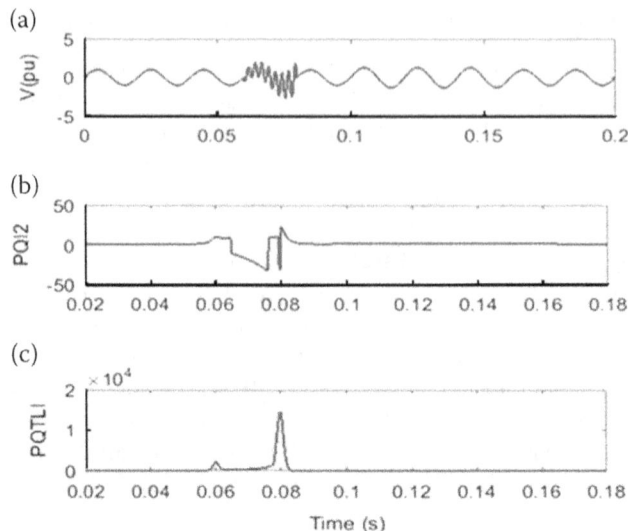

FIGURE 10.11   Voltage waveform with oscillatory transient and voltage swell: (a) voltage waveform, (b) PQ index, and (c) PQ time location index.

Figure 10.11(a), it is observed that magnitude of the voltage waveform has increased between the time 0.01 s to 0.16 s.

It is also observed from Figure 10.11(a) that the magnitude of the waveform oscillates for the duration of 0.08 s to 0.10 s, indicating that OT is associated with the voltage waveform. Figure 10.11(b) indicates that the magnitude of the PQI has increased between 0.01 s to 0.16 s, which indicates the presence of the voltage swell. Figure 10.11(b) also indicates that magnitude of the PQI changes abruptly during the time of 0.08 s to 0.10 s, which indicates the presence of the OT associated with the voltage signal. Figure 10.11(c) shows that the PQTLI has sharp magnitude peaks at the time instant of 0.10 s and 0.16 s, which indicates the start and end of the sag in voltage. Figure 10.11(c) also shows that the PQTLI has high magnitude peaks at the moments 0.08 s and 0.10 s, which localizes the OT by identifying the start of the OT and end of the OT. Hence, it is established that the proposed approach effectively identified and localized the PQ disturbance of swell and OT associated with the voltage signal.

10.3.6   VOLTAGE SIGNAL WITH IMPULSIVE TRANSIENT AND VOLTAGE SAG

The voltage signal with sag between the time 0.01 s to 0.16 s and impulsive transient between the time duration 0.065 s to 0.68 s is simulated and processed using the Stockwell transform and absolute values output matrix STMA is computed. A mesh plot of this matrix is obtained that represents the time on the x-axis, normalized frequency on the y-axis, and amplitude on the z-axis, which is shown in Figure 10.12. From this figure, it is observed that for the normalized frequency corresponding to the fundamental frequency of 50 Hz, the amplitude except

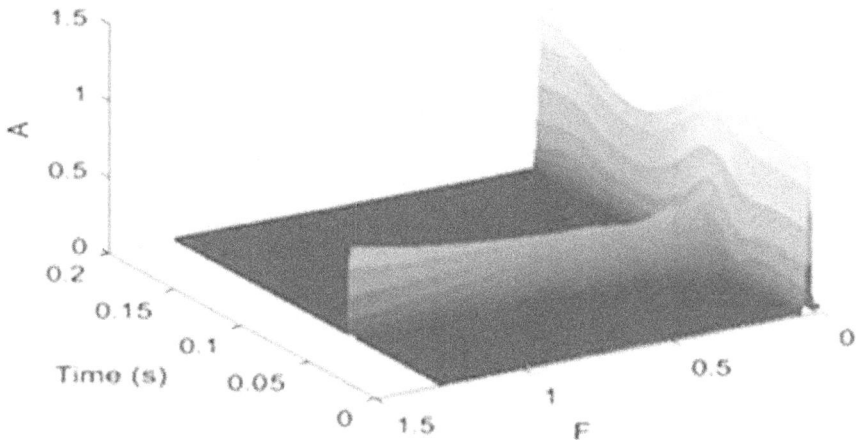

FIGURE 10.12 Mesh plot for the voltage waveform with impulsive transient and voltage sag.

between 0.01 s to 0.16 s is unity. During the time 0.01 s to 0.16 s, the magnitude corresponding to the fundamental frequency decreases, which indicates the presence of the voltage sag. Further, it is also observed that finite values are observed over the entire frequency range between the time duration 0.085 s to 0.88 s, indicating the presence of the IT. Hence, sag and IT associated with the voltage signal have been detected effectively. The mesh plot gives the visual inspection of the available voltage sag with IT.

The voltage signal with a PQ disturbance of sag between 0.01 s to 0.16 s and impulsive transient between the time duration 0.065 s to 0.068 s is processed using the Stockwell transform and proposed power quality index (PQI) and proposed power quality time location index (PQTLI) are computed. Voltage signal, PQI, and PQTLI are shown in Figure 10.13(a), (b), and (c), respectively. From Figure 10.13(a), it is observed that the magnitude of the voltage waveform has decreased between 0.01 s to 0.16 s.

It is also observed from Figure 10.13(a) that the magnitude of the waveform has a sharp magnitude peak for the duration of 0.065 s to 0.068 s, indicating that IT is associated with the voltage waveform. Figure 10.13(b) indicates that the magnitude of the PQI has decreased between 0.01 s to 0.16 s, which indicates the presence of the voltage sag. Figure 10.13(b) also indicates that the magnitude of the PQI has a sharp magnitude peak during the time of 0.065 s to 0.068 s, which indicates the presence of the IT associated with the voltage signal. Figure 10.13(c) shows that the PQTLI has a sharp magnitude peak during 0.065 s to 0.068 s, which indicates the presence of IT associated with the voltage signal. Figure 10.13(c) shows that magnitude of PQTLI has decreased between 0.01 s to 0.16 s, which indicates the presence of the voltage sag. Hence, it is established that the proposed approach effectively identified and localized the PQ disturbance of sag and IT associated with the voltage signal.

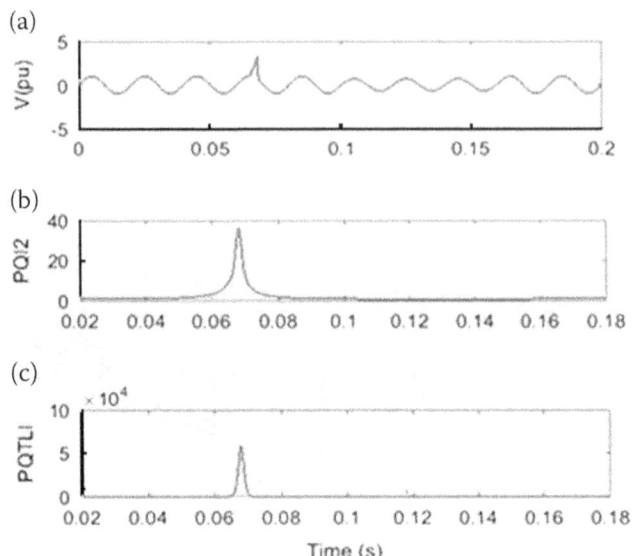

FIGURE 10.13 Voltage waveform with impulsive transient and voltage sag: (a) voltage waveform, (b) PQ index, and (c) PQ time location index.

10.3.7 VOLTAGE SIGNAL WITH IMPULSIVE TRANSIENT AND VOLTAGE SWELL

The voltage signal with a swell between 0.01 s to 0.16 s and impulsive transient between the time duration 0.065 s to 0.68 s is simulated and processed using the Stockwell transform and absolute values output matrix STMA is computed. A mesh plot of this matrix is obtained that represents the time on the x-axis, normalized frequency on the y-axis, and amplitude on the z-axis, which is shown in Figure 10.14. From this figure, it is observed that for the normalized frequency corresponding to the fundamental frequency of 50 Hz, the amplitude except between the time 0.01 s to 0.16 s is unity. During 0.01 s to 0.16 s, magnitude corresponding to the fundamental frequency increases, which indicates the presence of the voltage swell. Further, it is also observed that finite values are observed over the entire frequency range between 0.085 s to 0.88 s, indicating the presence of the IT. Hence, swell and IT associated with the voltage signal have been detected effectively. The mesh plot gives the visual inspection of the available voltage sag with IT.

The voltage signal with a PQ disturbance of swell between 0.01 s to 0.16 s and impulsive transient between 0.065 s to 0.068 s is processed using the Stockwell transform and proposed power quality index (PQI) and proposed power quality time location index (PQTLI) are computed. Voltage signal, PQI, and PQTLI are shown in Figure 10.15(a), (b), and (c), respectively. From Figure 10.15(a), it is observed that magnitude of the voltage waveform has increased between 0.01 s to 0.16 s.

It is also observed from Figure 10.15(a) that the magnitude of the waveform has a sharp magnitude peak for the duration of 0.065 s to 0.068 s, indicating that IT is associated with the voltage waveform. Figure 10.5(b) indicates that the magnitude

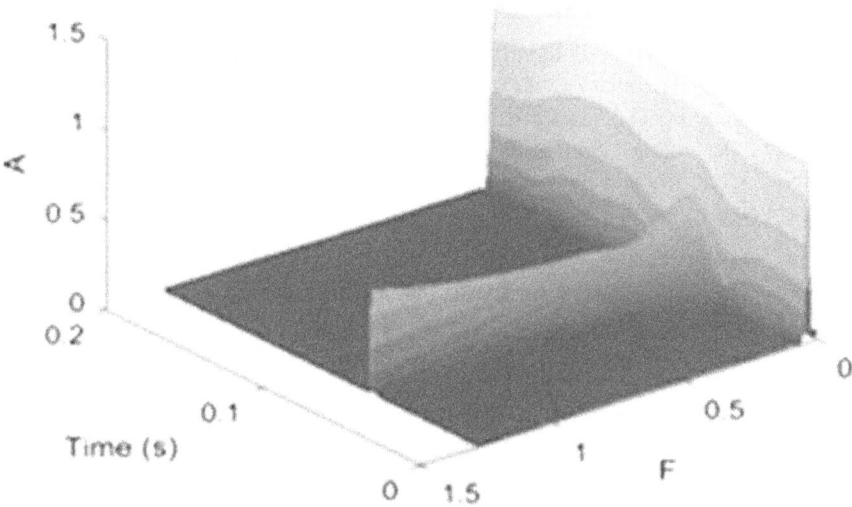

FIGURE 10.14 Mesh plot for the voltage waveform with impulsive transient and voltage sag.

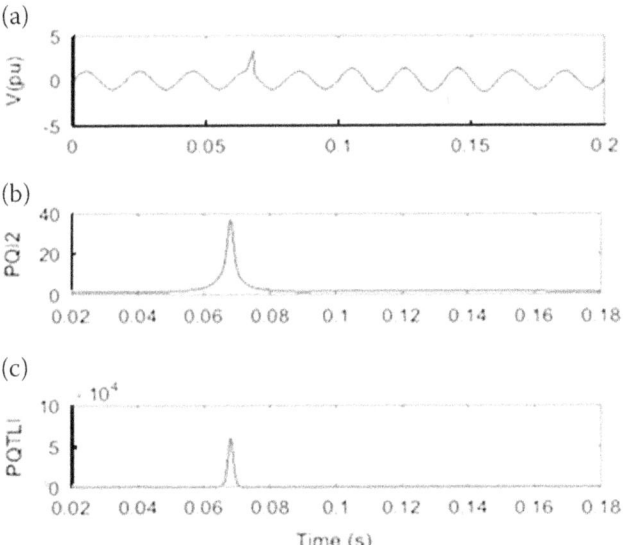

FIGURE 10.15 Voltage waveform with impulsive transient and voltage sag: (a) voltage waveform, (b) PQ index, and (c) PQ time location index.

of the PQI has increased between 0.01 s to 0.16 s, which indicates the presence of the voltage swell. Figure 10.15(b) also indicates that magnitude of the PQI has a sharp magnitude peak during 0.065 s to 0.068 s, which indicates the presence of the IT associated with the voltage signal. Figure 10.15(c) shows that the PQTLI has a sharp magnitude peak during the time duration of 0.065 s to 0.068 s, which indicates the presence of IT associated with the voltage signal. Figure 10.15(c) shows that the magnitude of the PQTLI has decreased between 0.01 s to 0.16 s, which indicates the presence of the voltage swell. Hence, it is established that the proposed approach effectively identified and localized the PQ disturbance of swell and IT associated with the voltage signal.

10.3.8 VOLTAGE SIGNAL WITH SIMULTANEOU OCCURRENCE OF VOLTAGE SWELL, OSCILLATORY TRANSIENT, AND HARMONICS

The voltage signal with a swell between 0.01 s to 0.16 s, oscillatory transient between 0.06 s to 0.08 s, and harmonics is simulated and processed using the Stockwell transform and absolute values output matrix STMA is computed. A mesh plot of this matrix is obtained that represents time on the x-axis, normalized frequency on the y-axis, and amplitude on the z-axis, which is shown in Figure 10.16. From this figure, it is observed that for the normalized frequency corresponding to the fundamental frequency of 50 Hz, the amplitude except between 0.01 s to 0.16 s is unity. During 0.01 s to 0.16 s, the magnitude corresponding to the fundamental frequency increases, which indicates the presence of the voltage swell. Further, it is also observed that a sharp magnitude peak is observed between the time duration

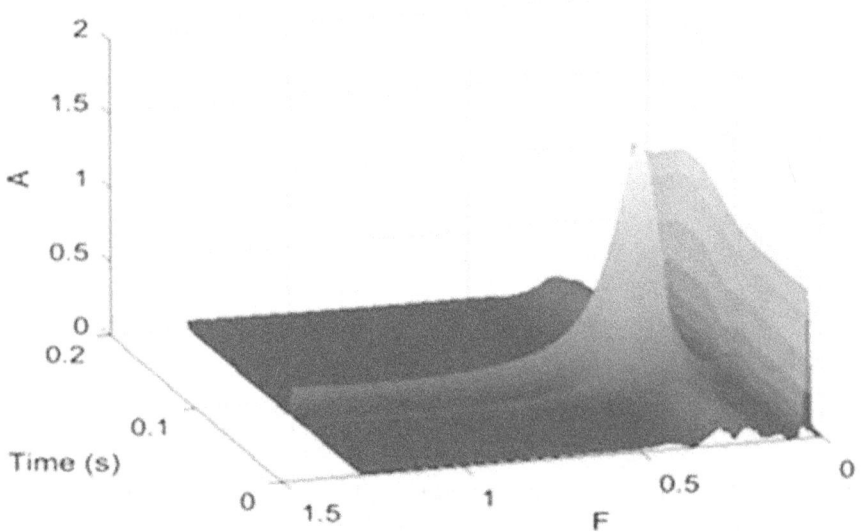

FIGURE 10.16 Mesh plot for the voltage waveform with simultaneous occurrence of voltage swell, oscillatory transient, and harmonics.

0.08 s to 0.10 s, indicating the presence of the OT. It is also observed that due to the presence of OT, low magnitude finite values are present throughout the time range, which indicates that all the frequencies are associated with the OT. Further, it is also observed that the amplitude for the frequency corresponding to the 3rd, 5th, and 7th harmonic components is also high for the entire time range. Hence, swell, OT, and harmonics associated with the voltage signal have been detected effectively. The mesh plot gives the visual inspection of the available voltage swell with OT and harmonics.

The voltage signal with a PQ disturbance of swell between 0.01 s to 0.16 s, OT between 0.06 s to 0.08 s, and harmonics over the entire time range is processed using the Stockwell transform and proposed PQI and proposed PQTLI are computed. Voltage signal, PQI, and PQTLI are shown in Figure 10.17(a), (b), and (c), respectively. From Figure 10.17(a), it is observed that magnitude of the voltage waveform has increased between 0.01 s to 0.16 s. It is also observed from Figure 10.17(a) that the magnitude of the waveform oscillates for the duration of 0.08 s to 0.10 s, indicating that OT is associated with the voltage waveform. It is also observed that the magnitude of the voltage waveform has continuous ripples over the entire time range, which indicates the presence of harmonics.

Figure 10.17(b) indicates that the magnitude of the PQI has increased between 0.01 s to 0.16 s, which indicates the presence of the voltage swell. Figure 10.17(b) also indicates that the magnitude of the PQI changes abruptly during 0.08 s to 0.10 s, which indicates the presence of the OT associated with the voltage signal. Figure 10.17(b) also indicates that the magnitude of the PQI has continuous ripples of high magnitude and fixed pattern over the entire time range, which indicates the

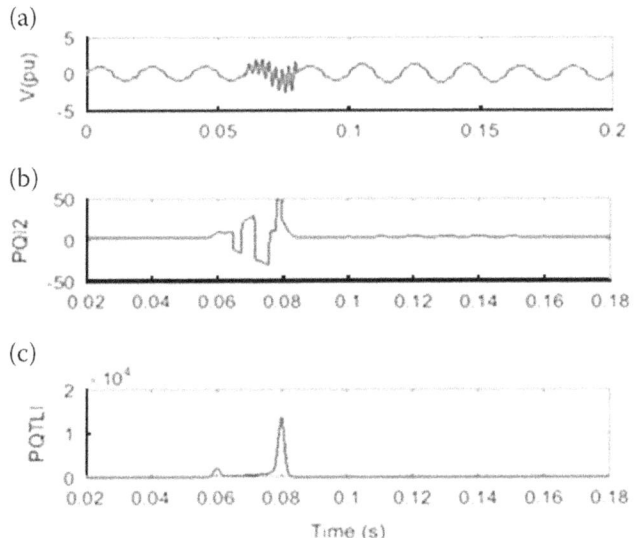

FIGURE 10.17 Voltage waveform with simultaneous occurrence of voltage swell, oscillatory transient, and harmonics: (a) voltage waveform, (b) PQ index, and (c) PQ time location index.

presence of the harmonics. Figure 10.17(c) shows that the PQTLI has sharp magnitude peaks at the time instant of 0.10 s and 0.16 s, which indicates the start and end of the sag in voltage. Figure 10.17(c) also shows that the PQTLI has high magnitude peaks at the moments 0.08 s and 0.10 s, which localizes the OT by identifying the start of the OT and end of the OT. Figure 10.17(c) also shows that the PQTLI has continuous ripples with sharp tips and a fixed pattern over the entire time range, which indicates the presence of harmonics with the signal. Hence, it is established that the proposed approach effectively identified and localized the PQ disturbance of the swell and OT associated with the voltage signal.

10.3.9 Voltage Signal with Simultaneous Occurrence of Voltage Sag, Oscillatory Transient, and Harmonics

The voltage signal with sag between 0.01 s to 0.16 s, oscillatory transient between 0.06 s to 0.08 s, and harmonics is simulated and processed using the Stockwell transform and absolute values output matrix STMA is computed. A mesh plot of this matrix is obtained, which represents the time on the x-axis, normalized frequency on the y-axis, and amplitude on the z-axis, which is shown in Figure 10.18. From this figure, it is observed that for the normalized frequency corresponding to the fundamental frequency of 50 Hz, the amplitude except between 0.01 s to 0.16 s is unity. During 0.01 s to 0.16 s, the magnitude corresponding to the fundamental frequency decreases, which indicates the presence of the voltage sag. Further, it is also observed that a sharp magnitude peak is observed between 0.08 s to 0.10 s,

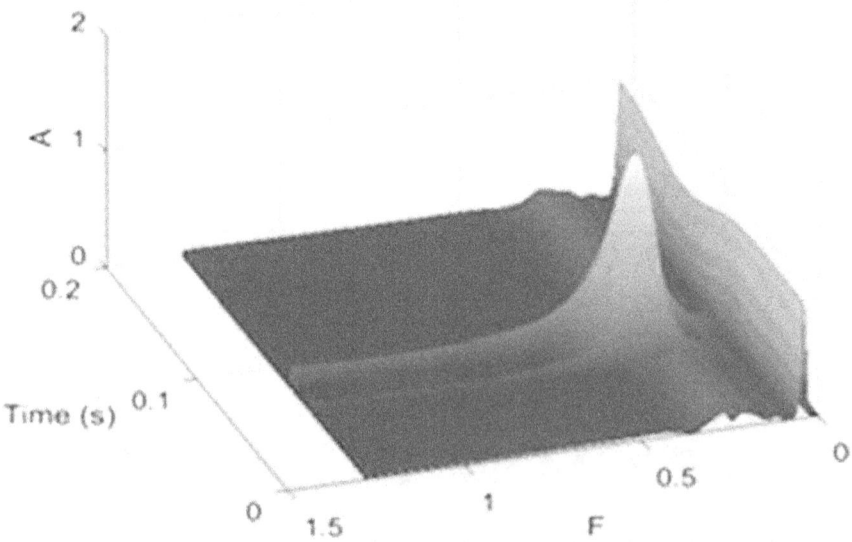

FIGURE 10.18 Mesh plot for the voltage waveform with simultaneous occurrence of voltage sag, oscillatory transient, and harmonics.

indicating the presence of the OT. It is also observed that due to the presence of OT, low magnitude finite values are present throughout the time range, which indicates that all the frequencies are associated with the OT. Further, it is also observed that the amplitude for the frequency corresponding to the 3rd, 5th, and 7th harmonic components is also high for the entire time range. Hence, sag, OT, and harmonics associated with the voltage signal have been detected effectively. The mesh plot gives the visual inspection of the available voltage sag with OT and harmonics.

The voltage signal with a PQ disturbance of sag between 0.01 s to 0.16 s, OT between 0.06 s to 0.08 s, and harmonics over the entire time range is processed using the Stockwell transform and proposed PQI and proposed PQTLI are computed. Voltage signal, PQI, and PQTLI are shown in Figure 10.19(a), (b), and (c), respectively. From Figure 10.19(a), it is observed that magnitude of the voltage waveform has decreased between 0.01 s to 0.16 s. It is also observed from Figure 10.19(a) that the magnitude of the waveform oscillates for the duration of 0.08 s to 0.10 s, indicating that OT is associated with the voltage waveform. It is also observed that the magnitude of the voltage waveform has continuous ripples over the entire time range, which indicates the presence of harmonics. Figure 10.19(b) indicates that the magnitude of the PQI has decreased between 0.01 s to 0.16 s, which indicates the presence of the voltage sag. Figure 10.19(b) also indicates that the magnitude of the PQI changes abruptly during 0.08 s to 0.10 s, which indicates the presence of the OT associated with the voltage signal. Figure 10.19(b) also indicates that magnitude of the PQI has continuous ripples of high magnitude and fixed pattern over the entire time range, which indicates the presence of the harmonics.

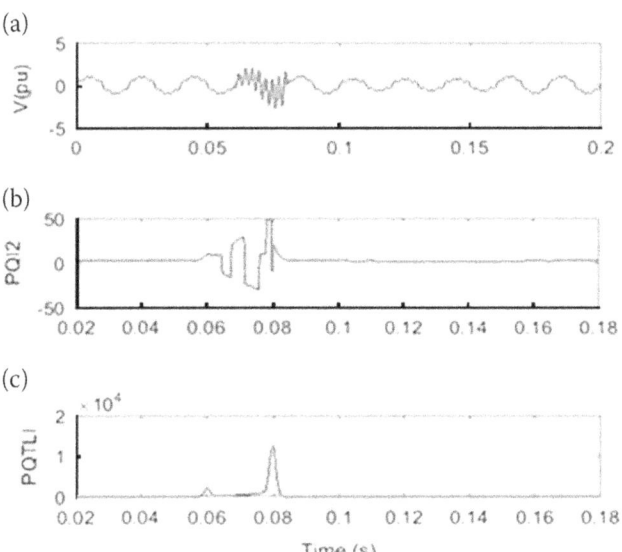

FIGURE 10.19 Voltage waveform with simultaneous occurrence of voltage sag, oscillatory transient, and harmonics: (a) voltage waveform, (b) PQ index, and (c) PQ time location index.

Figure 10.19(c) shows that the PQTLI has sharp magnitude peaks at the time instants of 0.10 s and 0.16 s, which indicates the start and end of the sag in voltage. Figure 10.19(c) also shows that the PQTLI has high magnitude peaks at the moments 0.08 s and 0.10 s, which localizes the OT by identifying the starting of the OT and end of the OT. Figure 10.19(c) also shows that the PQTLI has continuous ripples with sharp tips and fixed pattern over the entire time range, which indicates the presence of harmonics with the signal. Hence, it is established that the proposed approach effectively identified and localized the PQ disturbance of sag and OT associated with the voltage signal.

10.3.10 VOLTAGE SIGNAL WITH SIMULTANEOUS OCCURRENCE OF OSCILLATORY TRANSIENT, IMPULSIVE TRANSIENT, VOLTAGE SAG, AND HARMONICS

The voltage signal with sag between 0.01 s to 0.16 s, oscillatory transient between 0.06 s to 0.08 s, impulsive transient between 0.022 s to 0.025 s, and harmonics is simulated and processed using the Stockwell transform and absolute values output matrix STMA is computed. A mesh plot of this matrix is obtained, which represents the time on the x-axis, normalized frequency on the y-axis, and amplitude on the z-axis, which is shown in Figure 10.20. From this figure, it is observed that for the normalized frequency corresponding to the fundamental frequency of 50 Hz, the amplitude except between 0.01 s to 0.16 s is unity. During 0.01 s to 0.16 s, the magnitude corresponding to the fundamental frequency decreases, which indicates the presence of the voltage sag. Further, it is also observed that a sharp magnitude peak is observed between 0.08 s to 0.10 s, indicating the presence of the OT. It is

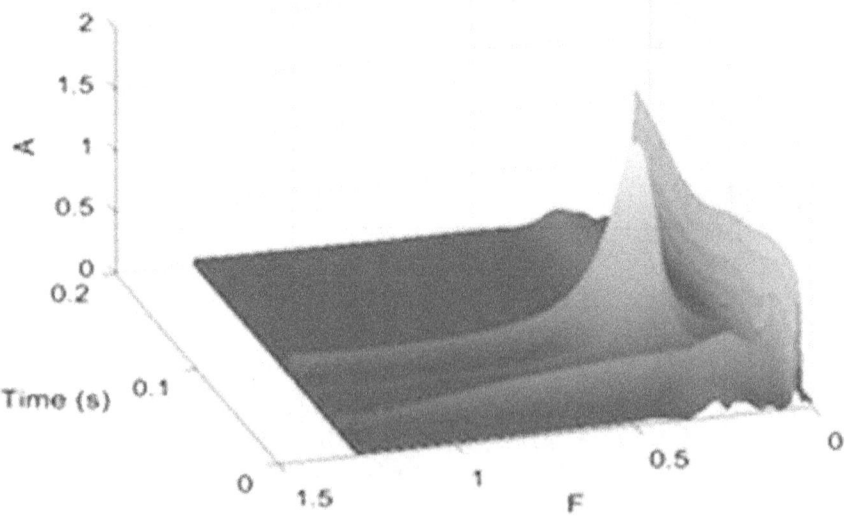

FIGURE 10.20 Mesh plot for the voltage waveform with simultaneous occurrence of oscillatory transient, impulsive transient, voltage sag, and harmonics.

also observed that due to the presence of OT, low magnitude finite values are present throughout the time range, which indicates that all the frequencies are associated with the OT. Further, it is also observed that the amplitude for the frequency corresponding to the 3rd, 5th, and 7th harmonic components is also high for the entire time range. Further, it is also observed that finite values are observed over the entire frequency range between 0.085 s to 0.88 s, indicating the presence of the IT. Hence, sag, OT, IT, and harmonics associated with the voltage signal have been detected effectively. The mesh plot gives the visual inspection of the available voltage sag with OT, IT, and harmonics.

The voltage signal with a PQ disturbance of sag between 0.01 s to 0.16 s, OT between 0.06 s to 0.08 s, impulsive transient between 0.065 s to 0.068 s, and harmonics over the entire time range is processed using the Stockwell transform and proposed PQI and proposed PQTLI are computed. Voltage signal, PQI, and PQTLI are shown in Figure 10.21(a), (b), and (c), respectively. From Figure 10.21(a), it is observed that the magnitude of the voltage waveform has decreased between 0.01 s to 0.16 s. It is also observed from Figure 10.21(a) that the magnitude of the waveform oscillates for the duration of 0.08 s to 0.10 s, indicating that OT is associated with the voltage waveform. It is also observed that the magnitude of the voltage waveform has continuous ripples over the entire time range, which indicates the presence of harmonics.

It is also observed from Figure 10.21(a) that the magnitude of the waveform has a sharp magnitude peak for the duration of 0.065 s to 0.068 s, indicating that IT is

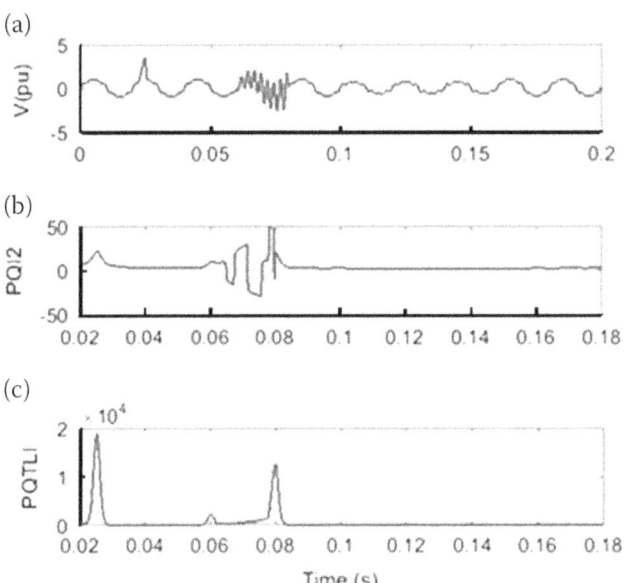

FIGURE 10.21 Voltage waveform with simultaneous occurrence of oscillatory transient, impulsive transient, voltage sag, and harmonics: (a) voltage waveform, (b) PQ index, and (c) PQ time location index.

associated with the voltage waveform. Figure 10.21(b) indicates that the magnitude of the PQI has decreased between 0.01 s to 0.16 s, which indicates the presence of the voltage sag. Figure 10.21(b) also indicates that the magnitude of the PQI changes abruptly during 0.08 s to 0.10 s, which indicates the presence of the OT associated with the voltage signal. Figure 10.21(b) also indicates that magnitude of the PQI has continuous ripples of high magnitude and a fixed pattern over the entire time range, which indicates the presence of the harmonics. Figure 10.21(b) also indicates that the magnitude of the PQI has a sharp magnitude peak during 0.065 s to 0.068 s, which indicates the presence of the IT associated with the voltage signal. Figure 10.21(c) shows that the PQTLI has sharp magnitude peaks at the time instant of 0.10 s and 0.16 s, which indicates the start and end of the sag in voltage. Figure 10.21(c) also shows that the PQTLI has high magnitude peaks at the moments 0.08 s and 0.10 s, which localizes the OT by identifying the start of the OT and end of the OT. Figure 10.21(c) also shows that the PQTLI has continuous ripples with sharp tips and a fixed pattern over the entire time range, which indicates the presence of harmonics with the signal. Figure 10.21(c) shows that the PQTLI has a sharp magnitude peak during the time duration of 0.065 s to 0.068 s, which indicates the presence of IT associated with the voltage signal. Hence, it is established that the proposed approach effectively identified and localized the PQ disturbance of sag and OT associated with the voltage signal.

10.4   FEATURE ESTIMATION AND CLASSIFICATION OF PQ EVENTS

Extraction of features from the PQI and PQTLI plots, which can used for the classification complex PQ issues, are discussed in this section. Classification of complex PQ disturbances is also discussed in this section.

10.4.1   FEATURE EXTRACTION

The features F1 to F6 extracted from the PQI and PQTLI using the statistical techniques are detailed in this section. These features are taken as input to the rule-based decision tree for classifying the PQ disturbances. Definitions of these features are detailed as follows:

> **F1:** Standard deviations of PQI
> **F2:** Standard deviations of PQTLI
> **F3:** Skewness of PQI
> **F4:** Skewness of PQTLI
> **F5:** Kurtosis of PQI
> **F6:** Kurtosis of PQTLI

The values of features F1 to F6 computed for the investigated PQ disturbances are provided in Table 10.1.

TABLE 10.1

Features Used for Classification of PQ Disturbances

S. No.	PQ Disturbance	Symbol	F1	F2	F3	F4	F5	F6
1	Voltage Sag + Harmonics	CP1	0.5448	8.9665	0.2332	8.6127	3.1600	83.5155
2	Voltage Swell + Harmonics	CP2	0.7047	9.1889	0.3465	7.9870	1.5883	74.5555
3	MI + Harmonics	CP3	1.0262	9.7427	−0.1551	7.2776	1.6308	63.7087
4	OT + Voltage Sag	CP4	6.0507	1.3190×10^3	−2.7833	8.3648	15.7671	78.2387
5	OT + Voltage Swell	CP5	6.0775	1.3621×10^3	−2.9531	8.3272	15.7943	77.1419
6	IT + Voltage Sag	CP6	3.6867	4.8817×10^3	6.5707	9.4923	51.3588	97.8559
7	IT + Voltage Swell	CP7	3.6984	5.0209×10^3	6.7189	9.4912	53.1899	97.8384
8	Voltage Swell + OT + Harmonics	CP8	7.3721	1.2430×10^3	0.7807	8.3024	19.8265	77.3574
9	Voltage Sag + OT+ Harmonics	CP9	7.3723	1.1768×10^3	0.9794	8.1639	20.0988	74.7869
10	OT + IT + Voltage Sag + Harmonics	CP10	7.7238	2.0821×10^3	0.7713	6.2404	16.8373	44.7586

10.4.2 CLASSIFICATION OF COMPLEX PQ DISTURBANCES

The complex PQ events have been classified using the decision-supported rules. These decision rules are driven by the feature values given in Table 10.1. Complex PQ events are grouped into different clusters using the features and decision rules. Further, the complex PQ issues are classified one by one from the decision-supported rules from these clusters one by one. A flow chart of classification of the complex PQ events is illustrated in Figure 10.22. In this graph, terminal nodes indicate the final response in terms of the categorized complex PQ events. It is observed that all the investigated complex PQ disturbances are classified effectively using the decision rules driven by the features F1 to F6. Classification accuracy of the approach is evaluated by testing the algorithm on a data set of 50 data computed by changing the various parameters used to define the complex PQ issues. The correctly identified and incorrectly identified complex PQ events are described in Table 10.2.

From Table 10.2, it is observed that the accuracy of the proposed approach is observed as high as 96.2% for classification of the complex PQ events.

10.5 PERFORMANCE VALIDATION

To establish the performance of the algorithm introduced for identification and classification of the PQ events, accuracy of this algorithm is compared with the accuracy of the algorithm reported in the reference [36]. This reference has proposed an

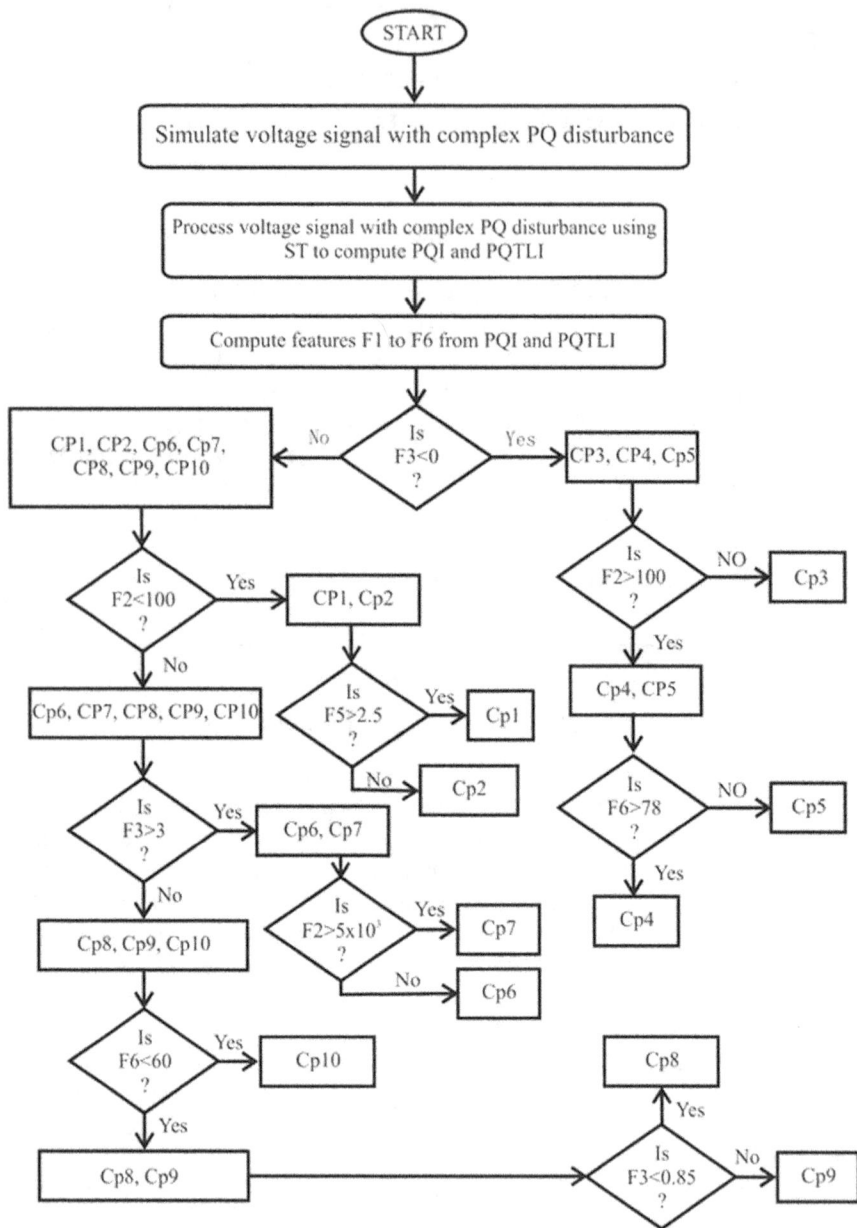

FIGURE 10.22   Classification tree of complex PQ events.

algorithm based on the discrete wavelet transform (DWT) and artificial neural network (ANN) and efficiency of 88% reported for this algorithm. However, the efficiency of the proposed algorithm based on ST and DT is 96.2%, which is much
higher compared to the algorithm reported in [36].

TABLE 10.2
Classification Results of Complex PQ Events

S. No.	PQ Disturbance	Symbol	Correctly Identified PQ Events (in numbers)	Incorrectly Identified PQ Events (in numbers)	Accuracy (%)
1	Voltage Sag + Harmonics	CP1	49	1	98
2	Voltage Swell + Harmonics	CP2	49	1	98
3	MI + Harmonics	CP3	50	0	100
4	OT + Voltage Sag	CP4	46	4	92
5	OT + Voltage Swell	CP5	47	3	94
6	IT + Voltage Sag	CP6	48	2	96
7	IT + Voltage Swell	CP7	48	2	96
8	Voltage Swell + OT + Harmonics	CP8	47	3	94
9	Voltage Sag + OT + Harmonics	CP9	48	2	96
10	OT + IT + Voltage Sag + Harmonics	CP10	49	1	98
Overall accuracy (%)					96.2%

10.6 CONCLUSION

A technique for the identification of complex PQ disturbances is proposed in this work. In the proposed technique, signals with a PQ disturbance are simulated using the mathematical models. These signals are processed using the Stockwell transform to compute an output matrix in the frequency domain. A power quality index (PQI) and a PQ time location index (PQTLI) are computed from this matrix and used for the identification of the different types of complex PQ issues. Six statistical features are computed from the PQI and PQTLI, which are considered as the input to the rule-based decision tree for classification of the PQ disturbances. It is concluded that the proposed technique based on Stockwell transform and rule-based decision is found to be effective in the identification of complex PQ issues. The classification accuracy of the complex PQ issues has been achieved as high as 96.2%. Performance of the algorithm is compared with the accuracy of the algorithm based on the discrete wavelet transform (DWT) and artificial neural network (ANN), which has an efficiency of 88%. Hence, the proposed algorithm classifies the complex PQ events with accuracy greater than the accuracy of DWT and ANN-based approach reported in literature. Simulation studies of the identification and classification of the PQ issues that are simple in nature as well as complex in nature are presented in this work.

However, before the application of the algorithm in the PQ monitoring devices, hardware-based validation can be considered as future work.

REFERENCES

[1] O. P. Mahela, B. Khan, H. Haes Alhelou and S. Tanwar, "Assessment of power quality in the utility grid integrated with wind energy generation," *IET Power Electronics*, vol. 13, no. 13, pp. 2917–2925, 2020.

[2] O. P. Mahela et al., "Recognition of power quality issues associated with grid integrated solar photovoltaic plant in experimental framework," *IEEE Systems Journal*, vol. 15, no. 3, pp. 3740–3748, September 2021, 10.1109/JSYST.2020.302 7203

[3] O. P. Mahela, A. G. Shaik, B. Khan, R. Mahla and H. H. Alhelou, "Recognition of complex power quality disturbances using S-transform based ruled decision tree," *IEEE Access*, vol. 8, pp. 173530–173547, 2020.

[4] O. P. Mahela, B.. Khan, H. H.. Alhelou, S. Tanwar, and S. Padmanaban, "Harmonic mitigation and power quality improvement in utility grid with solar energy penetration using distribution static compensator". *IET Power Electronics*, vol. 14, pp. 912–922, 2021.

[5] R. K. Pachauri et al., "Impact of partial shading on various PV array configurations and different modeling approaches: A comprehensive review," *IEEE Access*, vol. 8, pp. 181375–181403, 2020.

[6] R. K. Pachauri, O. P. Mahela, B. Khan, A. Kumar, S. Agarwal, H. Haes Alhelou, and J. Bai, "Development of arduino assisted data acquisition system for solar photovoltaic array characterization under partial shading conditions", *Computers & Electrical Engineering*, vol. 92, pp. 107175, 2021.

[7] R. K. Pachauri, et al., "Shade dispersion methodologies for performance improvement of classical total cross-tied photovoltaic array configuration under partial shading conditions". *IET Renew. Power Gener*, vol. 15, pp. 1796–1811, 2021.

[8] R. Kaushik et al., "Recognition of islanding and operational events in power system with renewable energy penetration using a stockwell transform-based method," *IEEE Systems Journal*, to be published, 10.1109/JSYST.2020.3020919

[9] S. P. Bihari et al., "A comprehensive review of microgrid control mechanism and impact assessment for hybrid renewable energy integration," *IEEE Access*, vol. 9, pp. 88942–88958, 2021.

[10] O. P. Mahela, Y. Sharma, S. Ali, B. Khan and S. Padmanaban, "Estimation of islanding events in utility distribution grid with renewable energy using current variations and stockwell transform," *IEEE Access*, vol. 9, pp. 69798–69813, 2021.

[11] O. P. Mahela, and A. Gafoor Shaik, "Power quality improvement in distribution network using DSTATCOM with battery energy storage system", *International Journal of Electrical Power and Energy Systems (Elsevier)*, vol. 83, pp. 229–240, December 2016, 10.1016/j.ijepes.2016.04.011.

[12] O. P. Mahela, and A. Gafoor Shaik, "Power quality recognition in distribution system with solar energy penetration using S-Transform and Fuzzy C-Means Clustering", *Renewable Energy (Elsevier)*, vol. 106, pp. 37–51, June 2017, 10.1016/j.renene.2016.12.098.

[13] O. P. Mahela, B. Khan, H. Haes Alhelou, and P. Siano, Power quality assessment and event detection in distribution network with wind energy penetration using stockwell transform and fuzzy clustering, *IEEE Transactions on Industrial Informatics*, vol. 16, no. 11, pp. 6922–6932, November 2020, 10.1109/TII.2020.2 971709.

[14] O. P. Mahela, B. Khan, H. Haes Alhelou, and S. Tanwar, Assessment of power quality in the utility grid integrated with wind energy generation, *IET Power Electronics*, vol. 13, pp. 2917–2925, January 2020, 10.1049/iet-pel.2019.1351.

[15] G. S. Chawda, A. Gafoor Shaik, O. P. Mahela, S. Padmanaban, and J. B. Holm-Nielsen, "Comprehensive review of distributed FACTS control algorithms for power quality enhancement in utility grid with renewable energy penetration," *IEEE Access*, vol. 8, pp. 107614–107634, 2020, 10.1109/ACCESS.2020.3000931.

[16] G. S. Chawda, A. Gafoor Shaik, M. Shaik, P. Sanjeevikumar, J. B. Holm-Nielsen, O. P. Mahela, and K. Palanisamy, "Comprehensive review on detection and classification of power quality disturbances in utility grid with renewable energy penetration," *IEEE Access*, vol. 8, pp. 146807–146830, 2020, 10.1109/ACCESS.2020.3014732.

[17] R. A. Flore, State of the art in the classification of power quality events, an overview. In: 10th international conference on harmonics and quality of power, vol. 1; 2002. p.17–20

[18] S. S. Santos, E. J. Powers, W. M. Grady and P. Hofmann, "Power quality assessment via wavelet transform analysis," *IEEE Transactions on Power Delivery*, vol. 11, no. 2, pp. 924–930, April 1996.

[19] A. M. Gaouda, M. M. A. Salama, M. K. Sultan, and A. Y. Chikhani, "Power quality detection and classification using wavelet-multiresolution signal decomposition," *IEEE Transactions on Power Delivery*, vol. 14, no. 4, pp. 1469–1476, October 1999.

[20] H.-T. Yang, and C.-C. Liao, "A de-noising scheme for enhancing wavelet-based power quality monitoring system," *IEEE Transactions on Power Delivery*, vol. 16, no. 3, pp. 353–360, July 2001.

[21] M. Karimi, H. Mokhtari, and M. R. Iravani, "Wavelet based on-line disturbance detection for power quality applications," *IEEE Transactions on Power Delivery*, vol. 15, no. 4, pp. 1212–1220, October 2000.

[22] P. K. Dash, B. K. Panigrahi, D. K. Sahoo, and G. Panda, "Power quality disturbance data compression, detection, and classification using integrated spline wavelet and S-transform," *IEEE Transactions on Power Delivery*, vol. 18, no. 2, pp. 595–600, April 2003.

[23] P. Kanirajan, and V. S. Kumar, "Power quality disturbance detection and classification using wavelet and RBFNN," *Applied Soft Computing*, vol. 35, pp. 470–481, 2015.

[24] H. S. Behera, P. K. Dash, and B. Biswal, "Power quality time series data mining using S-transform and fuzzy expert system," *Applied Soft Computing*, vol. 10, pp. 945–955, 2010.

[25] H. Eristi, O. Yildirim, B. Eristi, Y. Demir, "Automatic recognition system of underlying causes of power quality disturbances based on S-transform and extreme learning machine," *Electric Power Systems Research*, vol. 61, pp. 553–562, 2014.

[26] M. A. S. Masoum, S. Jamali, and N. Ghaffarzadeh, "Detection and classification of power quality disturbances using discrete wavelet transform and wavelet networks," *IET Science, Measurement and Technology*, vol. 4, no. 4, pp. 193–205, 2010.

[27] O. P. Mahela, and A. G. Shaik, "Recognition of power quality disturbances using S-transform and rule-based decision tree," in: 2016 IEEE First International Conference on Power Electronics, Intelligent Control and Energy Systems (ICPE-ICES 2016), New Delhi, India, July 4–6, 2016.

[28] O. P., Mahela, and A. G. Shaik, "Recognition of power quality disturbances using S-Transform and fuzzy C-Means clustering," in: IEEE International Conference and Utility Exhibition on Co-generation, small power plants and district energy (ICUE 2016), BITEC, Bang Na, Bangkok, Thailand, September 14–16, 2016.

[29] H. Shareef, A. Mohamed, A. A. Ibrahim, "An image processing based method for power quality event identification," *Electric Power Systems Research*, vol. 46, pp. 184–197, 2013.

[30] K. Kumari, A. K. Dadhich, and O. P. Mahela, "Detection of power quality disturbances in the utility grid with solar energy using S-Transform". In: IEEE 7th Power India International Conference on advances in signal processing (PIICON 2016), Bikaner, India, November 25–27, 2016, 10.1109/POWERI.2016.8077351

[31] A. K. Sharma, O. P. Mahela, and S. R. Ola, "Detection of Power Quality Disturbances in The Utility grid Using Stockwell Transform". In: IEEE 7th Power India International Conference on advances in signal processing (PIICON 2016), Bikaner, India, November 25–27, 2016, 10.1109/POWERI.2016.8077376

[32] N. Gupta, M. Khosravy, N. Patel, N. Dey, and O. P. Mahela, "Mendelian evolutionary theory optimization algorithm", *Springer Soft Computing*, vol. 24, 2020.

[33] N. Gupta, M. Khosravy, N. Patel, O. P. Mahela, and G. Varshney, "Plant genetics-inspired evolutionary optimization: A descriptive tutorial", In: Khosravy M., Gupta N., Patel N., Senjyu T. (eds), *Frontier Applications of Nature Inspired Computation*pp. 53–77. Springer, 2020.

[34] R. H. G. Tan and V. K. Ramachandaramurthy, "Numerical model framework of power quality events," *Numerical Journal of Scientific Research Research*, vol. 43, no. 1, pp. 30–47, June 2010.

[35] O. P. Mahela and A. G. Shaik, "Recognition of power quality disturbances using S-Transform based ruled decision tree and fuzzy C-means clustering classifiers", *Applied Soft Computing (Elsevier)*, Vol. 59, pp. 243–257, October 2017, 10.1016/j.asoc.2017.05.061

[36] Y.-Y. Hong, and C.-W. Wang, "Switching detection/classification using discrete wavelet transform and self-organizing mapping network," *IEEE Transactions on Power Delivery*, vol. 20, no. 2, pp. 1662–1668, April 2005.

11 Recognition of Simple Power Quality Disturbances Using Wavelet Packet-Based Fast Kurtogram and Ruled Decision Tree Algorithm

Om Prakash Mahela
Power System Planning Division, Rajasthan Rajya Vidyut
Prasarn Nigam Ltd., Jaipur, India

Anup Singh and Sunil Agarwal
Department of Electrical Engineering, Apex Institute of
Engineering and Technology, Jaipur, India

Baseem Khan
Department of Electrical and Computer Engineering,
Hawassa University, Ethiopia

CONTENTS

DOI: 10.1201/b22884-11

11.1   INTRODUCTION

The quality of electric power is becoming an issue of important concern for both the electric utilities as well as their customers. This has motivated researchers and academicians to make power quality as an active and important research area [1–10]. Power quality is degraded due to disturbances like sag in voltage, swell in voltage, harmonic components superimposed on voltage signal, notches, spikes, etc. These disturbances cause problems like malfunctions and instabilities in the operation of consumer equipment. These disturbances also lead to a short lifetime and failure of electrical equipment [11]. Detection and classification of these disturbances are important steps to find sources and causes of these PQ disturbances. This can be achieved by many techniques and methods. Signal processing methods are finding the top position for their use in the detection of PQ disturbances and machine learning intelligent methods find great applications for classification of the PQ disturbances. An algorithm supported by adaptive filtering and a support vector machine (SVM) with multi-class is observed for identification and classification of the PQ disturbances in [12]. The empirical wavelet transform supported adaptive filtering method has been implemented for extraction of features for the purpose of detection of PQ disturbances. Classification has been achieved using the multiclass SVM. In [13], the authors introduced a parameter supported by wavelet packet transform (WPT) for the identification of PQ disturbances with short durations. The investigated PQ disturbances like voltage sag, voltage swell, momentary interruption (MI), harmonics, and oscillatory transient (OT) are simulated in accordance with the standard IEEE 1159–2009. A multi-objective evolutionary algorithm for analysis of power quality disturbances with the help of two-dimensional orthonormal Stockwell transform and machine learning has been observed in [14]. Extraction of features is achieved using two-dimensional fast discrete orthonormal Stockwell transform (FDOST) and these features have been utilized for classification by the genetic algorithm (GA). In [15], the authors framed an approach to process the PQ waveforms of non-stationary nature using fast Stockwell transform (FST) based on a modified

Gaussian window for the generation of time frequency contours to extract suitable features to automatically recognize the PQ disturbance patterns. These features are given as input to the bacterial foraging optimization algorithm (BFOA) supported by the fuzzy decision tree for classification with improved efficiency.

11.1.1 RELATED WORK

The techniques based on mathematics and signal processing approaches implemented for power quality estimation and reported in the literature are comprehensively re-viewed and presented in this section. In [16], the authors described an algorithm for measuring the PQ parameters that is suitable for implementation in a measuring device known as an estimator analyzer for power quality. The main concern is to overcome shortcomings of PQ analyzers for currents. This is sensitive to fast fluc-tuations of input parameters of signals. This has been achieved by a complementary application of discrete wavelet transform (DWT) in collaboration with the Fourier transform (FT) and chirp z-transform (CZT). In [17], the authors proposed the de-composition of a voltage signal having power quality disturbances. This method was based on the independent component analysis (ICA). Power signals are decomposed into components from where specific information related to various natures of dis-turbances can be investigated. This method is effective in identification of the multiple disturbances. Pre-processing stages of the ICA method have been proposed using a filter bank. In [18], the authors introduced an approach for the identification and classification of disturbances in the power system network. The concept was based on the combination of the linear Kalman filter with DWT for extraction of parameters such as amplitude and slope from recorded voltage or current waveform. DWT, when implemented with the Kalman filter, has improved the performance of the algorithm. Recorded waveforms with distortions are passed through the DWT for determination of noise and covariance of noisy component, which is fed together to the Kalman filter. The fuzzy-expert system (FES) is implemented for classification purposes. An advanced technique for the classification of PQ disturbances with the help of the hidden Markov model (HMM) and wavelet transform (WT) is proposed in [19]. Energy distributions of signals are obtained using a WT at each level of decom-position level, which are considered for training the HMM. Statistical parameters of features are evaluated and used for initializing the training matrices of the HMM. This helped to maximize the accuracy of classification. Fifteen PQ disturbances of various natures are taken to train and evaluate method. The Dempster–Shafer algorithm is also incorporated to improve classification accuracy. The effect of noise is investigated and performance of the method for denoising of signals is also estimated. In [20], the author presented a method using wavelet norm entropy supported extraction of fea-tures of PQ disturbance for the classification purpose. A disturbance classification is performed using a wavelet neural network (WNN). This performs a feature extraction step and classification is achieved using a wavelet feature extractor supported by norm entropy and a classifier is based on a multi-layer perceptron. Seven types of PQ signals have been investigated. The performance of classification for various wavelet families is used for the proposed approach. Sensitivity of the WNN is also in-vestigated for different noise levels. The rate of accuracy of classification is achieved up to 92.5% even under noisy conditions. In [21], the authors introduced an integrated algorithm based on the DWT and fast Fourier transform (FFT) for PQ recognition. Detection of PQ disturbance is achieved by processing the signal using the DWT. Discrete wavelet coefficients are utilized for the calculation of average energy entropy from squared detailed coefficients and considered as a feature. Different PQ disturbances are first detected and in subsequent steps these are classified into four groups of sag, swell, interruption, and harmonics with the help of this feature. Subsequent classification of every main group is achieved using FFT-based features. In [22], the authors proposed an algorithm for detection of attack on a power system. This is based on the machine learning technique that can be trained with the help of information and logged data collected by phasor measurement units (PMUs). Initially, the features are constructed and then data are sent to various machine learning steps in which a random forest is considered a simple classifier of AdaBoost. The model is evaluated using open source simulation consisting of 37 power system events that have been used for evaluation of the model. It is demonstrated that the proposed model achieved an accuracy rate of 93.91% and detection rate of 93.6%. In [23], an approach to detect and classify the single and combined nature of PQ disturbance is proposed. This is based on a combination of fuzzy logic and particle swarm optimization (PSO). In the proposed method, effective features of PQ disturbance are extracted from parameters evaluated using the Fourier and wavelet transform-based decomposition of signal. The fuzzy system classified different types of PQ disturbances using the above-mentioned features. Subsequently, the PSO is utilized for accurate determination of parameters of membership function for a fuzzy system. Investigated disturbances include impulse, MI, swell, sag, notches, transients, harmonic, and flicker. Performance of the algorithm is also evaluated in the presence of noise. In [24], the authors discussed an approach using decision rules to classify the PQ disturbances. A signal with a PQ disturbance is first characterized with the help of multi-resolution analysis supported by S-transform, which is considered as a tool for feature extraction [25,26–35]. Subsequently, a simple and robust rule-based algorithm for the identification of PQ disturbances; this algorithm used linear and parabolic rules for classification of patterns and decision boundaries are set by a heuristic search. Classification is achieved using a modular structure where each module works separately to identify specific disturbances [36–45].

A detailed analysis of the survey of literature is presented in this chapter, and it is established that the present methods implemented for power quality disturbances identification are using simple features with signal processing techniques. However, an advance signal may be introduced that can be used for the simple nature as well multiple nature power quality disturbances simultaneously. This is considered in this chapter for further research using a wavelet packet supported fast kurtogram, which is found to be effective in identification and classification of power quality disturbances in the recent scenario.

11.1.2 Contribution of the Proposed Work

The research work considered is focused to design an algorithm using wavelet packet supported fast kurtogram and decision rules for identification and classification of the simple nature power quality disturbances. The contributions of this research are as follows:

- An algorithm based on a wavelet packet supported fast kurtogram and decision rules has been introduced for the identification as well as classification of the simple nature of PQ disturbances.
- Performance of the algorithm is evaluated by testing the algorithm on the standard PQ disturbances that are generated with the help of mathematical relations.
- The study is performed with the help of MATLAB® software.

11.1.3 ORGANIZATION OF THIS CHAPTER

This chapter is organized as follows: Section 11.2 discusses the methodology utilized in this research. Section 11.3 presents the results and discussion computed in this research. A comparative analysis of the proposed technique with existing literature is also presented in this section. Section 11.4 presents the conclusion of the proposed research.

11.2 METHODOLOGY

Formulation of power quality disturbances with the aid of mathematical relations have been introduced in this section. The generation of simple nature PQ disturbances is elaborated. The algorithm based on a wavelet packet is supported by a fast kurtogram and decision rules proposed for the identification as well as classification of the simple nature PQ disturbances have been included in the section.

11.2.1 FORMULATION OF PROBLEM

Recently, the structure of power is in a dynamic state and has continuously been changing due to the integration of power electronic converter supported loads and equipment that are non-linear. Deployment of smart grid topologies have and large integration of renewable energy have also affected the structure of power networks. Above-mentioned technologies have resulted in the introduction of power quality disturbances of various nature. Specifically, multiple nature PQ disturbances are being observed. Existing approaches for identification of power quality disturbances are not well acquainted with fast technological developments in the power sector. Hence, advance methods are required that may be used for the identification and classification of power quality disturbances in the recent scenario.

11.2.2 FORMULATION AND GENERATION OF SIMPLE NATURE PQ DISTURBANCES

Power quality disturbances superimposed on the voltage signals have been formulated using the standards set by the IEEE-1159 standard with the help of MATLAB software.

These voltage signals (50 Hz) are generated using the mathematical relation reported in [46]. The boundary parameters of the signals reported in [46] have been used in the present study. Mathematical modeling of the simple nature PQ disturbances investigated in this study is given below. Here, the symbols used are are A: amplitude, f: frequency, V: voltage, T: time period, τ: time constant, ω: angular frequency, and u (t): unit step function.

1. Voltage signal without any disturbance:

$$V(t) = A \sin(\omega t)$$

2. Voltage signal with sag disturbance:

$$V(t) = [1 - \alpha(u(t - t_1) - u(t - t_2))]\sin(\omega t)$$

3. Voltage signal with swell disturbance:

$$V(t) = [1 + \alpha(u(t - t_1) - u(t - t_2))]\sin(\omega t)$$

4. Voltage signal with momentary interruption disturbance:

$$V(t) = [1 - \alpha(u(t - t_1) - u(t - t_2))]\sin(\omega t)$$

5. Voltage signal with harmonics disturbance:

$$V(t) = \alpha_1 \sin(\omega t) + \alpha_3 \sin(3\omega t) + \alpha_5 \sin(5\omega t) + \alpha_7 \sin(7\omega t)$$

6. Voltage signal with oscillatory transient disturbance:

$$V(t) = \sin(\omega t) + \alpha e^{\frac{(t-t_1)}{\tau}} \sin(\omega_n (t - t_1)(u(t_2 - u(t_1))))$$

7. Voltage signal with impulsive transient disturbance:

$$V(t) = \sin(\omega t) + \alpha e^{\frac{(t-t_1)}{\tau}} - \alpha e^{\frac{(t-t_1)}{\tau}} (u(t_2 - u(t_1)))$$

8. Voltage signal with notches disturbance:

$$V(t) = \sin(\omega t) - sign(\sin(\omega t))$$
$$\times \left[\sum_{n=0}^{9} K \times \{u(t - (t_1 + 0.02n)) - u(t - (t_2 + 0.02n))\}\right]$$

9. Voltage signal with spikes disturbance:

$$V(t) = \sin(\omega t) - sign(\sin(\omega t))$$

$$\times \left[\textstyle\sum_{n=0}^{9} K \times \{u(t - (t_1 + 0.02n)) - u(t - (t_2 + 0.02n))\} \right]$$

11.2.3 WAVELET PACKET SUPPORTED FAST KURTOGRAM AND DECISION RULES–BASED ALGORITHM FOR IDENTIFICATION AND CLASSIFICATION OF SIMPLE NATURE PQ DISTURBANCES

The algorithm proposed in this work that can be implemented for identification and classification of the power quality disturbance is simple in nature using the wavelet packet supported fast kurtogram and decision rules can be described with the following steps:

- Generate the simple nature power quality disturbances using a mathematical formulation.
- Compute the fast kurtogram of a voltage signal containing a PQ disturbance up to the first level of decomposition using a fast decimated filter bank tree supported by wavelet packet transform. The sampling frequency of a signal considered is 3.2 kHz. Analysis of the fast kurtogram gives availability of transients.
- Transient signals have been filtered out from the kurtogram with optimal carrier frequency of 1,600 Hz up to the first level of decomposition.
- An envelope of filtered signal and amplitude spectrum of squared envelope is obtained. An analysis of these plots gives the availability of the PQ disturbances.

11.3 RESULTS AND DISCUSSION

The results of the identification of simple nature power quality disturbances using the MATLAB-based simulation studies have been described in detail in this section. Classification results using the various decision rules are also included in this section. Performance of the proposed approach and its comparison with an algorithm already reported in literature has also been presented in this section. Descriptions of the results to point out the final outcome have been provided in different sections according to the relevance of the results.

11.3.1 VOLTAGE SIGNAL WITHOUT PQ DISTURBANCE

The voltage signal without any type of disturbance has been formulated using the mathematical relation given in Section 11.2.2 and simulated in the MATLAB software. A fast kurtogram of a voltage signal containing a PQ disturbance has been

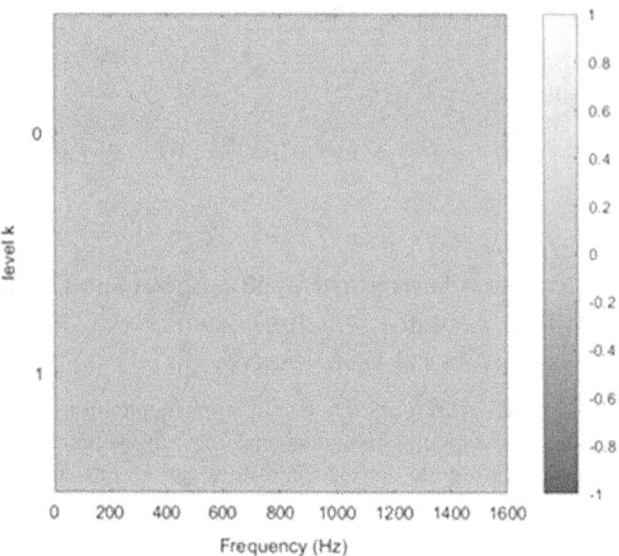

FIGURE 11.1   Spectrum of fast kurtogram for pure sine wave voltage signal.

computed using the fast decimated filter bank tree supported by the wavelet packet transform and shown in Fig. 5.1. The sampling frequency of a signal is considered as 3.2 kHz. Analysis of the fast kurtogram gives availability of transients. Filter bandwidth for the fast kurtogram of the pure voltage signal has been observed as 1,600 Hz and the cutoff frequency is 800 Hz. Since high-frequency transients are not associated with the signal, hence the only kurtogram has included the frequency on the x-axis up to 1,600 Hz and range of Δf between -1 to $+1$ on the y-axis, as illustrated in Figure 11.1. It is observed from Figure 11.1 that at zero level and first level of decomposition only the fundamental frequency of 50 Hz is observed in the kurtogram. High transient frequency components have not been observed in the kurtogram representation of the signal of voltage without any type of disturbance. The difference of observed value of kurtosis and maximum value is observed as zero, which is expected for the pure sinusoidal nature signal of voltage. The fast kurtogram of voltage signal shown in Figure 11.1 will be considered as the reference spectrum for the purpose of identification of other types of disturbances associated with the voltage signals for further analysis of the simple nature power quality disturbances.

The transient signals have been filtered out from the wavelet packet supported fast kurtogram shown in Figure 11.1. The optimal carrier frequency of 1,600 Hz has been used and filtering has been done up to the first level. A voltage signal without any type of disturbance is shown in Figure 11.2(a). The nvelope of the filtered signal is shown in Figure 11.2(b) where the red line gives the threshold for the transient disturbance. The amplitude of spectrum of the squared envelope is detailed in Figure 11.2(c). The cutoff frequency for filtering the signal has been observed to be 1,200 Hz with a value of kurtosis as -0.5.

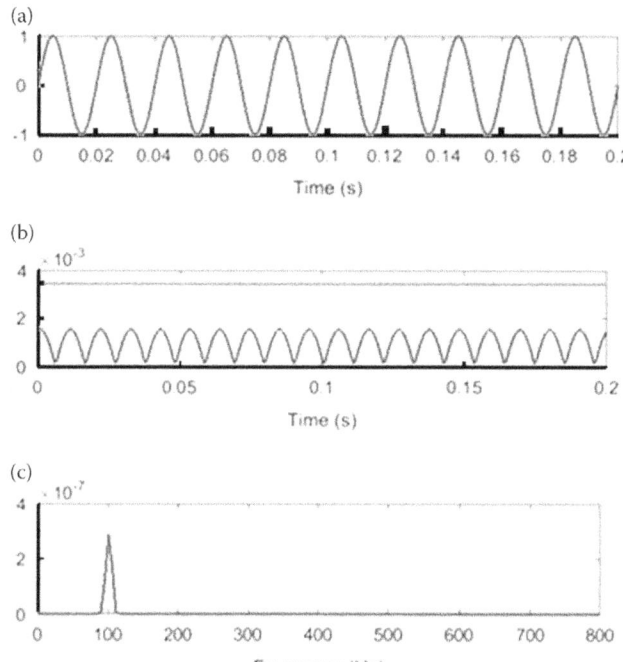

FIGURE 11.2 Pure sine wave: (a) voltage signal, (b) envelope of filtered signal, and (c) amplitude spectrum of squared envelope.

It is observed from the envelope of filtered signal depicted in Figure 11.2(b) that ripples of very low magnitude correspond to the pure voltage signal of 50 Hz frequency. The magnitude of these ripple components is below the threshold value, represented by a red line indicating that any type of transient has not been associated with the voltage signal. This is inferred from the amplitude spectrum of squared envelope, shown in Figure 11.2(c), that only one sharp magnitude peak is observed that corresponds to the fundamental frequency of 50 Hz. Hence, no disturbance is associated with the waveform and it is pure in nature.

11.3.2 VOLTAGE SIGNAL WITH SAG DISTURBANCE

The voltage signal with sag disturbance has been formulated using the mathematical relation given in Section 11.2.2 and simulated in the MATLAB software. The fast kurtogram of a voltage signal containing a sag PQ disturbance has been computed using the fast decimated filter bank tree supported by the wavelet packet transform and shown in Figure 11.3. The sampling frequency of the signal is considered as 3.2 kHz.

An analysis of the fast kurtogram gives the availability of transients. The filter bandwidth for the fast kurtogram of the voltage signal with sag has been observed as 800 Hz and the cutoff frequency is 1,200 Hz. The kurtogram has included the frequency on the x-axis up to 1,600 Hz and range of Δf between 0 to 3 on the y-axis,

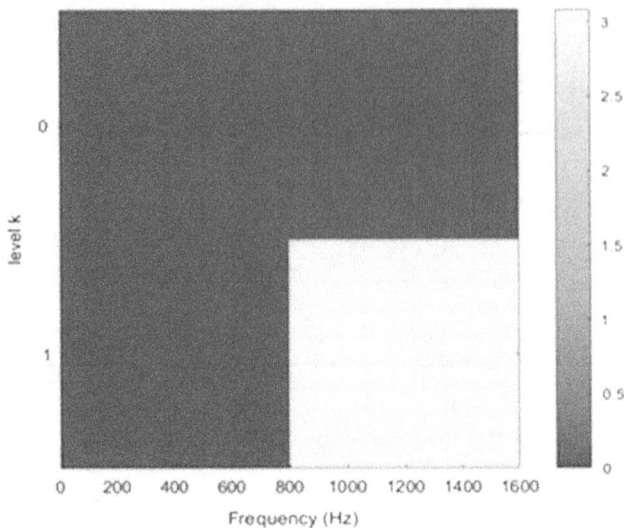

FIGURE 11.3   Spectrum of fast kurtogram for voltage signal with sag disturbance.

as illustrated in Figure 11.3. As observed from Figure 11.3, the zero level and first level of decomposition has been shown in the kurtogram. High transient frequency components have been observed in the kurtogram representation of the signal of voltage with a sag between the frequencies of 800 Hz to 1,600 Hz. This indicates due to the voltage sag initiation and termination, the high-frequency transients have been introduced in the signal. The difference of observed value of kurtosis and maximum value of kurtosis are observed as 3. On comparing the fast kurtogram of Figure 11.3 with that shown in Figure 11.1, it is observed that due to the voltage sag initiation, end of the high-frequency transients are introduced in the signal.

The transient signals have been filtered out from the wavelet packet supported fast kurtogram shown in Figure 11.4. The optimal carrier frequency of 1,600 Hz has been used and filtering has been done up to the first level. The voltage signal with sag disturbance is shown in Figure 11.4(a). The envelope of the filtered signal is shown in Figure 11.4(b) where the red line gives the threshold for the transient disturbance. The amplitude of spectrum of the squared envelope is detailed in Figure 11.4(c). The cutoff frequency for filtering the signal has been observed to be 1,200 Hz with a value of kurtosis as 3.

It is observed from the envelope of filtered signal depicted in Figure 11.4(b) that ripples of very low magnitude correspond to the 50 Hz frequency. The magnitude of these ripple components is below the threshold value. However, the sharp magnitude peaks observed at 0.06 s and 0.14 s indicate the initiation and end of the voltage sag. Further, the magnitude of the curve also decreases between the 0.06 s to 0.14 s, indicating the presence of sag associated with the voltage signal. Hence, analysis of the envelope of filtered signal effectively detects and localizes the sag disturbance. This is inferred from the amplitude spectrum of the squared envelope shown in Figure 11.4(c) that the sharp magnitude peak is observed that corresponds to the

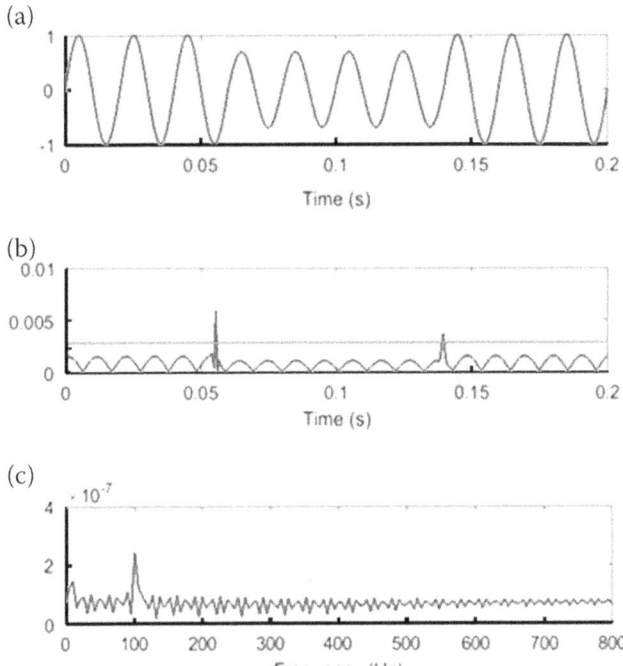

FIGURE 11.4 Voltage signal with sag disturbance: (a) voltage signal, (b) envelope of filtered signal, and (c) amplitude spectrum of squared envelope.

fundamental frequency of 50 Hz. However, low magnitude transient components are also introduced in the signal due to the initiation and end of the sag.

11.3.3 VOLTAGE SIGNAL WITH SWELL DISTURBANCE

The voltage signal with sag disturbance has been formulated using the mathematical relation given in Section 11.2.2 and simulated in the MATLAB software. A fast kurtogram of voltage signal containing swell PQ disturbance has been computed using the fast decimated filter bank tree supported by the wavelet packet transform and shown in Figure 11.5. The sampling frequency of the signal is considered 3.2 kHz. Analysis of the fast kurtogram gives availability of transients. The filter bandwidth for the fast kurtogram of voltage signal with swell has been observed as 800 Hz and the cutoff frequency is 1,200 Hz. The kurtogram has included the frequency on the x-axis up to 1,600 Hz and range of Δf between 0 to 1.3 on the y-axis, as illustrated in Figure 11.5. It is observed from Figure 11.5 that at zero level and first level of decomposition has been shown in the kurtogram. High transient frequency components have been observed in the kurtogram representation of the signal of voltage with a swell between the frequencies of 800 Hz to 1,600 Hz. This indicates that due to the voltage swell initiation and termination, the high-frequency transients have been introduced in the signal. The difference of

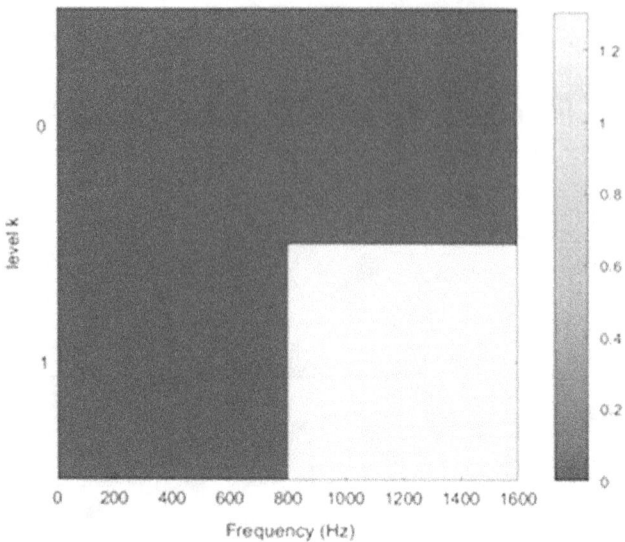

FIGURE 11.5   Spectrum of fast kurtogram for voltage signal with swell disturbance.

observed value of kurtosis and maximum value of kurtosis is observed as 1.3. On comparing the fast kurtogram of Figure 11.5 with that shown in Figure 11.1, it is observed that due to the voltage swell initiation and end, the high-frequency transients are introduced in the signal.

The transient signals have been filtered out from the wavelet packet supported fast kurtogram shown in Figure 11.6. The optimal carrier frequency of 1,600 Hz has been used and filtering has been done up to the first level. The voltage signal with swell disturbance is shown in Figure 11.6(a). The envelope of the filtered signal is shown in Figure 11.6(b) where the red line gives the threshold for the transient disturbance. The amplitude of spectrum of squared envelope is detailed in Figure 11.6(c). The cutoff frequency for filtering the signal has been observed to be 1,200 Hz, with a value of kurtosis as 1.3.

It is observed from the envelope of filtered signal depicted in Figure 11.6(b) that ripples of very low magnitude correspond to the 50 Hz frequency. The magnitude of these ripple components is below the threshold value. However, sharp magnitude peaks observed at 0.06 s and 0.14 s indicate the initiation and end of the swell in the voltage signal. Further, the magnitude of the curve also increases between 0.06 s to 0.14 s, indicating the presence of a swell associated with the voltage signal. Hence, analysis of the envelope of the filtered signal effectively detects and localizes the swell disturbance. This is inferred from the amplitude spectrum of the squared envelope shown in Figure 11.6(c) in that a sharp magnitude peak is observed that corresponds to the fundamental frequency of 50 Hz. However, low magnitude transient components are also introduced in the signal due to the initiation and end of the swell. A kurtosis of 1.3 of the envelope of the filtered signal has been observed.

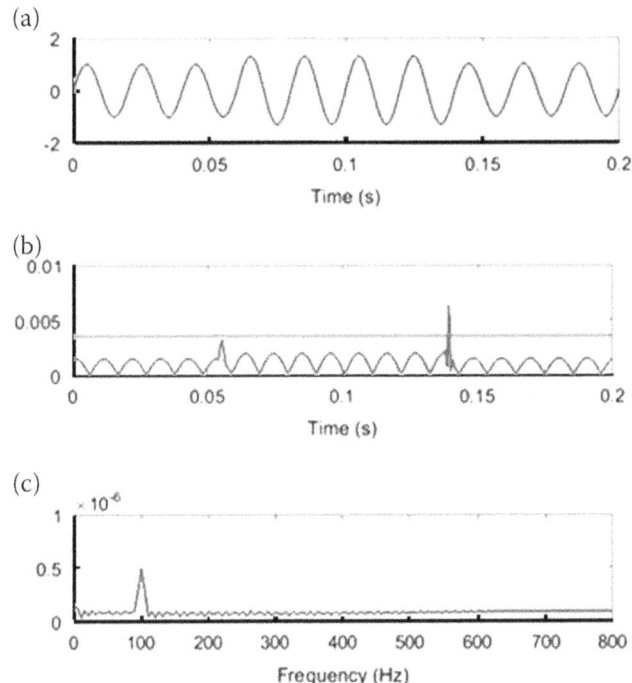

FIGURE 11.6 Voltage signal with swell disturbance: (a) voltage signal, (b) envelope of filtered signal, and (c) amplitude spectrum of squared envelope.

11.3.4 VOLTAGE SIGNAL WITH MOMENTARY INTERRUPTION DISTURBANCE

The voltage signal with sag disturbance has been formulated using the mathematical relation given in Section 11.2.2 and simulated in the MATLAB software. A fast kurtogram of a voltage signal containing a momentary interruption PQ disturbance has been computed using the fast decimated filter bank tree supported by the wavelet packet transform and shown in Figure 11.7. The sampling frequency of the signal is considered as 3.2 kHz.

Analysis of the fast kurtogram gives the availability of transients. The filter bandwidth for the fast kurtogram of voltage signal with momentary interruption has been observed as 800 Hz and the cutoff frequency is 1,200 Hz. The kurtogram has included the frequency on the x-axis up to 1,600 Hz and range of Δf between 0 to 1.3 on the y-axis, as illustrated in Figure 11.7. As observed from Figure 11.7 a zero level and first level of decomposition have been shown in the kurtogram. High transient frequency components have been observed in the kurtogram representation of the signal of voltage with a swell between the frequencies of 800 Hz to 1,600 Hz. This indicates that due to the voltage swell initiation and termination, the high-frequency transients have been introduced in the signal. The difference of observed value of kurtosis and maximum value of kurtosis is observed as 61.4. On comparing the fast kurtogram of Figure 11.7 with that shown in Figure 11.1, it is observed that

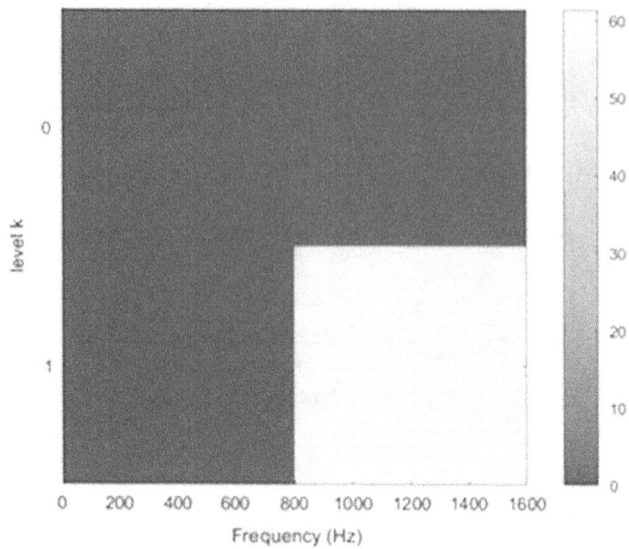

FIGURE 11.7   Spectrum of fast kurtogram for voltage signal with swell momentary interruption disturbance.

due to the voltage momentary interruption initiation and end, the high-frequency transients are introduced in the signal.

The transient signals have been filtered out from the wavelet packet supported fast kurtogram shown in Figure 11.8. The optimal carrier frequency of 1,600 Hz has been used and filtering has been done up to the first level. The voltage signal with momentary interruption disturbance is shown in Figure 11.8(a). The envelope of the filtered signal is shown in Figure 11.8(b) where the red line gives the threshold for the transient disturbance. The amplitude of spectrum of the squared envelope is detailed in Figure 11.8(c). The cutoff frequency for filtering the signal has been observed to be 1,200 Hz with a value of kurtosis as 61.3.

It is observed from the envelope of filtered signal depicted in Figure 11.8(b) that ripples of very low magnitude correspond to the 50 Hz frequency. The magnitude of these ripple components is below the threshold value. However, sharp magnitude peaks observed at 0.06 s and 0.14 s indicate the initiation and end of the momentary interruption in the voltage signal. Further, the magnitude of the curve also decreases to approximately zero (below 10% of normal amplitude) between 0.06 s to 0.14 s, indicating the presence of a momentary interruption associated with the voltage signal. Hence, analysis of the envelope of the filtered signal effectively detects and localizes the momentary interruption disturbance. This is inferred from the amplitude spectrum of the squared envelope shown in Figure 11.8(c) that transient components are observed throughout the frequency range due to the end and initiation of the momentary interruption. A kurtosis of 1.3 of the envelope of filtered signal has been observed.

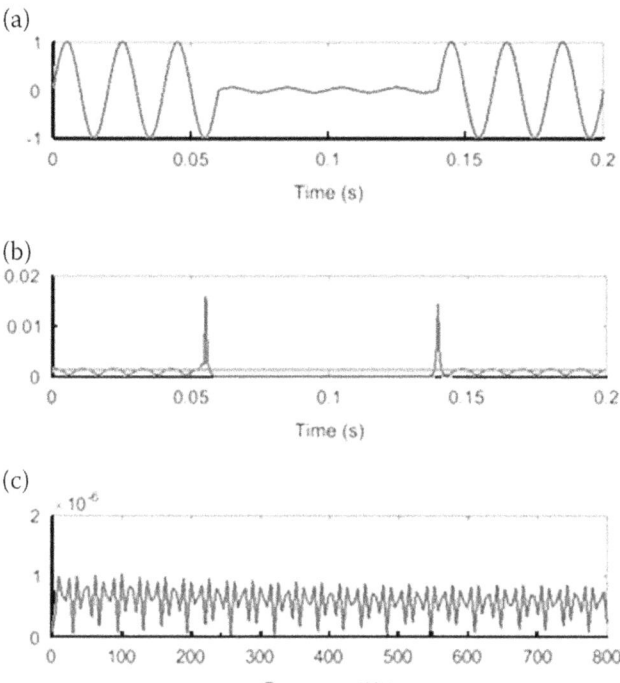

FIGURE 11.8 Voltage signal with momentary interruption disturbance: (a) voltage signal, (b) envelope of filtered signal, and (c) amplitude spectrum of squared envelope.

11.3.5 Voltage Signal with Harmonic Disturbance

The voltage signal with harmonic disturbances has been formulated using the mathematical relation given in Section 11.2.2 and simulated in the MATLAB software. A fast kurtogram of a voltage signal containing harmonics PQ disturbance has been computed using the fast decimated filter bank tree supported by the wavelet packet transform and shown in Figure 11.9. The sampling frequency of the signal is considered as 3.2 kHz. Analysis of the fast kurtogram gives availability of the transients. Filter bandwidth for the fast kurtogram of voltage signal with harmonics has been observed as 1,600 Hz and the cutoff frequency is 800 Hz. The kurtogram has included the frequency on the x-axis up to 1,600 Hz and range of Δf between −1 to 1 on the y-axis, as illustrated in Figure 11.9. It is observed from Figure 11.9 that zero level and first level of decomposition has been shown in the kurtogram. High transient frequency components have not been observed in the kurtogram representation of the signal of voltage with harmonics. This indicates that due to the voltage swell initiation and termination, the high-frequency transients have not been introduced in the signal. However, harmonic frequency components are introduced. The difference of observed value of kurtosis and maximum value of kurtosis is observed as 0. On comparing the fast kurtogram of Figure 11.9 with that shown in Figure 11.1, it is observed that due to the harmonics, the high-frequency transients are not introduced in the signal.

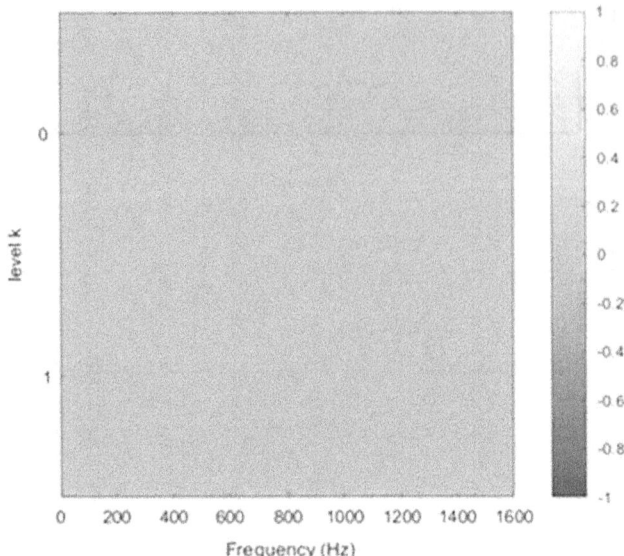

FIGURE 11.9   Spectrum of fast kurtogram for voltage signal with harmonics disturbance.

The transient signals have been filtered out from the wavelet packet supported fast kurtogram shown in Figure 11.9. The optimal carrier frequency of 1,600 Hz has been used and filtering has been done up to the first level. The voltage signal with harmonics disturbance is shown in Figure 11.10(a). The envelope of the filtered signal is shown in Figure 11.10(b) where the red line gives the threshold for the transient disturbance. The amplitude of the spectrum of the squared envelope is detailed in Figure 11.10(c). The cutoff frequency for filtering the signal has been observed to be 1,200 Hz with a value of kurtosis as –0.8.

It is observed from the envelope of filtered signal depicted in Figure 11.10(b) that ripple frequency components of harmonic frequencies have been superimposed over the magnitude of power frequency of 50 Hz. The magnitude of these ripple components is below the threshold value. The pattern of these ripple magnitudes have been modified compared to the respective curve of pure sinusoidal curve. Hence, analysis of envelope of filtered signal effectively detects harmonic components of 3rd, 5th and 7th harmonics superimposed with the signal. It is inferred from the amplitude spectrum of the squared envelope shown in Figure 11.10(c) that harmonic components have been detected with the sharp magnitude peaks observed in addition to the peak corresponding to the fundamental frequency components. Kurtosis equal to –0.2 of the envelope of filtered signal has been observed.

11.3.6   VOLTAGE SIGNAL WITH OSCILLATORY TRANSIENT DISTURBANCE

The voltage signal with an oscillatory transient disturbance has been formulated using the mathematical relation given in Section 11.2.2 and simulated in the MATLAB software. A fast kurtogram of a voltage signal containing oscillatory

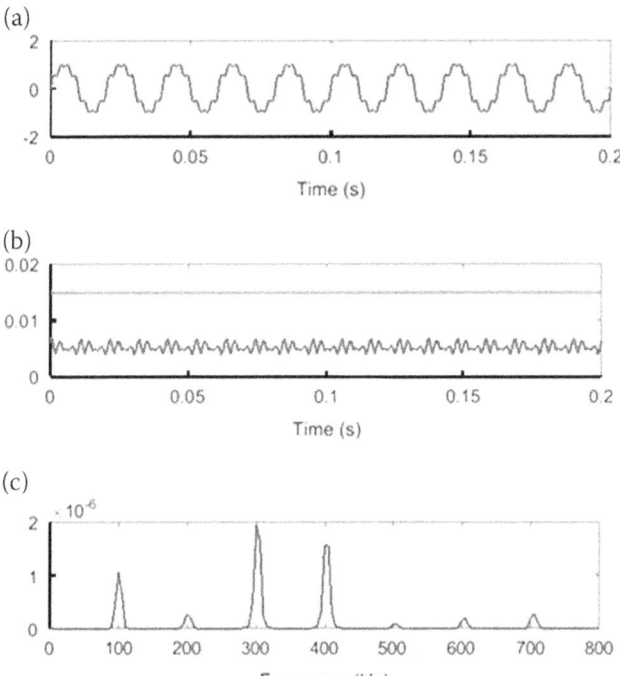

FIGURE 11.10 Voltage signal with harmonics disturbance: (a) voltage signal, (b) envelope of filtered signal, and (c) amplitude spectrum of squared envelope.

transient PQ disturbance has been computed using the fast decimated filter bank tree supported by the wavelet packet transform and shown in Figure 11.11. The sampling frequency of the signal is considered as 3.2 kHz.

The analysis of the fast kurtogram gives the availability of transients. The filter bandwidth for the fast kurtogram of a voltage signal with oscillatory transient has been observed as 800 Hz and the cutoff frequency is 1,200 Hz. The kurtogram has included the frequency on the x-axis up to 1,600 Hz and range of Δf between 0 to 29.5 on the y-axis, as illustrated in Figure 11.11. It is observed from Figure 11.11 that zero level and first level of decomposition have been shown in the kurtogram. High transient frequency components have been observed in the kurtogram representation of the signal of voltage with oscillatory transient between the frequencies of 800 Hz to 1,600 Hz. This indicates that due to the oscillatory transient initiation and termination, the high-frequency transients have been introduced in the signal. The difference of observed value of kurtosis and maximum value of kurtosis is observed as 29.5. On comparing the fast kurtogram of Figure 11.11 with that shown in Figure 11.1, it is observed that due to the oscillatory transient initiation and end, the high-frequency transients are introduced in the signal.

The transient signals have been filtered out from the wavelet packet supported fast kurtogram shown in Figure 11.11. The optimal carrier frequency of 1,600 Hz

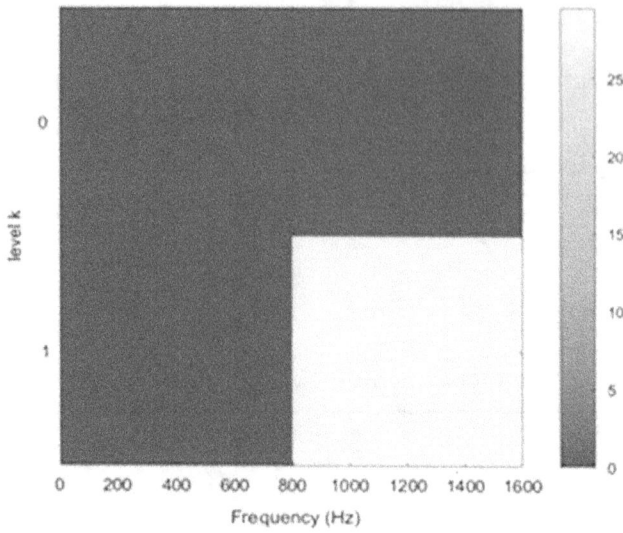

FIGURE 11.11 Spectrum of fast kurtogram for voltage signal with oscillatory transient disturbance.

has been used and filtering has been done up to the first level. The voltage signal with oscillatory transient disturbance is shown in Figure 11.12(a). The envelope of the filtered signal is shown in Figure 11.12(b) where the red line gives the threshold for the transient disturbance. The amplitude of the spectrum of the squared envelope is detailed in Figure 11.12(c). The cutoff frequency for filtering the signal has been observed to be 1,200 Hz with a value of kurtosis as 29.4.

It is observed from the envelope of the filtered signal depicted in Figure 11.12(b) that a high magnitude projection above the threshold values has been observed between 0.08 s to 0.1 s, indicating the presence of oscillatory transient. Small magnitude ripples below the threshold on this curve represent the availability of power frequency of 50 Hz. The magnitude of these ripple components is below the threshold value. Hence, the presence of oscillatory transient has been detected effectively from the envelope of the filtered signal. This is inferred from the amplitude spectrum of the squared envelope shown in Figure 11.12(c) that ripples are observed on the surface throughout the time range indicating the presence of oscillatory transient. Kurtosis of the squared envelope of the filtered signal has been observed equal to 29.4.

11.3.7 Voltage Signal with Impuslive Transient Disturbance

The voltage signal with impulsive transient disturbance has been formulated using the mathematical relation given in Section 11.2.2 and simulated in the MATLAB® software. A fast kurtogram of voltage signal containing impulsive transient PQ disturbance has been computed using the fast decimated filter bank tree supported by the wavelet packet transform and shown in Figure 11.13. The sampling frequency of the signal is considered as 3.2 kHz.

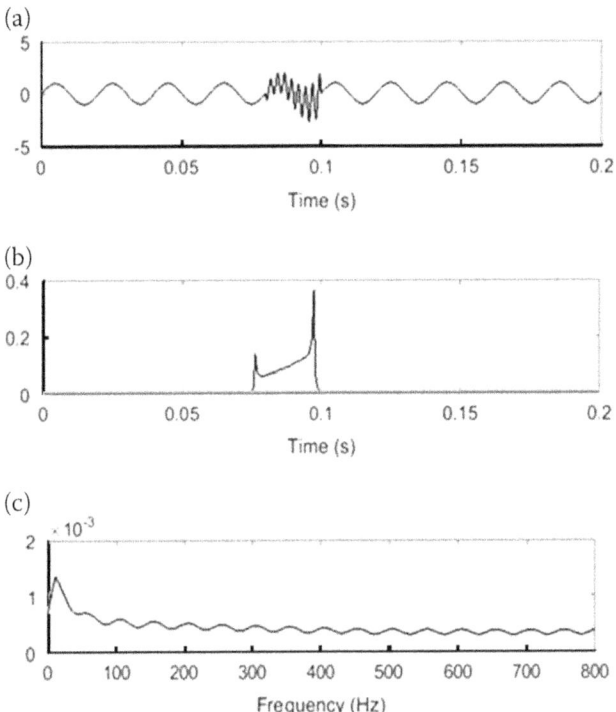

FIGURE 11.12 Voltage signal with oscillatory transient disturbance: (a) voltage signal, (b) envelope of filtered signal, and (c) amplitude spectrum of squared envelope.

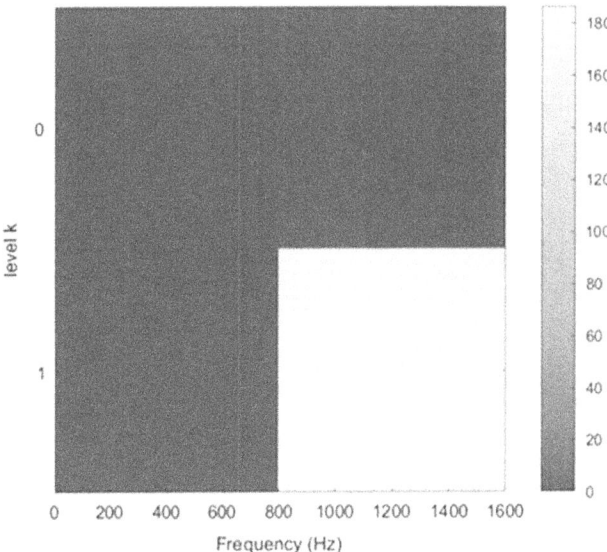

FIGURE 11.13 Spectrum of fast kurtogram for voltage signal with impulsive transient disturbance.

An analysis of the fast kurtogram gives the availability of transients. The filter bandwidth for the fast kurtogram of voltage signal with impulsive transient has been observed as 800 Hz and the cutoff frequency is 1,200 Hz. The kurtogram has included the frequency on the x-axis up to 1,600 Hz and range of Δf between 0 to 186.5 on the y-axis, as illustrated in Figure 11.13. It is observed from Figure 11.13 that zero level and first level of decomposition have been shown in the kurtogram. High transient frequency components have been observed in the kurtogram representation of the signal of voltage with impulsive transient between the frequencies of 800 Hz to 1,600 Hz. This indicates that due to the impulsive transient initiation and termination, the high-frequency transients have been introduced in the signal. The difference of observed value of kurtosis and maximum value of kurtosis is observed as 186.5. On comparing the fast kurtogram of Figure 11.13 with that shown in Figure 11.1, it is observed that due to the impulsive transient initiation and end, the high-frequency transients are introduced in the signal.

The transient signals have been filtered out from the wavelet packet supported fast kurtogram shown in Figure 11.13. The optimal carrier frequency of 1,600 Hz has been used and filtering has been done up to the first level. The voltage signal with impulsive transient disturbance is shown in Figure 11.14(a). The envelope of the filtered signal is shown in Figure 11.14(b) where the red line gives the threshold for the transient disturbance. The amplitude of the spectrum of the squared envelope is detailed in Figure 11.14(c). The cutoff frequency for filtering the signal has been observed to be 1,400 Hz with a value of kurtosis as 186.5.

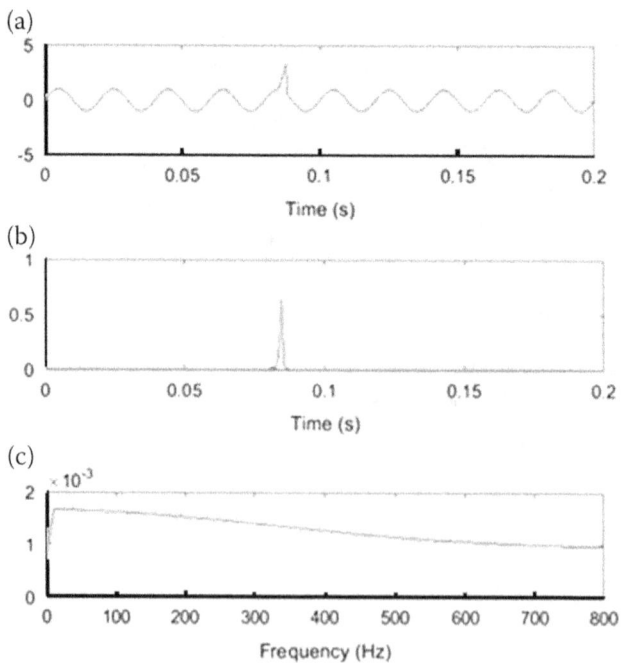

FIGURE 11.14   Voltage signal with impulsive transient disturbance: (a) voltage signal, (b) envelope of filtered signal, and (c) amplitude spectrum of squared envelope.

It is observed from the envelope of the filtered signal depicted in Figure 11.14(b) that a sharp high magnitude projection above the threshold values has been at 0.08 s, which indicates the presence of impulsive transient. Small magnitude ripples below the threshold on this curve represent the availability of a power frequency of 50 Hz. The magnitude of these ripple components is below the threshold value. Hence, the presence of impulsive transient has been detected effectively from the envelope of the filtered signal. This is inferred from the amplitude spectrum of the squared envelope shown in Figure 11.14(c) that ripples are observed on the surface throughout the time range, indicating the presence of impulsive transient. Kurtosis of the squared envelope of the filtered signal has been observed equal to 186.5.

11.3.8 VOLTAGE SIGNAL WITH NOTCHES DISTURBANCE

The voltage signal with notches disturbance has been formulated using the mathematical relation given in Section 11.2.2 and simulated in the MATLAB software. A fast kurtogram of voltage signal containing notches PQ disturbance has been computed using the fast decimated filter bank tree supported by the wavelet packet transform and shown in Figure 11.15. The sampling frequency of the signal is considered as 3.2 kHz.

The analysis of the fast kurtogram gives the availability of transients. Filter bandwidth for the fast kurtogram of voltage signal with notches transient has been observed as 800 Hz and the cutoff frequency is 1,200 Hz. The kurtogram has been included the frequency on the x-axis up to 1,600 Hz and range of Δf between 0 to

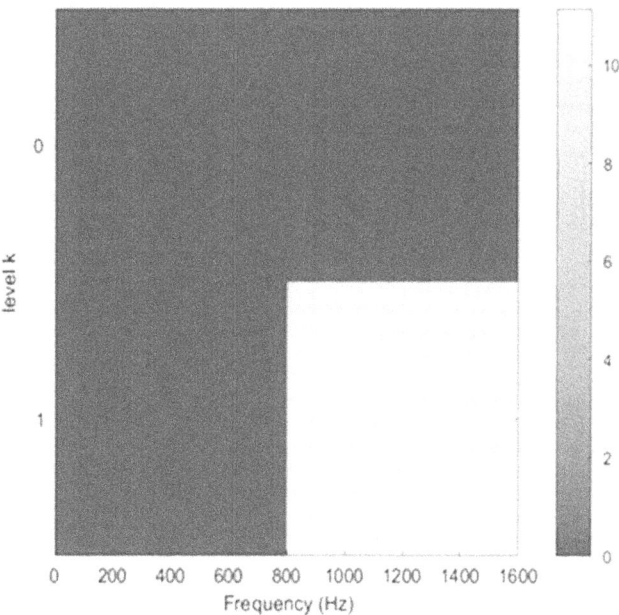

FIGURE 11.15   Spectrum of fast kurtogram for voltage signal with notches disturbance.

11.1 on the y-axis, as illustrated in Figure 11.15. It is observed from Figure 11.15 that zero level and first level of decomposition have been shown in the kurtogram. High transient frequency components have been observed in the kurtogram representation of the signal of voltage with notches between the frequencies of 800 Hz to 1,600 Hz. This indicates that due to the notches initiation and termination, the high-frequency transients have been introduced in the signal. The difference of observed value of kurtosis and maximum value of kurtosis is observed as 11.1. On comparing a fast kurtogram of Figure 11.15 with that shown in Figure 11.1, it is observed that due to the notches transient initiation and end, the high-frequency transients are introduced in the signal.

The transient signals have been filtered out from the wavelet packet supported fast kurtogram shown in Figure 11.15. The optimal carrier frequency of 1,600 Hz has been used and filtering has been done up to the first level. The voltage signal with notches disturbance is shown in Figure 11.16(a). The envelope of the filtered signal is shown in Figure 11.16(b) where the red line gives the threshold for the transient disturbance. The amplitude of the spectrum of the squared envelope is detailed in Figure 11.16(c). The cutoff frequency for filtering the signal has been observed to be 1,200 Hz with a value of kurtosis as 11.1.

It is observed from the envelope of the filtered signal depicted in Figure 11.16(b) that sharp high magnitude peaks above the threshold have been observed at regular

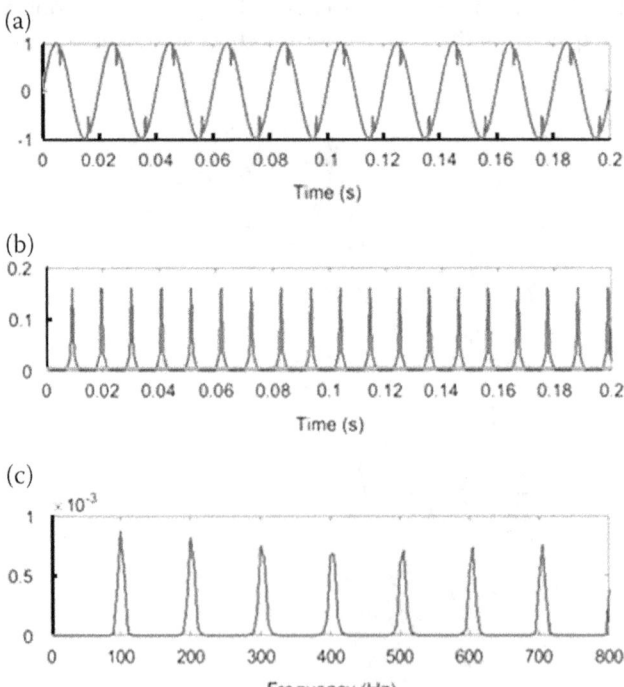

FIGURE 11.16 Voltage signal with notches disturbance: (a) voltage signal, (b) envelope of filtered signal, and (c) amplitude spectrum of squared envelope.

patterns, which indicate the presence of notches. Small magnitude ripples below the threshold on this curve represent the availability of power frequency of 50 Hz. Hence, the presence of notches has been detected effectively from the envelope of the filtered signal. This is inferred from the amplitude spectrum of the squared envelope shown in Figure 11.16(c) that ripples are observed on the surface throughout the time range, indicating the presence of notches transient. Kurtosis of the squared envelope of the filtered signal has been observed equal to 11.5.

11.3.9 VOLTAGE SIGNAL WITH SPIKES DISTURBANCE

The voltage signal with spikes disturbance has been formulated using the mathematical relation given in Section 11.2.2 and simulated in the Matlab software. A fast kurtogram of a voltage signal containing spikes PQ disturbance has been computed using the fast decimated filter bank tree supported by the wavelet packet transform and shown in Figure 11.17. The sampling frequency of the signal is considered as 3.2 kHz.

The analysis of the fast kurtogram gives the availability of transients. Filter bandwidth for the fast kurtogram of voltage signal with notches transient has been observed as 800 Hz and the cutoff frequency is 1,200 Hz. The kurtogram has been included in the frequency on the x-axis up to 1,600 Hz and range of Δf between 0 to 13.3 on the y-axis, as illustrated in Figure 11.17. It is observed from Figure 11.17 that zero level and first level of decomposition have been shown in the kurtogram. High transient frequency components have been observed in the kurtogram representation of the signal of voltage with spikes between the frequencies of 800 Hz to 1,600 Hz. This indicates that due to the spikes initiation and termination, the high-frequency transients have been introduced in the signal. The difference of

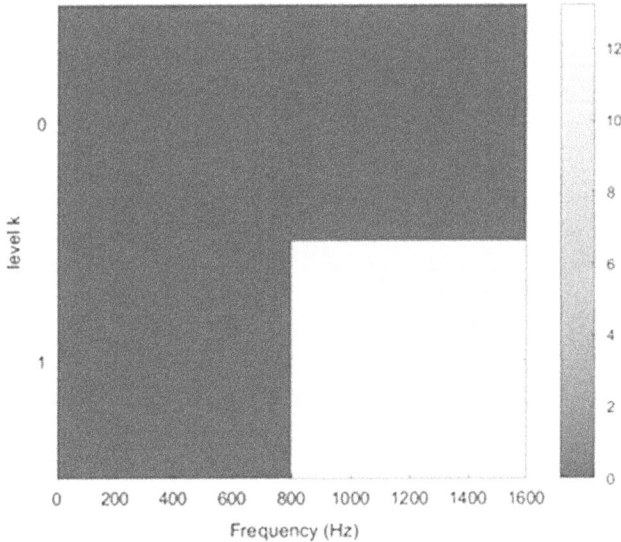

FIGURE 11.17 Spectrum of fast kurtogram for voltage signal with spikes disturbance.

observed value of kurtosis and maximum value of kurtosis is observed as 13.3. On comparing the fast kurtogram of Figure 11.17 with that shown in Figure 11.1, it is observed that due to the spikes, transient initiation and end, the high-frequency transients are introduced in the signal.

The transient signals have been filtered out from the wavelet packet supported fast kurtogram shown in Figure 11.17. The optimal carrier frequency of 1,600 Hz has been used and filtering has been done up to the first level. The voltage signal with spikes disturbance is shown in Figure 11.18(a). The envelope of the filtered signal is shown in Figure 11.18(b) where the red line gives the threshold for the transient disturbance. The amplitude of the spectrum of the squared envelope is detailed in Figure 11.18(c). The cutoff frequency for filtering the signal has been observed to be 1,200 Hz with a value of kurtosis as 13.2.

It is observed from the envelope of the filtered signal depicted in Figure 11.18(b) that sharp high magnitude peaks above the threshold have been observed at regular patterns, which indicate the presence of spikes. Small magnitude ripples below the threshold on this curve represent the availability of a power frequency of 50 Hz. Hence, the presence of spikes has been detected effectively from the envelope of the filtered signal. This is inferred from the amplitude spectrum of the squared envelope shown in Figure 11.18(c) that ripples are observed on the surface throughout the time range, indicating the presence of spikes transient. Kurtosis of the squared envelope of filtered signal has been observed equal to 13.2.

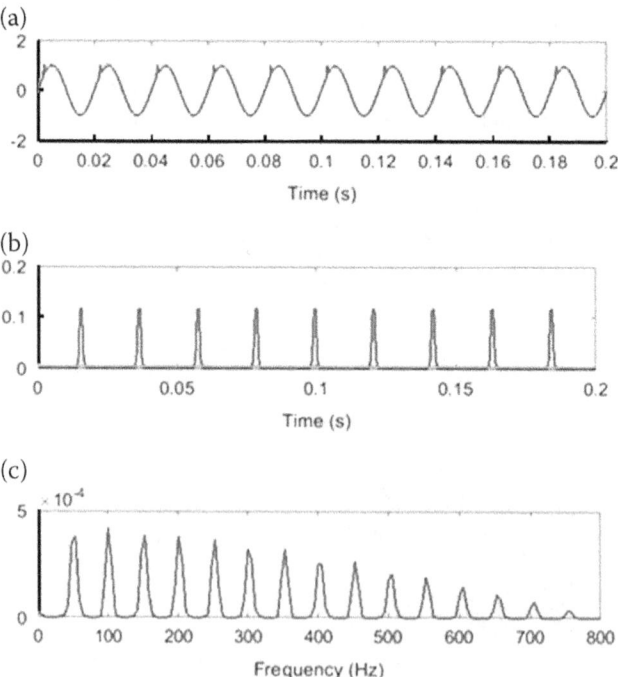

FIGURE 11.18 Voltage signal with spikes disturbance: (a) voltage signal, (b) envelope of filtered signal, and (c) amplitude spectrum of squared envelope.

11.3.10 CLASSIFICATION OF POWER QUALITY DISTURBANCES

Classification of the power quality disturbances is evaluated using the decision rules–based tree. An input data set for the decision rules is framed using the peak values, standard deviation, variance, and kurtosis of the envelope of the filtered signal. Here, the kurtosis is calculated using the MATLAB function *kurtosis*. The complete data set is shown in Table 11.1. It is observed that the kurtosis of the envelope of the filtered signal is found to be a key factor for classification of the PQ disturbances. However, other parameters such as peak values, standard deviation, and variance will be effective in the condition of the conditions.

A decision tree showing the classification of the simple nature power quality disturbances is shown in Figure 11.19. Classification is started using the kurtosis values. Disturbances with K > 50 are momentary interruption (MI) and impulsive transient (IT). These are further classified using the kurtosis value. IT has K > 125 and a voltage signal with momentary interruption (MI) has K < 125.

Pure sine wave voltage signal, sag, swell, harmonics, oscillatory transient (OT), notch, and spike have values of kurtosis K < 50. These disturbances are further classified. OT has values K > 28 and further voltage sag is classified by the rule K > 20. Subsequently, the disturbances are divided into two groups using the peak values. Notches and spikes have PV > 0.1, whereas the pure voltage signal, swell, and harmonics have PV < 0.1. These are further classified using the PV values. The

TABLE 11.1
Data Set for Classification of Simple Nature PQ Disturbances

S. No.	Event	Measured Values			
		Peak Value (PV)	Standard Deviation (SD)	Variance (V)	Kurtosis (K)
1	Voltage signal with pure sine wave	0.0015	4.6339e-04	2.1473e-07	1.8634
2	Voltage signal with sag	0.0059	5.5654e-04	3.0974e-07	24.2799
3	Voltage signal with swell	0.0063	6.4190e-04	4.1204e-07	15.2717
4	Voltage signal with interruption	0.0157	0.0013	1.7941e-06	87.3171
5	Voltage signal with superimposed harmonics	0.0067	8.1347e-04	6.6173e-07	2.2936
6	Voltage signal with oscillatory transient	0.3634	0.0370	0.0014	32.5630
7	Voltage signal with impulsive transient	0.6413	0.0417	0.0017	189.2185
8	Voltage signal with notches	0.1613	0.0387	0.0015	12.0247
9	Voltage signal with spikes	0.1167	0.0273	7.4513e-04	14.4034

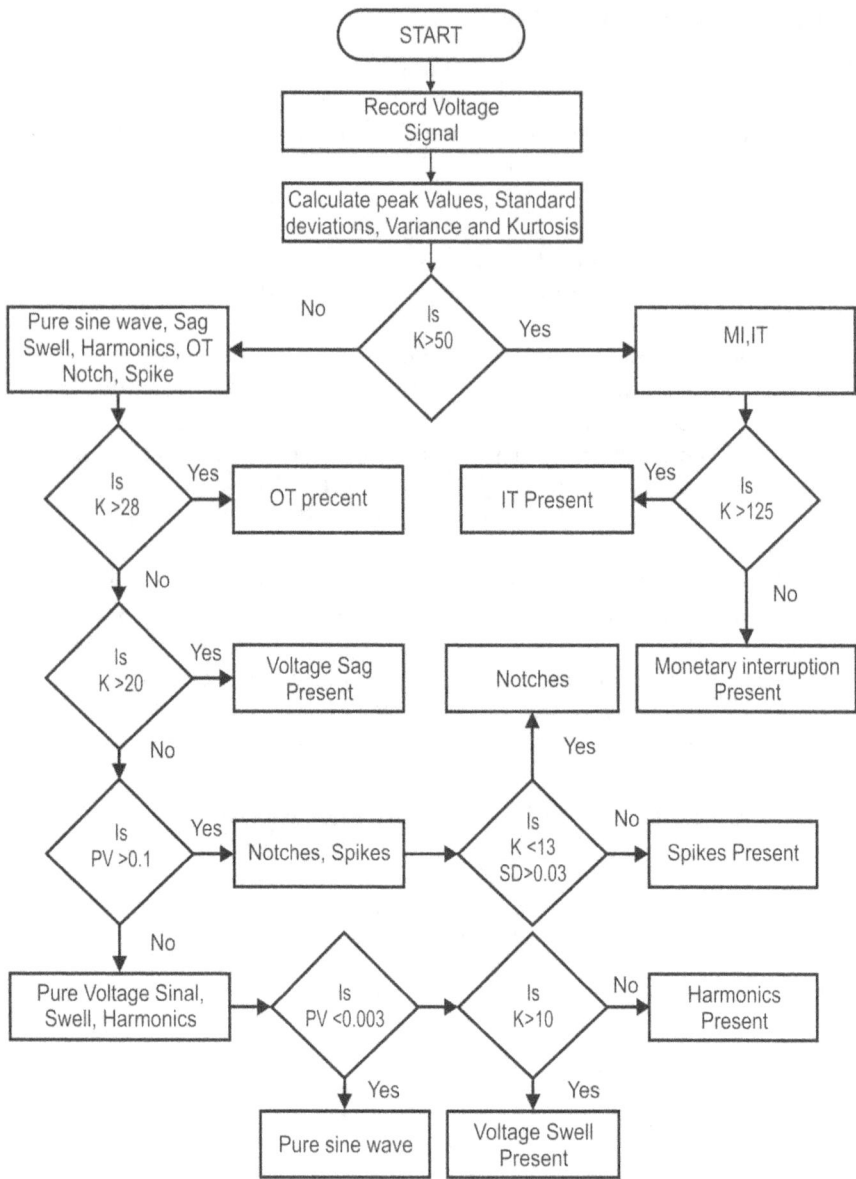

FIGURE 11.19 Decision rules supported tree for classification of simple nature power quality disturbances.

pure sine wave has PV < 0.003, whereas the voltage swell and harmonics have values PV > 0.003. These are further classified using kurtosis values. A voltage swell has K > 10 and harmonics has K < 10. Hence, various types of simple nature PQ disturbances have been classified effectively. The decision rules used for classification purposes have also been provided in Figure 11.19.

11.3.11 PERFORMANCE OF PROPOSED ALGORITHM

The performance of the proposed algorithm has been evaluated by testing the algorithm on 50 sets of each type of signal, which are obtained by changing the parameters such as magnitude, frequency, time of incidence of the disturbance, duration of disturbance, etc. Accurately classified and inaccurately classified disturbances are used to rate the performance of the proposed algorithm. The performance of the algorithm is given in Table 11.2.

It is observed from Table 11.2 that the average classification efficiency of 98.67% has been achieved using the proposed algorithm.

11.3.12 COMPARISON OF PERFORMANCE OF PROPOSED ALGORITHM
WITH REPORTED TECHNIQUES

The performance of the proposed algorithm has been compared with the discrete wavelet transform and fuzzy C-means clustering approach reported in the [47]. Fuzzy C-means clusters using the features extracted with the help of discrete

TABLE 11.2
Performance of Algorithm for Recognition of Simple Nature PQ Disturbances

S. No.	Event	Number of Disturbances with Accurate Classification	Number of Disturbances with Inaccurate Classification	Percentage Accurate Classification
1	Voltage signal with pure sine wave	50	0	100%
2	Voltage signal with sag	50	0	100%
3	Voltage signal with swell	50	0	100%
4	Voltage signal with momentary interruption	50	0	100%
5	Voltage signal with superimposed harmonics	47	3	94%
6	Voltage signal with oscillatory transient	50	0	100%
7	Voltage signal with impulsive transient	50	0	100%
8	Voltage signal with notches	48	2	96%
9	Voltage signal with spikes	49	1	98%
Average classification accuracy				98.67%

wavelet transform are found to overlap in the [47] due to which the accuracy of the classification is achieved approximately up to 96% only, whereas the proposed algorithm using the kurtosis values of the envelope of the filtered signal obtained from the fast kurtogram based on the wavelet packet transform based features for classification using decision rules has been found to be more accurate. In the proposed algorithm, efficiency up to 98% has been achieved. Hence, the method proposed in this study is found to be efficient compared to the method reported in the reference [47].

11.4 CONCLUSION

This research presents an algorithm using the wavelet packet supported fast kurtogram and decision rules that can be implemented for identification and classification of the power quality disturbance. The proposed algorithm is based on the features extracted from a fast kurtogram and envelope of the filtered signal and amplitude spectrum of the squared envelope. The performance of the proposed algorithm has been evaluated by testing the algorithm on 50 sets of each type of the signal, which are obtained by changing the parameters such as magnitude, frequency, time of incidence of the disturbance, duration of disturbance, etc. Accurately classified and inaccurately classified disturbances are used to rate the performance of the proposed algorithm. The performance of the algorithm has been tested using the performance of the proposed algorithm compared with the discrete wavelet transform and fuzzy C-means clustering approach. Further, the algorithm is also effectively tested for the recognition of multiple power quality disturbances.

It is concluded that the proposed algorithm based on the wavelet packet supported a fast kurtogram and decision rules are found to be effective in the detection and classification of the simple nature power quality disturbances. This algorithm is effective in the detection of the disturbances with an efficiency up to 98%, which is more efficient compared to the discrete wavelet transform and fuzzy C-means clustering approach.

REFERENCES

[1] O. P. Mahela, B. Khan, H. Haes Alhelou, and S. Tanwar, "Assessment of Power Quality in the Utility Grid Integrated with Wind Energy Generation," *IET Power Electronics*, vol. 13, no. 13, pp. 2917–2925, 2020.
[2] O. P. Mahela et al., "Recognition of Power Quality Issues Associated With Grid Integrated Solar Photovoltaic Plant in Experimental Framework," *IEEE Systems Journal*, vol. 15, no. 3, pp. 3740–3748, September 2021, 10.1109/JSYST.2020.302 7203
[3] O. P. Mahela, A. G. Shaik, B. Khan, R. Mahla, and H. H. Alhelou, "Recognition of Complex Power Quality Disturbances Using S-Transform Based Ruled Decision Tree," *IEEE Access*, vol. 8, pp. 173530–173547, 2020.
[4] O. P. Mahela, Khan, B., Alhelou, H. H., Tanwar, S., and Padmanaban, S., "Harmonic Mitigation and Power Quality Improvement in Utility Grid with Solar Energy Penetration Using Distribution Static Compensator," *IET Power Electron*, vol. 14, pp. 912–922, 2021.

[5] R. K. Pachauri et al., "Impact of Partial Shading on Various PV Array Configurations and Different Modeling Approaches: A Comprehensive Review," in *IEEE Access*, vol. 8, pp. 181375–181403, 2020.

[6] R. K. Pachauri, O. P. Mahela, B. Khan, A. Kumar, S. Agarwal, H. H. Alhelou, and J. Bai, "Development of Arduino Assisted Data Acquisition System for Solar Photovoltaic Array Characterization Under Partial Shading Conditions," *Computers & Electrical Engineering*, vol. 92, pp. 107175, 2021.

[7] R. K. Pachauri et al., "Shade Dispersion Methodologies for Performance Improvement of Classical Total Cross-tied Photovoltaic Array Configuration Under Partial Shading Conditions." *IET Renewable Power Generation*, vol. 15, pp. 1796–1811, 2021.

[8] R. Kaushik et al., "Recognition of Islanding and Operational Events in Power System With Renewable Energy Penetration Using a Stockwell Transform-Based Method," *IEEE Systems Journal*, 10.1109/JSYST.2020.3020919

[9] S. P. Bihari et al., "A Comprehensive Review of Microgrid Control Mechanism and Impact Assessment for Hybrid Renewable Energy Integration," *IEEE Access*, vol. 9, pp. 88942–88958, 2021.

[10] O. P. Mahela, Y. Sharma, S. Ali, B. Khan, and S. Padmanaban, "Estimation of Islanding Events in Utility Distribution Grid With Renewable Energy Using Current Variations and Stockwell Transform," *IEEE Access*, vol. 9, pp. 69798–69813, 2021.

[11] D. D. Ferreira, J. M. de Seixa, and A. S. Cerqueira, "A Method Based on Independent Component Analysis for Single and Multiple Power Quality Disturbance Classification," *Electric Power Systems Research*, vol. 119, pp. 425–431, 2015.

[12] K. Thirumala, S. Pal, T. Jain, and A. C. Umarikar, "A Classification Method for Multiple Power Quality Disturbances Using EWT Based Adaptive Filtering and Multiclass SVM," *Neurocomputing*, vol. 334, pp. 265–274, 2019.

[13] C. A. Naik, and P. Kundu, "Wavelet Packet Transform Based Parameter for the Analysis of Short Duration Power Quality Disturbances," *IFAC (Elsevier)*, vol. 48, pp. 485–489, 2015.

[14] S. Karasu, and Z. Saraç, "Investigation of Power Quality Disturbances by Using 2D Discrete Orthonormal S-transform, Machine Learning and Multi-objective Evolutionary Algorithms," *Swarm and Evolutionary Computation*, vol. 44, pp. 1060–1072, 2019.

[15] B. Biswal, H. S. Behera, R. Bisoi, and P. K. Dash, "Classification of Power Quality Data Using Decision Tree and Chemotactic Differential Evolution Based Fuzzy Clustering," *Swarm and Evolutionary Computation*, vol. 4, pp. 12–24, 2012.

[16] T. Tarasiuk, "Estimator-Analyzer of Power Quality: Part I–Methods and Algorithms", *Measurement*, vol. 44, pp. 238–247, 2011.

[17] M. A. A. Lima, A. S. Cerqueira, D. V. Coury, and C. A. Duque, "A Novel Method for Power Quality Multiple Disturbance Decomposition Based on Independent Component Analysis," *Electrical Power and Energy Systems*, vol. 42, pp. 593–604, 2012.

[18] A. A. Abdelsalam, A. A. Eldesouky, and A. A. Sallam, "Classification of Power System Disturbances Using Linear Kalman Filter and Fuzzy-expert System," *Electrical Power and Energy Systems*, vol. 43, pp. 688–695, 2012.

[19] H. Dehghani, B. Vahidi, R. A. Naghizadeh, and S. H. Hosseinian, "Power Quality Disturbance Classification Using a Statistical and wavelet-based Hidden Markov Model with Dempster–shafer Algorithm," *Electrical Power and Energy Systems*, vol. 47, pp. 368–377, 2013.

[20] M. Uyar, S. Yildirim, and M. T. Gencoglu, "An Effective wavelet-based Feature Extraction Method for Classification of Power Quality Disturbance Signals," *Electric Power Systems Research*, vol. 78, pp. 1747–1755, 2008.

[21] S. A. L. M. Waghmare, "Integrated DWT–FFT Approach for Detection and Classification of Power Quality Disturbances," *Electrical Power and Energy Systems*, vol. 61, pp. 594–605, 2014.

[22] D. Wang, X. Wang, Y. Zhang, and L. Jin, "Detection of Power Grid Disturbances and Cyber-attacks Based on Machine Learning," *Journal of Information Security and Applications*, vol. 46, pp. 42–52, 2019.

[23] R. Hooshmand, and A. Enshaee, "Detection and Classification of Single and Combined Power Quality Disturbances Using Fuzzy Systems Oriented by Particle Swarm Optimization Algorithm," *Electric Power Systems Research*, vol. 80, pp. 1552–1561, 2010.

[24] A. Rodríguez, J. A. Aguado, F. Martín, J. J. López, F. Muñoz, and J. E. Ruiz, "Rule-based Classification of Power Quality Disturbances Using S-transform," *Electric Power Systems Research*, vol. 86, pp. 113–121, 2012.

[25] R. Kapoor, and M. K. Saini, "Detection and Tracking of Short Duration Variations of Power System Disturbances Using Modified Potential Function," *Electrical Power and Energy Systems*, vol. 47, pp. 394–401, 2013.

[26] P. Kanirajan, and V. S. Kumar, "Power Quality Disturbance Detection and Classification Using wavelet and RBFNN," *Applied Soft Computing*, vol. 35, pp. 470–481, 2015.

[27] L. C.M. Andrade, M. Oleskovicza, and R. A. S. Fernandes, "Adaptive Threshold Based on wavelet Transform Applied to the Segmentation of Single and Combined Power Quality Disturbances," *Applied Soft Computing*, vol. 38, pp. 967–977, 2016.

[28] U. Singh, and S. N. Singh, "A New Optimal Feature Selection Scheme for Classification of Power Quality Disturbances Based on Ant Colony Framework," *Applied Soft Computing Journal*, vol. 74, pp. 216–225, 2019.

[29] Z. Moravej, S. A. Banihashemi, and M. H. Velayati, "Power Quality Events Classification and Recognition Using a Novel Support Vector Algorithm," *Energy Conversion and Management*, vol. 50, pp. 3071–3077, 2009.

[30] S. Wang, and H. Chen, "A Novel Deep Learning Method for the Classification of Power Quality Disturbances Using Deep Convolution Neural Network," *Applied Energy*, vol. 235, pp. 1126–1140, 2019.

[31] J. G. M. S. Decanini, M. S. Tonelli-Neto, F. C. V. Malange, and C. R. Minussi, "Detection and Classification of Voltage Disturbances Using a Fuzzy-ARTMAP-wavelet Network," *Electric Power Systems Research*, vol. 81, pp. 2057–2065, 2011.

[32] Z. Liu, Q. Zhang, Z. Han, and G. Chen, "A New Classification Method for Transient Power Quality Combining Spectral Kurtosis with Neural Network," *Neurocomputing*, vol. 125, pp. 95–101, 2014.

[33] T. Chakravorti, L. Priyadarshini, P. K. Dash, and B. N. Sahu, "Islanding and Non-islanding Disturbance Detection in Microgrid Using Optimized Modes Decomposition Based Robust Random Vector Functional Link Network," *Engineering Applications of Artificial Intelligence*, vol. 85, pp. 122–136, 2019.

[34] R. Moreno, N. Visairo, C. Núñez, and E. Rodríguez, "A Novel Algorithm for Voltage Transient Detection and Isolation for Power Quality Monitoring," *Electric Power Systems Research*, vol. 114, pp. 110–117, 2014.

[35] A. G. Shaik, and O. P. Mahela, "Power Quality Assessment and Event Detection in Hybrid Power System," *Electric Power Systems Research*, vol. 161, pp. 26–44, 2018.

[36] M. Karimi, H. Mokhtari, and M. Reza Iravani, "wavelet Based On-Line Disturbance Detection for Power Quality Applications," *IEEE Transactions on Power Delivery*, vol. 15, no. 4, pp. 1212–1220, October 2000.

[37] H.-T., Yang, and C.-C. Liao, "A De-Noising Scheme for Enhancing wavelet-Based Power Quality Monitoring System," *IEEE Transactions on Power Delivery*, vol. 16, no. 3, pp. 353–360, July 2001.

[38] H., Mokhtari, M. Karimi-Ghartemani, and M. Reza Iravani, "Experimental Performance Evaluation of a wavelet-Based On-Line Voltage Detection Method for Power Quality Applications," *IEEE Transactions on Power Delivery*, vol. 17, no. 1, pp. 161–172, January 2002.

[39] P. K. Dash, B. K. Panigrahi, D. K. Sahoo, and G. Panda, "Power Quality Disturbance Data Compression, Detection, and Classification Using Integrated Spline wavelet and S-Transform," *IEEE Transactions on Power Delivery*, vol. 18, no. 2, pp. 595–600, April 2003.

[40] I. W. C. Lee, and P. K. Dash, "S-Transform-Based Intelligent System for Classification of Power Quality Disturbance Signals," *IEEE Transactions on Industrial Electronics*, vol. 50, no. 4, pp. 800–805, August 2003.

[41] H. Zhang, P. Liu, and O. P. Malik, "Detection and Classification of Power Quality Disturbances in Noisy Conditions," *IEE Proceedings – Generation, Transmission and Distribution*, vol. 150, no. 5, pp. 567–572, September 2003.

[42] D. G. Ece, and Ö. N. Gerek, "Power Quality Event Detection Using Joint 2-D-wavelet Subspaces," *IEEE Transactions on Instrumentation and Measurement*, vol. 53, no. 4, pp. 1040–1046, August 2004.

[43] C.-H. Lin, and M.-C. Tsao, "Power Quality Detection with Classification Enhancible wavelet-probabilistic Network in a Power System," *IEE Proceedings – Generation, Transmission and Distribution*, vol. 152, no. 6, pp. 969–976, November 2005.

[44] C.-H. Lin, and C.-H. Wang, "Adaptive wavelet Networks for Power-Quality Detection and Discrimination in a Power System," *IEEE Transactions on Power Delivery*, vol. 21, no. 3, pp. 1106–1113, July 2006.

[45] S. Mishra, C. N. Bhende, and B. K. Panigrahi, "Detection and Classification of Power Quality Disturbances Using S-Transform and Probabilistic Neural Network," *IEEE Transactions on Power Delivery*, vol. 23, no. 1, pp. 280–287, January 2008.

[46] O. P. Mahela, and A. G. Shaik, "Recognition of Power Quality Disturbances Using S-transform Based Ruled Decision Tree and Fuzzy C-means Clustering Classifiers," *Applied Soft Computing*, vol. 59, pp. 243–257, 2017.

[47] O. P. Mahela, U. K. Sharma, and T. Manglani, "Recognition of Power Quality Disturbances Using Discrete wavelet Transform and Fuzzy C-means Clustering," Power Indian International Conference (PIICON 2018), NIT Kurukshetra, India, December, 2018.

12 Identification of Transmission Line Faults Using Voltage-Based Stockwell Transform Features and Decision Rules Supported Fault Classification

Om Prakash Mahela
Power System Study Division, Rajasthan Rajya Vidyut Prasarn Nigam Ltd., Jaipur, India

Vishnu Dutt Sharma and Sunil Agarwal
Department of Electrical Engineering, Apex Institute of Engineering and Technology, Jaipur, India

Baseem Khan
Department of Electrical and Computer Engineering, Hawassa University, Ethiopia

CONTENTS

DOI: 10.1201/b22884-12

12.1   INTRODUCTION

The complexity of the power networks has raised the requirement of improved/ advanced schemes for protection of these complex grids that require fast and accurate estimation of the faults. Fault estimation included the identification and to classify the various faults. This also includes the discrimination of phases that are faulty from the healthy phases [1]. Techniques using the approaches of signal processing techniques have been reported for identification, classification, and localization of faults incident on the lines used for the transmission of power from generating plants to the grid substations (GSS) [2]. In [3], the authors introduced an algorithm to identify faulty event incidents on the transmission line. This is achieved by the use of pilot impedances and synchronized data. This resulted in the improvement of sensitivity as well as reliability of protection scheme. In [4], the author proposed a method to identify, classify, and localize the faulty conditions on the both the overhead transmission line and underground cable. This method used the concept of entropy computed by hybrid combination of the fast discrete orthogonal ST (FDOST) to extract the features to identify the faults. Classification of these faulty events has been achieved using a support vector machine (SVM) and support vector regression (SVR). In [5], the authors introduced an algorithm that uses the wavelet transform and alienation coefficients to design a protection scheme for the multi-terminal transmission line (MTTL). This is achieved by the application of WT-supported approximate coefficients of voltage and current signals computed over a time period of a quarter cycle to identify, classify, and localize the faults on MTTL. In [6], the authors introduced an approach for estimating the location of fault incident point on the power transmission lines. Voltage measurements performed using the wide area measurement systems (WAMS) have been used to implement the proposed protection scheme. A network bus admittance matrix is also computed to recognize the faults on the transmission line. Recognition of a faulted line and fault location is achieved with high accuracy in a single step. A review of literature related to the fault recognition techniques is detailed in the below subsection.

12.1.1 RELATED WORK

A detailed comprehensive literature review focused on the transmission line protection schemes is presented in this work. A literature review included in this section mainly focused on the identification, classification, and localization of faults on a transmission line. Application of the signal processing approaches in the field of transmission line protection is also presented in this section. A review of literature is focused on transmission line protection is detailed in this section. In [7], the authors introduced an adaptive transmission line protection approach that has used the synchronized phasor measurement units. Positive-sequence voltage and current phasors recorded at both ends of the transmission line are used to determine transmission line parameters and fault location on a transmission line. A proposed scheme is found to be effective to provide protection to both the single circuit and double circuit transmission lines. This is also effective for detecting faults even in the presence of power swing conditions. A proposed scheme provides fast operation and accurate transmission line protection. In [8], the authors introduced a protection scheme for a low voltage direct current (LVDC) ring-bus microgrid network. A scheme has been designed by the use of a multi-criterion system (MCS) and neural network (NN). A proposed approach is aimed at high-speed detection of line-to-ground (LG) and line-to-line (LL) faults, which involve low impedance. The criteria of definite threshold for differential current have been eliminated by the use of specific rules and a multi-criterion system. A proposed MCS protection scheme has high speed and accuracy in comparison to differential protection. A NN estimated the location of the fault on the line. A method for location of faults on a transmission line using impedance has been reported in [9]. Mathematical formulation is based on the use of a generalized equation for fault distance estimation. This is independent of the type of fault and the least squares method. The voltage as well as currents measured at two terminals of transmission line and transmission line parameters are considered. The proposed method gives an effective performance for fault detection. This is also effective in locating the high and low impedance faults. The proposed method has a high accuracy for location of faults on the transmission line. In [10], a technique for detecting the high impedance transmission line faults has been introduced. Stockwell transform is used for extraction of phase angle of third harmonic current measured at grid substation. The parameters have been continuously monitored using a moving standard deviation. Values lower than the standard deviation based self adaptive threshold indicates the presence of faults. The proposed method is efficient for fast identification of high impedance faults in various types of contact surfaces and fault locations. It is also effective in differentiating the high impedance faults from the operational disturbances in the network. Performance of the proposed technique is not affected by fault location, fault inception angle, system loading etc. In [11], a method using cut-in point for fault detection is introduced. This is based on dividing transient zero sequence currents (TZSC) in steady-state components (SSCs) and transient (TCs) components. The proposed method has a high fault detection accuracy and takes minimum time. Accuracy and applicability of the proposed method have been tested on a radial distribution network of IEEE-13 nodes. In [12], the authors presented a single-end traveling wave (TW) fault location approach. Three recorders of TW are mounted at local, middle, and remote locations of the transmission line. This method eliminates the requirement of accurate synchronous sampling. In this proposed method, polarity coefficients of current TW and absolute difference of time differences of arrival (TDOAs) of TWs at both terminals of transmission lines have been utilized to detect fault occurrence on the transmission line. The fault location has been determined using a ratio of TDOAs of TWs recorded on both ends of a transmission line. A decision tree supported traveling wave (TW)–based algorithm for estimation of a faulty section of transmission line and fault location algorithm have been proposed for multi-terminal transmission lines (MTTL) in [13]. The difference between fault incidence instant and traveling wave reaching time measured is considered travel time (TT) and utilized for identification of a faulty section. A decision is made by comparing the measured TT with the actual TT along the individual transmission line. Once the faulty section is identified, two adjacent arrival times of a faulty section have been used for the estimation of the fault location. In [14], the authors introduced an approach for prediction of faulty events in the power system network. This used an environmental attributes-based framework for spatial temporal distribution. Distribution of anticipated faulty events has also been predicted using forecasted information of environmental attributes. Relative weights have been introduced to distinguish diverse influence of each environmental element on the reliability of a power system network. Efficiency is established by investigating the impact of each single fault cause using an empirical study. Flexibility of the proposed framework in real-time applications has been demonstrated. Various techniques implemented for recognition of faults is detailed in [15–35].

The technique using signal processing approaches for detection, classifying, and localizing the faults on transmission lines are observed in literature. However, this method uses single features that have their demerits. A recent focus is for designing smart protection schemes that are fast, accurate, and reliable. This can be achieved by combining several features. This has been taken as the key factor for the research undertaken in this chapter, which is mainly focused on developing an algorithm by the combination of different features of voltage and current signals for detecting and classifying the faults.

12.1.2 CONTRIBUTION OF THE WORK

Research work included in this chapter is aimed to design an algorithm that can be implemented for identification and classification of faults on power transmission lines using the voltage features. The main contributions are as follows:

- This research work investigates to design an algorithm based on processing of voltage signals using the Stockwell transform to identify faults on a power transmission line.
- The classification of the faults has been achieved using the rule-based decision tree.
- The proposed study is performed using MATLAB® software in a Simulink environment.

12.1.3 ORGANIZATION OF THIS CHAPTER

The manuscript is organized as follows: Section 12.2 discusses the methodology utilized in this work. Section 12.4 presents the results with detailed discussion. Section 12.5 presents the conclusion of this chapter.

12.2 METHODOLOGY AND TEST SYSTEM

The test system implemented for implementation of the proposed study of fault estimation on the transmission line is described in this section. The data of the test system are detailed in this section. The algorithm proposed for the estimation of faults using the Stockwell transform and rule-based decision tree is detailed in this section. Brief descriptions of the Stockwell transform and rule-based decision tree are also illustrated in this section.

12.2.1 PROBLEM FORMULATION

Presently, the complexity level of the utility transmission network is continuously increasing. Transmission lines are uprated in terms of the voltage levels as well as the power transmission capacities. Further, the capacities of the power transformers are also increasing day by day. The dynamic components, such as thyristor controlled reactors (TCR), static synchronous compensator (STATCOM), etc. are also being continuously deployed in the network of the transmission. In a recent scenario, the renewable energy (RE) is also integrated to the utility network in terms of the large-sized RE generators. The above-mentioned dynamic changes have resulted in the following issues/problems in the power system network that needs to be addressed:

- Protection systems deployed needs upgradation. The signal processing approaches may lead to the design of the advanced protection schemes.
- Stability of the power system network is at threat. Suitable practices and design parameters need to be deployed to maintain the stability of the utility grid.
- Power quality is deteriorating. Suitable techniques need to be designed for the recognition and mitigation of the PQ issues in the present-day power system network.

This study has considered the issues of the power system protection. A suitable technique has been designed that can be used for the estimation of faults for designing the protection schemes for complex power system network.

12.2.2 TEST SYSTEM USED FOR THE STUDY

The test network consisting of a transmission line with two terminals is used to perform the proposed study. This test transmission line is described in Figure 12.1. The transmission line is integrated in between bus-1 and bus-2, as detailed in Figure 12.1. A circuit breaker (CB) is placed near bus-1, which is used to trip the transmission line during the faulty event. Generator sources 1 and 2 are used to simulate the large area

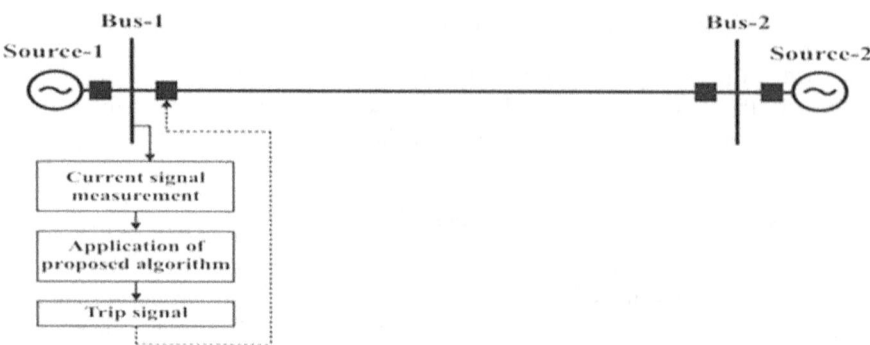

FIGURE 12.1 Test system with transmission line integrated between bus-1 and bus-2.

TABLE 12.1
Test System Parameters

Attribute	Value of Parameter
Total Length of line	230 km
Voltage of generator-1	$500 \angle 20°$ kV
Voltage of generator-2	$500 \angle 0°$ kV
Equivalent impedance of generator-1	$17.177 + j45.529\ \Omega$
Equivalent impedance of generator-2	$15.31 + j45.925\ \Omega$
Positive sequence impedance of transmission line	$4.983 + j117.83\ \Omega$
Zero sequence impedance of transmission line	$12.682 + j364.196\ \Omega$
Positive sequence admittance of transmission line	$j1.468$ m℧
Zero sequence admittance of transmission line	$j1.099$ m℧
Transmission line capacity	$433.63 + j294.52$ MVA

utility networks. In the proposed study, bus-1 is treated as the sending terminal of power transmission line and bus-2 is taken as the receiving terminal of transmission line. Parameters of the test line and test system are included in Table 12.1 [30].

12.2.3 VOLTAGE SUPPORTED ALGORITHM USED FOR THE ESTIMATION OF FAULT CONDITIONS

The proposed voltage supported algorithm used for recognition of the fault conditions on the test transmission line can be implemented using the following steps:

- Simulate a faulty condition on the test transmission line and measure voltage on the sending end of the transmission line.
- Process voltage signals using the Stockwell transform to obtain an output matrix and designate this as the STM, which is a complex valued matrix.
- Compute absolute values of each element of STM and designate it as the ASTM.

- Calculate the median intermediate fault index (MIFI) from the ASTM matrix using the following MATLAB commands corresponding to all phases. The subscripts 1, 2, and 3 are used for the phases A, B, and C, respectively.

  A1 = median(ASTM1);
  A2 = median(ASTM2);
  A3 = median(ASTM3);

- Calculate the maximum value intermediate fault index (MVIFI) from the ASTM matrix using the following MATLAB commands corresponding to all phases. The subscripts 1, 2, and 3 are used for the phases A, B, and C, respectively.

  B1 = max(ASTM1);
  B2 = max(ASTM2);
  B3 = max(ASTM3);

- Calculate the summation intermediate fault index (SIFI) from the ASTM matrix using the following MATLAB commands corresponding to all phases. The subscripts 1, 2, and 3 are used for the phases A, B, and C, respectively.

  C1 = sum(ASTM1);
  C2 = sum(ASTM2);
  C3 = sum(ASTM3);

- Calculate the co-variance fault index (CFI) from the ASTM matrix using the following MATLAB commands corresponding to all phases. The subscripts 1, 2, and 3 are used for the phases A, B, and C, respectively.

  D1 = cov(max(ASTM1))
  D2 = cov(max(ASTM2))
  D3 = cov(max(ASTM2))

- Calculate the proposed hybrid fault index (HFI) using the following MATLAB commands corresponding to all phases. The subscripts 1, 2, and 3 are used for the phases A, B, and C, respectively. Here the weight factor (WF) is considered equal to 10^6.

  FI1 = A1.*B1.*C1.*D1.*WF;
  FI2 = A2.*B2.*C2.*D2.*WF;
  FI3 = A3.*B3.*C3.*D3.*WF;

- Select a threshold value for the hybrid fault index (HFIT) equal to 25 for the proposed HFI to estimate the faulty condition from the healthy conditions. This HFIT is selected by the testing algorithm for the different conditions such as different values of the fault location, fault impedance, fault incidence angle, reverse power flow, etc.

12.3 RESULTS AND DISCUSSION

Results for the estimation of the fault conditions on test transmission using voltage-based method are detailed in this section. A discussion of the simulation results are included in detail. The results for estimation of fault condition such as phase-a to ground (ag), phase-a and phase-b fault (ab), phase-a and phase-b to ground fault (abg), three-phase fault (abc), and three phases to ground fault (abcg) are described in this section. Further, the results of fault estimation during different operating scenarios have also been discussed in this section.

12.3.1 FAULT ON PHASE-A AND INVOLVEMENT OF GROUND

A phase-A to ground (AG) fault is simulated at the center of the test transmission line (115 km from node-1). Voltage pertaining to all phases is recorded on node-1 of the test line. Voltages associated with all phases are shown in Figure 12.2(a). These voltages associated to all phases are processed by the use of a Stockwell transform for computing the STM matrix. The absolute value matrix (ASTM) is computed from the STM. The proposed MIFI, MVIFI, SIFI, and HFI are computed from the ASTM using the proposed algorithm and detailed in Figure 12.2(b), (c), (d), and (e) in respective order.

Figure 12.2(a) details the voltage associated with the phase-A decreases after the incidence of the AG fault, whereas values of voltage pertaining with phases-B and C follows the sinusoidal nature same as that recorded in the healthy period. Figure 12.2(b) details the values of voltage-based MIFI increase after the incidence of fault condition for all phases. Figure 12.2(c) details the values of voltage-based MVIFI associated with the faulty phase-A changes and deviates from the straight-line nature. However, the values of MVIFI pertaining to healthy phases-B and C are the same in the post-fault condition as that observed in the pre-fault condition. Figure 12.2(d) details the values of voltage-based SIFI increase after incidence of a faulty condition for all phases. However, an increase in the values of SIFI corresponding to the faulty phase-A are higher in comparison to healthy phases-B and C. Figure 12.2(e) describes the values of voltage-based HFI associated with the faulty phase-A increase after the incidence of the fault condition. However, the values of HFI associated to healthy phases-B and C are same in the post-fault condition as that observed in the pre-fault condition, which is zero. The values of voltage-based HFI have crossed the threshold HFIT (25), indicating that the phase-A is faulty. This index for the healthy phases-B and C is below the HFIT. Hence, the phase-A to ground (AG) fault has been recognized effectively by using the voltage-based method.

The magnitudes of voltage supported the co-variance fault index (CFI) for all the phases in the event of the phase-A to ground fault described in Table 12.2. It is observed that the CFI for the phase-A has high magnitudes, whereas its value associated to phases-B and C is nearly equal to zero.

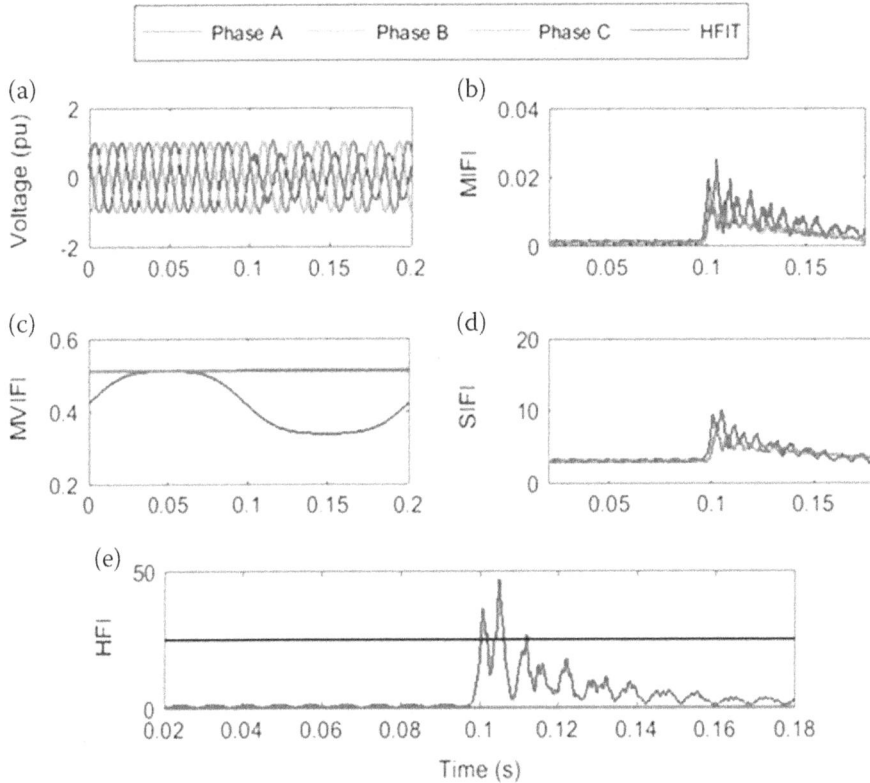

FIGURE 12.2 Voltage-based estimation of fault on phase-A and involvement of ground: (a) voltage of all phases, (b) median-based intermediate fault index, (c) maximum value–based intermediate fault index, (d) summation-based intermediate fault index, and (e) proposed hybrid fault index.

TABLE 12.2

Magnitude of Voltage-Based Covariance Fault Index During Faulty Conditions

Type of Fault	Covariance Fault Index		
	Phase-A	**Phase-B**	**Phase-C**
AG	0.0047	1.2933e-07	2.4815e-06
AB	0.0067	0.0026	2.2766e-08
ABG	0.0081	0.0075	1.7373e-06
ABC	0.0092	0.0091	0.0092
ABCG	0.0092	0.0091	0.0092

12.3.2 FAULT ON PHASES-A AND B WITHOUT INVOLVEMENT OF GROUND

A fault on phases-A and B without involving ground (AB fault) has been simulated at the center of test transmission line (115 km from node-1). The voltage pertaining to all phases is recorded on node-1 of the test line. The voltages associated with all the phases are shown in Figure 12.3(a). These voltages associated to all phases are processed by the use of a Stockwell transform for computing the STM matrix. The absolute value matrix (ASTM) is computed from the STM. The proposed MIFI, MVIFI, SIFI, and HFI are computed from the ASTM using the proposed algorithm and detailed in Figure 12.3(b), (c), (d), and (e) in respective order.

Figure 12.3(a) details the voltage associated with the phases-A and B decreases after the incidence of AB fault, whereas values of voltage associated with the phase-C follow the sinusoidal nature same as that recorded in the healthy period. Figure 12.3(b) details the values of voltage-based MIFI increase after the incidence of fault condition for phases. However, an increase in magnitude associated to phases-A and B is high in comparison to phase-A. Figure 12.3(c) details the values of voltage-based MVIFI associated with the faulty phases-A and B changes and deviates from

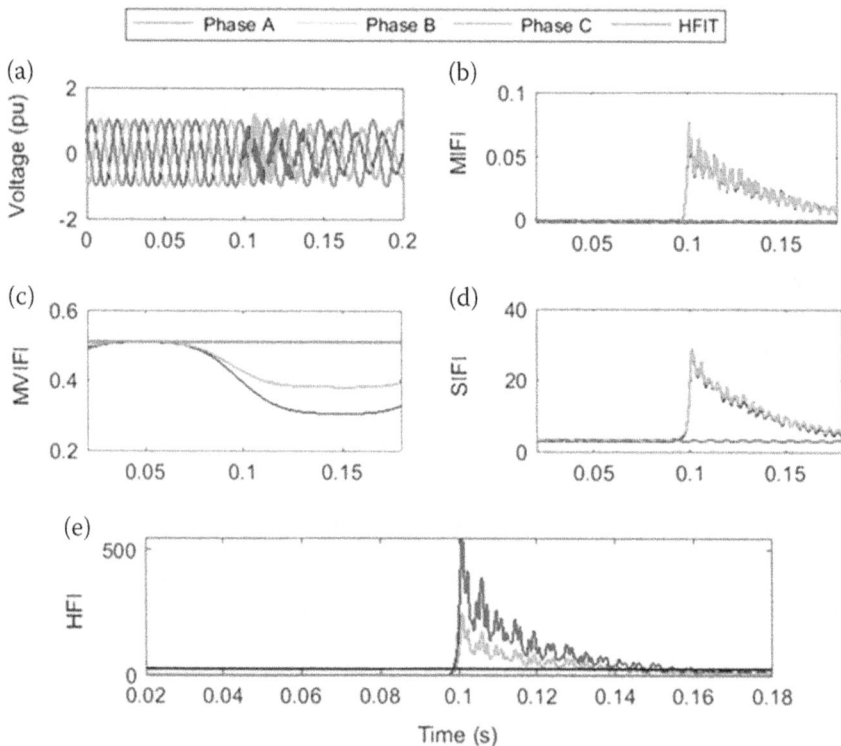

FIGURE 12.3 Voltage-based estimation fault on phase-A and phase-B fault: (a) voltage of all phases, (b) median-based intermediate fault index, (c) maximum value–based intermediate fault index, (d) summation-based intermediate fault index, and (e) proposed hybrid fault index.

the straight-line nature. However, the values of MVIFI associated to the healthy phase-C are the same in the post-fault condition as that observed in the pre-fault condition. Figure 12.3(d) details the values of a voltage-based SIFI increase corresponding to the phases-A and B after the incidence of the faulty condition. However, its value associated to healthy phase-C is nearly zero. Figure 12.3(e) details the values of voltage-based HFI associated with the faulty phases-A and B increased after incidence of fault condition. However, the values of HFI associated to the healthy phase-C is the same in the post-fault condition as that observed in the pre-fault condition, which is zero. The values of voltage-based HFI have crossed the threshold HFIT (25), indicating that the phases-A and B are faulty. This index for the healthy phase-C is below the HFIT. Hence, the phases-A and B fault without involvement of ground has been recognized effectively by the use of voltage-based method.

It is observed that the CFI for the phases-A and B has high values, whereas its value pertaining to the phase-C is very low.

12.3.3 Fault on Phases-A and B with Involvement of Ground

A fault on phases-A and B with involvement of ground (ABG fault) is simulated at the center of the test transmission line (115 km from node-1). Voltages with all phases are recorded on bus-1 of the test line. The voltage signals associated with all the phases are shown in Figure 12.4(a). These voltages pertaining to all phases are processed by the use of a Stockwell transform for computing the STM matrix. The absolute value matrix (ASTM) is computed from the STM. The proposed MIFI, MVIFI, SIFI, and HFI are computed from the ASTM using the proposed algorithm and detailed in Figure 12.4(b), (c), (d), and (e) in respective order.

Figure 12.4(a) details the voltage associated with the phases-A and B decrease after the incidence of the ABG fault, whereas values of voltage associated with the phase-C follows the sinusoidal nature the same as that recorded in the healthy period. Figure 12.4(b) details the values of a voltage-based MIFI increase after the incidence of fault condition for phases. However, an increase in magnitude associated with phases-A and B is high in comparison to phase-A. Figure 12.4(c) details the values of voltage-based MVIFI associated with the faulty phases-A and B changes and deviates from the straight-line nature. However, the values of MVIFI associated with healthy phase-C are the same in the post-fault condition as that observed in the pre-fault condition. Figure 12.4(d) details the values of voltage-based SIFI increase corresponding to the phases-A and B after the incidence of the fault condition. However, its value corresponding to the healthy phase-C is nearly zero. Figure 12.4(e) details the values of voltage-based HFI associated with the faulty phases-A and B increase after the incidence of the fault condition. However, the values of HFI associated to the healthy phase-C is same in the post-fault condition as that observed in the pre-fault condition, which is zero. The values of voltage-based HFI have crossed the threshold HFIT (25), indicating that the phases-A and B are faulty. This index for the healthy phase-C is below the HFIT. Hence, the phases-A and B fault with involvement of ground (ABG fault) has been recognized effectively by the use of a voltage-based method.

It is observed that the CFI for the phases-A and B has high values, whereas its value associated to the phase-C is very low.

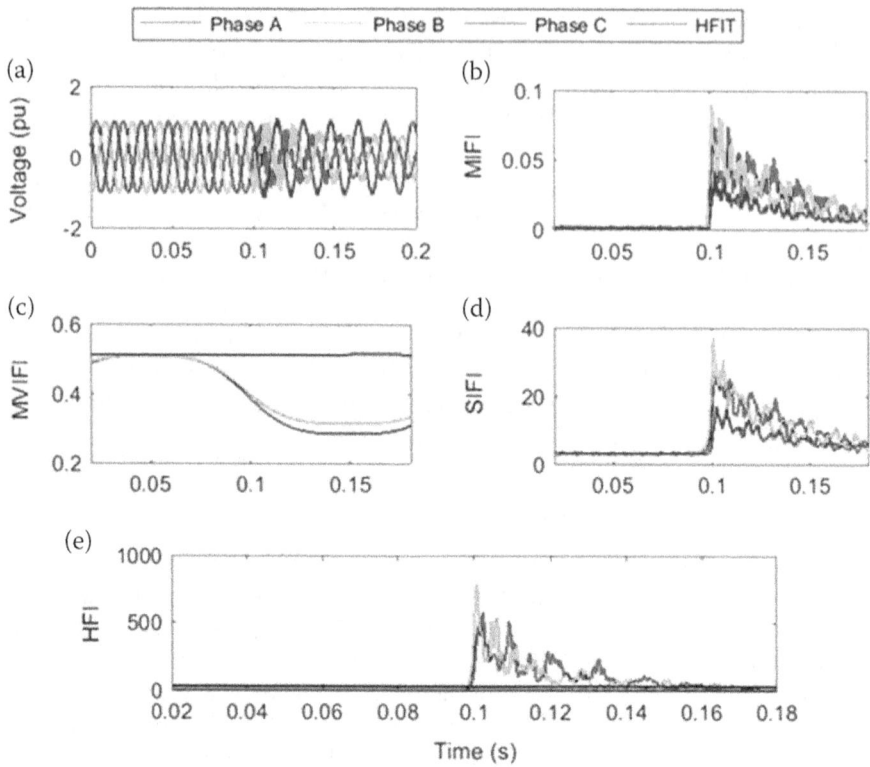

FIGURE 12.4   Voltage-based estimation fault on phases-A and B involving ground: (a) voltage of all phases, (b) median-based intermediate fault index, (c) maximum value–based intermediate fault index, (d) summation-based intermediate fault index, and (e) proposed hybrid fault index.

12.3.4   FAULT ON ALL THE PHASES WITHOUT INVOLVEMENT OF GROUND

A fault on all the phases without involving the ground (ABC fault) is simulated at the center of the test transmission line (115 km from node-1). Voltage pertaining to all phases is recorded on node-1 of the test line. The voltages associated with all the phases are shown in Figure 12.5(a). These voltages pertaining to all the phases are processed by the use of a Stockwell transform for computing the STM matrix. The absolute value matrix (ASTM) is computed from the STM. The proposed MIFI, MVIFI, SIFI, and HFI are computed from the ASTM using the proposed algorithm and detailed in Figure 12.5(b), (c), (d), and (e) in respective order.

   Figure 12.5(a) details the voltage associated with all phases decreases after the incidence of ABC fault. Figure 12.5(b) details the values of a voltage-based MIFI increase after the incidence of fault condition for all phases. Figure 12.5(c) details the values of voltage-based MVIFI associated with all the faulty phase changes and deviates from the straight-line nature. Figure 12.5(d) details the values of voltage-based SIFI increase pertaining to all the phases after the incidence of the fault condition. Figure 12.5(e) details the values of voltage-based HFI associated with all

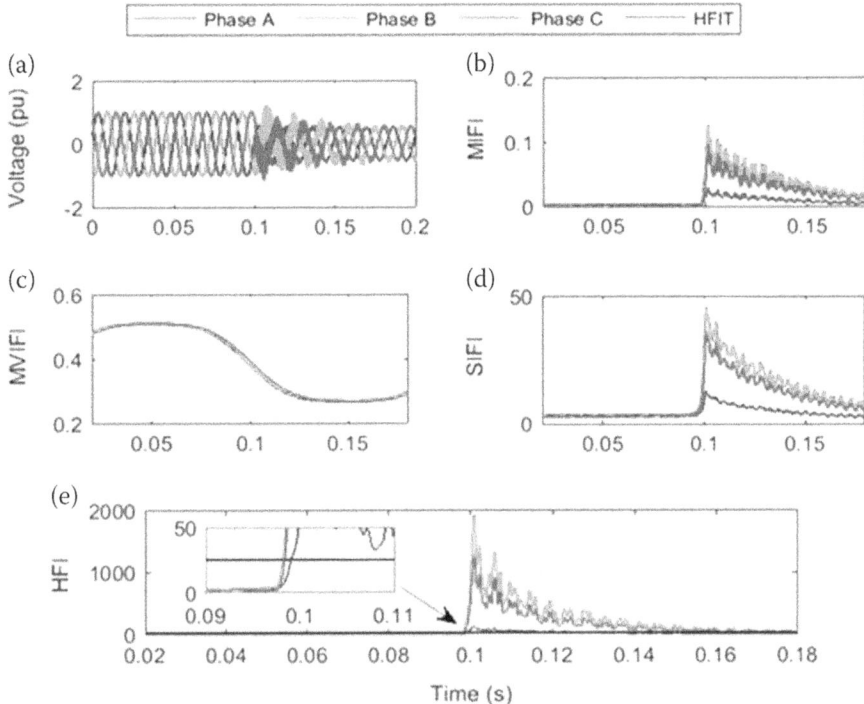

FIGURE 12.5 Voltage-based estimation fault on all phases without involving ground (a) voltage of all phases, (b) median-based intermediate fault index, (c) maximum value–based intermediate fault index, (d) summation-based intermediate fault index, and (e) proposed hybrid fault index.

the faulty phases increase after the incidence of the fault condition. The values of voltage-based HFI have crossed the threshold HFIT (25), indicating that all the phases are faulty in nature. Hence, the fault on all the phases without involving the ground (ABC fault) has been recognized effectively by the use of a voltage-based method. It is observed that the CFI for all the phases has high values.

12.3.5 FAULT ON ALL THE PHASES WITH INVOLVEMENT OF GROUND

A fault on all the phases with involving the ground (ABCG fault) is simulated at the center of the test transmission line (115 km from node-1). Voltages to all phases are recorded on bus-1 of the test line. Voltage signals pertaining to all the phases are shown in Figure 12.6(a). These voltages pertain to all the phases are processed using the Stockwell transform to obtain the STM matrix. The absolute value matrix (ASTM) is computed from the STM. The proposed MIFI, MVIFI, SIFI, and HFI are computed from the ASTM using the proposed algorithm and detailed in Figure 12.6(b), (c), (d), and (e) in respective order.

Figure 12.6(a) details the voltages associated with all phases decreased after the incidence of the ABCG fault. Figure 12.6(b) details the values of a voltage-based

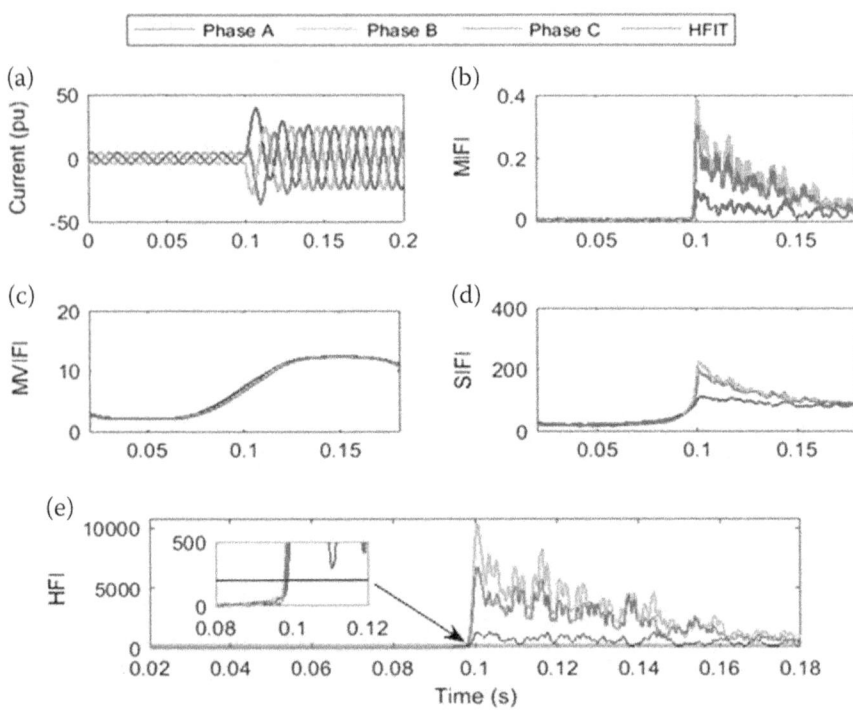

FIGURE 12.6 Voltage-based estimation fault on all phases and ground: (a) voltage of all phases, (b) median-based intermediate fault index, (c) maximum value–based intermediate fault index, (d) summation-based intermediate fault index, and (e) proposed hybrid fault index.

MIFI increase after the incidence of the fault condition for all phases. Figure 12.6(c) details the values of voltage-based MVIFI associated with all the faulty phase changes and deviates from the straight-line nature. Figure 12.6(d) details the values of a voltage-based SIFI increase pertaining to all the phases after the incidence of the fault condition. Figure 12.6(e) details the values of voltage-based HFI pertaining to all faulty phases increase after the incidence of the fault condition. The values of voltage-based HFI have crossed the threshold HFIT (25), indicating that all phases are faulty. Hence, fault on all phases involving the ground (ABCG fault) is recognized effectively by the use of a voltage-based method.

This is inferred that the CFI for all the phases has high values.

12.3.6 Impact of Variations in Fault Incidence Angle

For investigating the effect of variations in a fault incidence angle, a fault on phases-A and B involving the ground (ABG fault) is simulated at the center of the test transmission line (115 km from the bus-1) with fault incidence angle equal to 45°. Voltage associated with all the phases is recorded on node-1 of the test line. The voltages associated with all phases are shown in Figure 12.7(a). These voltages pertaining to all phases are processed using a Stockwell transform for computing the STM matrix. The absolute value matrix (ASTM) is computed from the STM. The proposed MIFI,

MVIFI, SIFI, and HFI are computed from the ASTM using the proposed algorithm and detailed in Figure 12.7(b), (c), (d), and (e) in respective order.

Figure 12.7 details the voltage associated with the phases-A and B decreases after the incidence of the ABG fault, whereas values of voltage associated with the phase-C follow the sinusoidal nature, the same as that recorded in the healthy period. Figure 12.7(b) details the values of a voltage-based MIFI increase after the incidence of fault condition for phases. However, an increase in magnitude pertaining to phases-A and B is high in comparison to phase-A. Figure 12.7(c) details the values of voltage-based MVIFI associated with the faulty phases-A and B changes and deviates from the straight-line nature. However, the values of MVIFI pertaining to the healthy phase-C are the same in the post-fault condition as that observed in the pre-fault condition. Figure 12.7(d) details the values of a voltage-based SIFI increase corresponding to phases-A and B after the incidence of the fault condition. However, its value corresponding to healthy phase-C is nearly zero. Figure 12.7(e) details the values of voltage-based HFI associated with the faulty phases-A and B increase after incidence of the fault condition. However, the values of HFI pertaining to healthy phase-C is the same in the post-fault condition as that observed in the pre-fault condition, which is zero. The values of voltage-based HFI have crossed the threshold HFIT (25), indicating that phases-A

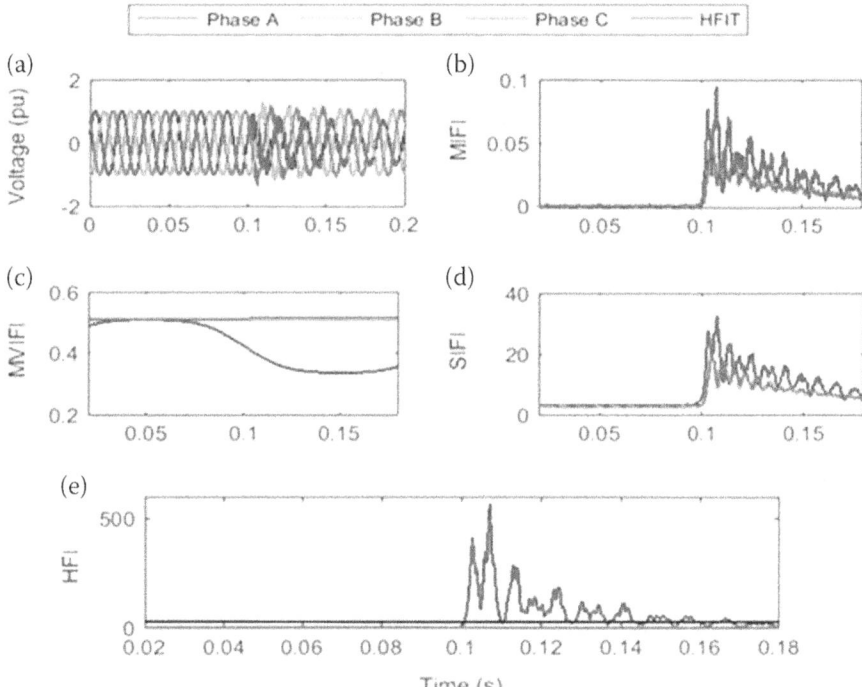

FIGURE 12.7 Voltage-based estimation of fault on phase-A and involvement of ground incident at an angle of 45°: (a) voltage of all phases, (b) median-based intermediate fault index, (c) maximum value–based intermediate fault index, (d) summation-based intermediate fault index, and (e) proposed hybrid fault index.

and B are faulty. This index for healthy phase-C is below the HFIT. Hence, the phases-A and B fault with involvement of the ground (ABG fault) with a fault incidence angle equal to 45° has been recognized effectively by the use of a voltage-based method.

For investigating the effect of variations in the fault incidence angle, a fault on phases-A and B with involvement of the ground (ABG fault) is simulated at the center of test transmission line (115 km from node-1) with a fault incidence angle equal to 90°. The voltage associated with all the phases is recorded on node-1 of the test line. The voltages pertaining to all phases are shown in Figure 12.8(a). These voltage signals corresponding to all the phases are processed using the Stockwell transform for computing the STM matrix. The absolute value matrix (ASTM) is computed from the STM. The proposed MIFI, MVIFI, SIFI, and HFI are computed from the ASTM using the proposed algorithm and detailed in Figure 12.8(b), (c), (d), and (e) in respective order.

Figure 12.8(a) details the voltage associated with the phases-A and B decreases after the incidence of the ABG fault, whereas the values of voltage associated with phase-C follows the sinusoidal nature, the same as that recorded in the healthy period. Figure 12.8(b) details the values of a voltage-based MIFI increase after the incidence

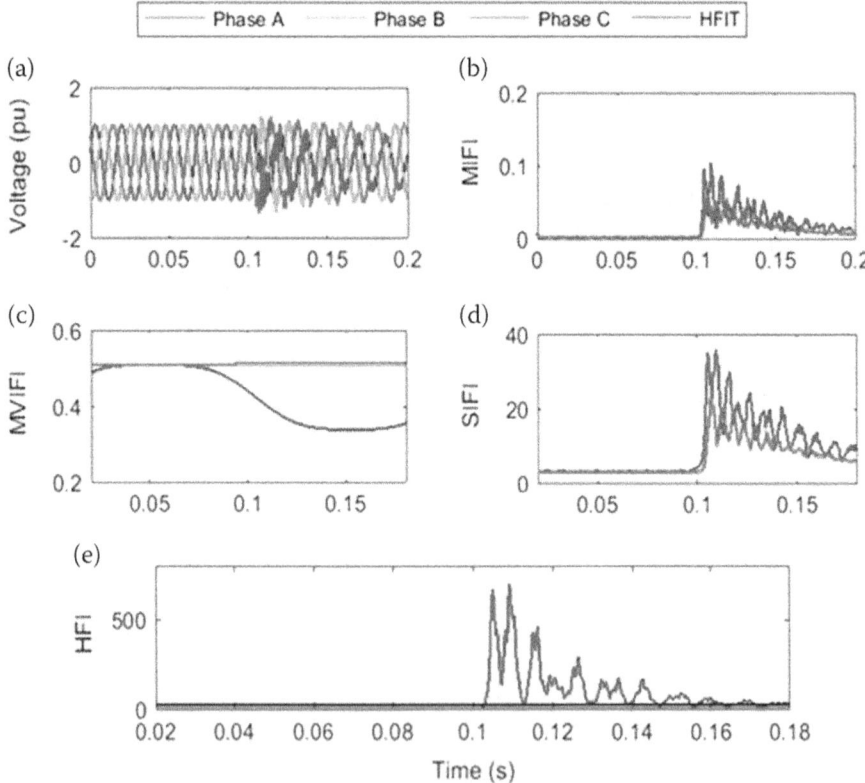

FIGURE 12.8 Voltage-based estimation of fault on phase-A and involvement of ground incident at an angle of 90°: (a) voltage of all phases, (b) median-based intermediate fault index, (c) maximum value–based intermediate fault index, (d) summation-based intermediate fault index, and (e) proposed hybrid fault index.

of fault condition for phases. However, an increase in magnitude pertaining to phases-A and B is high compared to phase-A. Figure 12.8(c) details the values of voltage-based MVIFI associated with the faulty phases-A and B changes and deviates from the straight-line nature. However, the values of MVIFI pertaining to healthy phase-C are the same in the post-fault condition as that observed in the pre-fault condition. Figure 12.8(d) details the values of voltage-based SIFI increase corresponding to the phases-A and B after incidence of the fault condition. However, its value corresponding to the healthy phase-C is nearly zero. Figure 12.8(e) details the values of voltage-based HFI associated with the faulty phases-A and B increase after the incidence of the fault condition. However, the values of HFI pertaining to healthy phase-C is the same in the post-fault condition as that observed in the pre-fault condition, which is zero. The values of voltage-based HFI have crossed the threshold HFIT (25), indicating that phases-A and B are faulty. This index for the healthy phase-C is below the HFIT. Hence, the phases-A and B fault with involvement of the ground (ABG fault) with the fault incidence angle equal to 90° has been recognized effectively by the use of the voltage-based method.

12.3.7 IMPACT OF VARIATIONS IN FAULT IMPEDANCE

For investigating the effect of variations in a fault incidence angle, a fault on phases-A and B with involvement of the ground (ABG fault) is simulated at the center of the test transmission line (115 km from node-1) with a fault impedance equal to 5°. Voltage associated with all the phases is recorded on node-1 of the test line. The voltages pertaining to all phases are shown in Figure 12.9(a). These voltage signals pertaining to all the phases are processed by the use of a Stockwell transform for computing the STM matrix. The absolute value matrix (ASTM) is computed from the STM. The proposed MIFI, MVIFI, SIFI, and HFI are computed from the ASTM using the proposed algorithm and detailed in Figure 12.9(b), (c), (d), and (e) in respective order.

Figure 12.9(a) details the voltage associated with phases-A and B decreases after the incidence of the ABG fault, whereas the values of voltage associated with phase-C follow the sinusoidal nature, the same as that recorded in the healthy period. Figure 12.9(b) details the values of voltage-based MIFI increase after the incidence of the fault condition for phases. However, an increase in magnitude associated to phases-A and B is high in comparison to phase-A. Figure 12.9(c) details the values of voltage-based MVIFI associated with the faulty phases-A and B changes and deviates from the straight-line nature. However, the values of MVIFI pertaining to the healthy phase-C are same in the post-fault condition as that observed in the pre-fault condition. Figure 12.9(d) details the values of voltage-based SIFI increase pertaining to the phases-A and B after incidence of the fault condition. However, its value corresponding to the healthy phase-C is nearly zero. Figure 12.9(e) details the values of voltage-based HFI associated with the faulty phases-A and B increase after the incidence of the fault condition. However, the values of HFI pertaining to the healthy phase-C is the same in the post-fault condition as that observed in the pre-fault condition, which is zero. The values of voltage-based HFI have crossed the threshold HFIT (25), indicating that the phases-A and B are faulty. This index for the healthy phase-C is below the HFIT. Hence, the phases-A and B fault with involvement of the

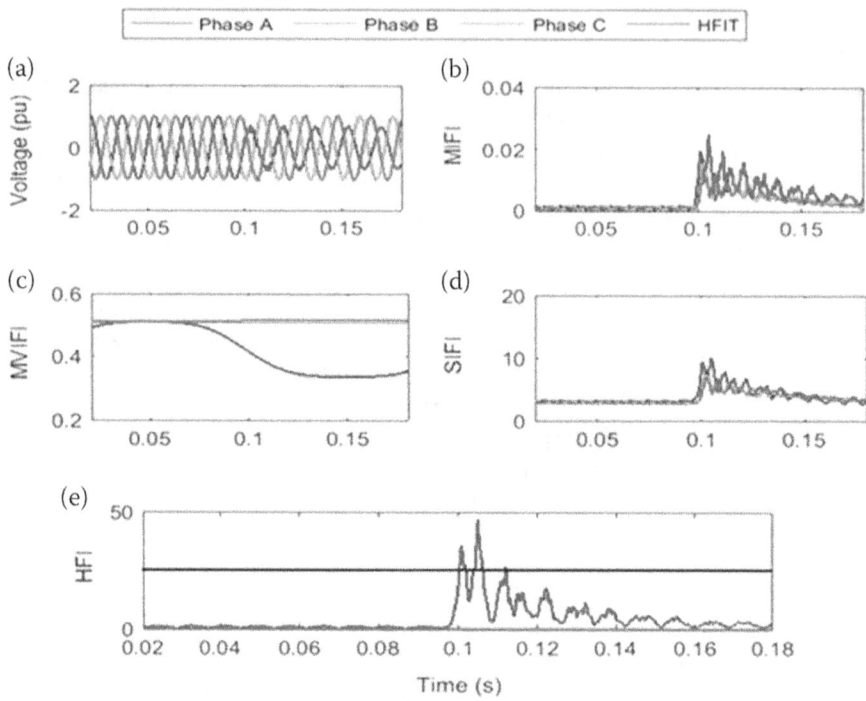

FIGURE 12.9 Voltage-based estimation of fault on phase-A and involvement of ground with fault impedance of 5Ω: (a) voltage of all phases, (b) median-based intermediate fault index, (c) maximum value–based intermediate fault index, (d) summation-based intermediate fault index, and (e) proposed hybrid fault index.

ground (ABG fault) with the fault impedance equal to 5Ω is recognized effectively by the use of the voltage-based method.

For investigating the effect of variations in a fault incidence angle, a fault on phases-A and B with involvement of the ground (ABG fault) is simulated at the center of the test transmission line (115 km from node-1) with a fault impedance equal to 10Ω. Voltage associated with all the phases is recorded on node-1 of the test line. Voltage signals associated with all the phases are shown in Figure 12.10(a). These voltages pertaining to all phases are processed using the Stockwell transform to obtain the STM matrix. The absolute value matrix (ASTM) is computed from the STM. The proposed MIFI, MVIFI, SIFI, and HFI are computed from the ASTM using the proposed algorithm and detailed in Figure 12.10(b), (c), (d), and (e) in respective order.

Figure 12.10(a) details the voltage associated with the phases-A and B decreases after the incidence of the ABG fault, whereas the values of voltage associated with the phase-C follows the sinusoidal nature, the same as that recorded in the healthy period. Figure 12.10(b) details the values of voltage-based MIFI increase after the incidence of fault condition for phases. However, an increase in magnitude pertaining to phases-A and B is high in comparison to phase-A. Figure 12.10(c) details the values of voltage-based MVIFI associated with the faulty phases-A and B changes and deviates from the straight-line nature. However, the values of MVIFI pertaining to the healthy phase-C are the same

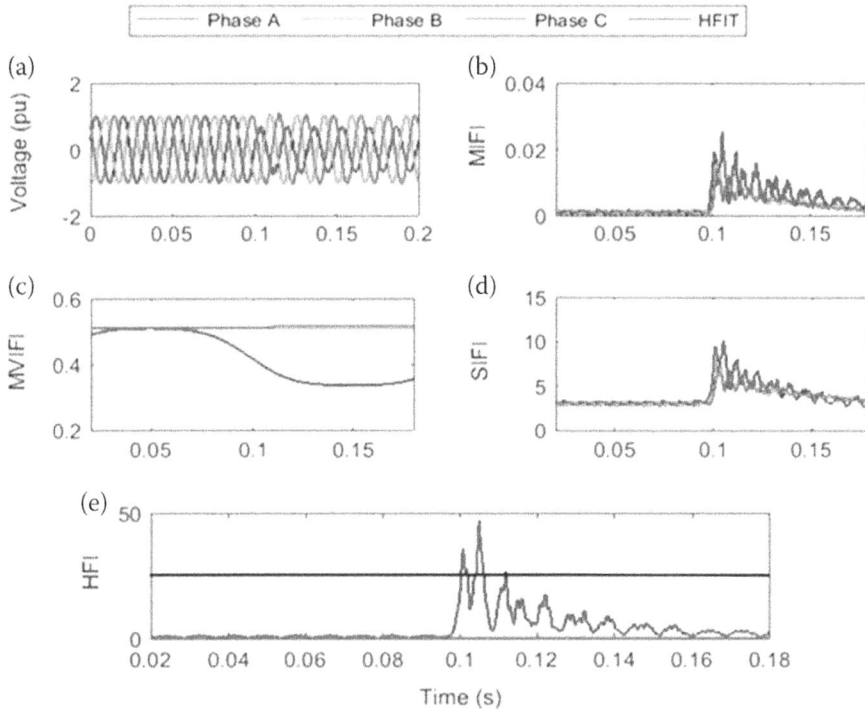

FIGURE 12.10 Voltage-based estimation of fault on phase-A and involvement of ground with fault impedance of 10Ω: (a) voltage of all phases, (b) median-based intermediate fault index, (c) maximum value–based intermediate fault index, (d) summation-based intermediate fault index, and (e) proposed hybrid fault index.

in the post-fault condition as that observed in the pre-fault condition. Figure 12.10(d) details the values of voltage-based SIFI increase pertaining to the phases-A and B after the incidence of the fault condition. However, its value corresponding to the healthy phase-C is nearly zero. Figure 12.10(e) details the values of voltage-based HFI associated with the faulty phases-A and B increase after an incidence of the fault condition. However, the values of HFI pertaining to the healthy phase-C is the same in the post-fault condition as that observed in the pre-fault condition, which is zero. The values of voltage-based HFI have crossed the threshold HFIT (25), indicating that the phases-A and B are faulty. This index for the healthy phase-C is below the HFIT. Hence, the phases-A and B fault with involvement of the ground (ABG fault) with a fault impedance equal to 10Ω has been recognized effectively by the use of a voltage-based method.

12.3.8 IMPACT OF VARIATIONS IN FAULT LOCATION

For investigating the effect of variations in fault location, a fault on phases-A and B with involvement of the ground (ABG fault) is simulated at a distance of 30 km from node-1. Voltage associated with all the phases is recorded on bus-1 of the test line. Voltages associated with all the phases are shown in Figure 12.11(a). These voltages

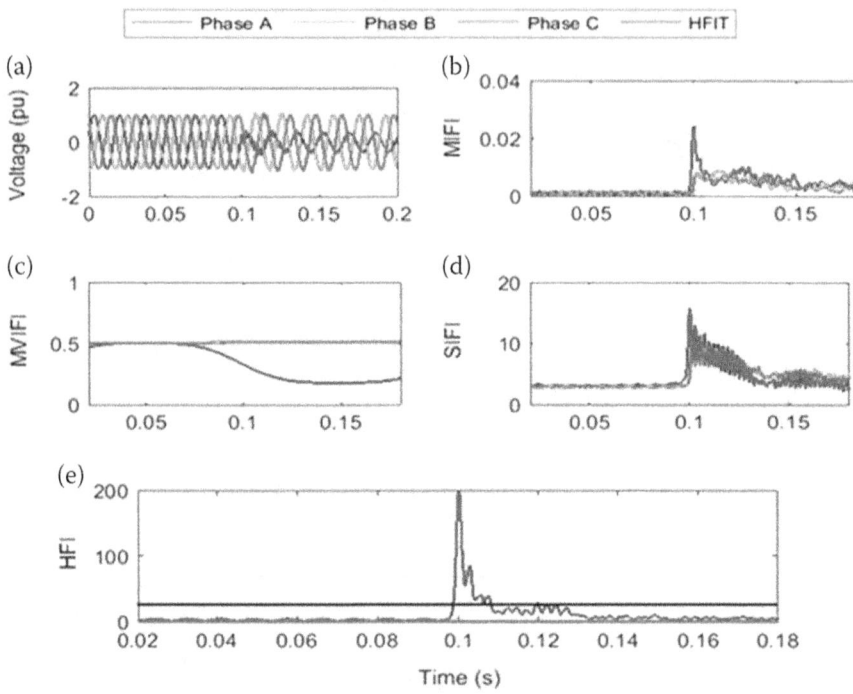

FIGURE 12.11 Voltage-based estimation of fault on phase-A and involvement of ground at a location of 30 km from the sending terminal of transmission line: (a) voltage of all phases, (b) median-based intermediate fault index, (c) maximum value–based intermediate fault index, (d) summation-based intermediate fault index, and (e) proposed hybrid fault index.

pertaining to all the phases are processed using the Stockwell transform for computing to the STM matrix. The absolute value matrix (ASTM) is computed from the STM. The proposed MIFI, MVIFI, SIFI, and HFI are computed from the ASTM using the proposed algorithm and detailed in Figure 12.11(b), (c), (d), and (e) in respective order.

Figure 12.11(a) details the voltage associated with the phases-A and B decreases after the incidence of the ABG fault, whereas values of voltage associated with the phase-C follow the sinusoidal nature, the same as that recorded in the healthy period. Figure 12.11(b) details the values of voltage-based MIFI increase after the incidence of the fault condition for phases. However, an increase in magnitude pertaining to phases-A and B is high compared to phase-A. Figure 12.11(c) details the values of voltage-based MVIFI associated with the faulty phases-A and B changes and deviates from the straight-line nature. However, the values of MVIFI pertaining to the healthy phase-C are the same in the post-fault condition as that observed in the pre-fault condition. Figure 12.11(d) details the values of voltage-based SIFI increase corresponding to the phases-A and B after incidence of the fault condition. However, its value pertaining to the healthy phase-C is nearly zero. Figure 12.11(e) details the values of voltage-based HFI associated with the faulty phases-A and B increase after incidence of the fault condition. However, the values of HFI corresponding to the healthy phase-C are same in the post-fault condition as that observed in the pre-fault condition, which is zero. The values of voltage-based HFI have crossed the threshold HFIT (25), indicating that phases-A and B are faulty. This index for the healthy phase-C is below the HFIT. Hence, the phases-A and B fault with involvement of the ground (ABG fault) incident at a distance of 30 km from sending terminal of transmission line is recognized effectively by the use of a voltage-based method.

For investigating the effect of variations in a fault location, a fault on phases-A and B with involvement of the ground (ABG fault) is simulated at a distance of 200 km from bus-1. Voltage associated with all the phases is recorded on node-1 of the test line. The voltage signals associated with all the phases are shown in Figure 12.12(a). These voltages pertaining to all the phases are processed using the Stockwell transform to obtain the STM matrix. The absolute value matrix (ASTM) is computed from the STM. The proposed MIFI, MVIFI, SIFI, and HFI are computed from the ASTM using the proposed algorithm and detailed in Figure 12.12(b), (c), (d), and (e) in respective order.

Figure 12.12(a) details the voltage associated with the phases-A and B decrease after the incidence of the ABG fault, whereas the values of voltage associated with the phase-C follow the sinusoidal nature, the same as that recorded in the healthy period. Figure 12.12(b) details the values of voltage-based MIFI increase after the incidence of fault condition for phases. However, an increase in magnitude pertaining to phases-A and B is high compared to phase-A. Figure 12.12(c) details the values of voltage-based

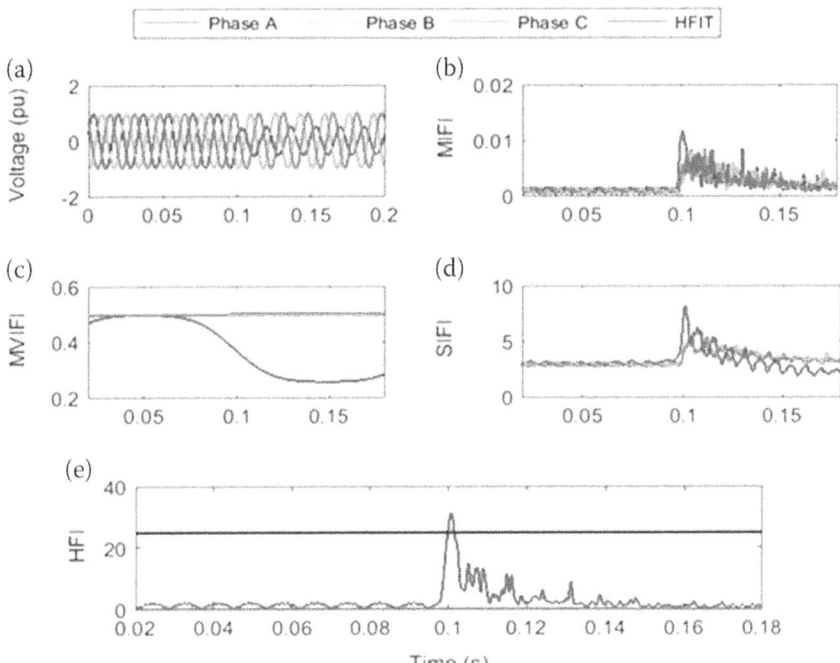

FIGURE 12.12 Voltage-based estimation of fault on phase-A and involvement of ground at a location of 200 km from the sending end of the transmission line: (a) voltage of all phases, (b) median-based intermediate fault index, (c) maximum value–based intermediate fault index, (d) summation-based intermediate fault index, and (e) proposed hybrid fault index.

MVIFI associated with the faulty phases-A and B changes and deviates from the straight-line nature. However, the values of MVIFI pertaining to the healthy phase-C are same in the post-fault condition as that observed in the pre-fault condition.

Figure 12.12(d) details the values of voltage-based SIFI increase corresponding to the phases-A and B after the incidence of a faulty condition. However, its value corresponding to the healthy phase-C is nearly zero. Figure 12.12(e) details the values of voltage-based HFI associated with the faulty phases-A and B increase after an incidence of the fault condition. However, the values of HFI pertaining to the healthy phase-C is the same in the post-fault condition as that observed in the pre-fault condition, which is zero. The values of voltage-based HFI have crossed the threshold HFIT (25), indicating that phases-A and B are faulty in nature. This index for the healthy phase-C is below the HFIT. Hence, the phases-A and B fault with involvement of the ground (ABG fault) incident at a distance of 200 km from the sending terminal of a transmission line is recognized effectively by the use of a voltage-based method.

12.3.9 CLASSIFICATION OF THE FAULTS USING VOLTAGE-BASED FEATURES

The classification of various types of fault has been achieved using the decision supported by the rules. This rule-based decision tree algorithm classified different types of the faulty events using magnitudes of voltage-supported covariance fault index (VCFI). The VCFI is computed by multiplying the CFI by the weight factor.

The values of VCFI detailed in Table 12.2 are used for classification of the faults. A threshold VTH1 is used to classify the faults using a number of faulty phases. The value of VTH1 is set equal to 100 for classification faults the category based on the number of faulty phases. If VCFI > 100 for all the phases, then the fault is a three-phase fault (ABC or ABCG). Furthermore, if VCFI > 100 for one phase, only then the type of fault is phase to ground fault (AG). For the two-phase fault, the values of CFI is greater than 1 for two phases. The threshold TH2 is used to classify the AB and ABG faults. A threshold value of 700 is set for the threshold TH2. If VCFI > TH2, then there is a ABG fault else AB fault.

The classification of the different faults using the rule-based decision tree using the CFI is detailed in Figure 12.3 (Table 12.3 and Figure 12.13).

TABLE 12.3
Magnitude of VCFI During Faulty Conditions

Type of Fault	Voltage-Based Covariance Fault Index (VCFI)		
	Phase-A	Phase-B	Phase-C
AG	470	0.012933	0.024815
AB	670	260	0.0022766
ABG	810	750	0.17373
ABC	920	910	920
ABCG	920	910	920

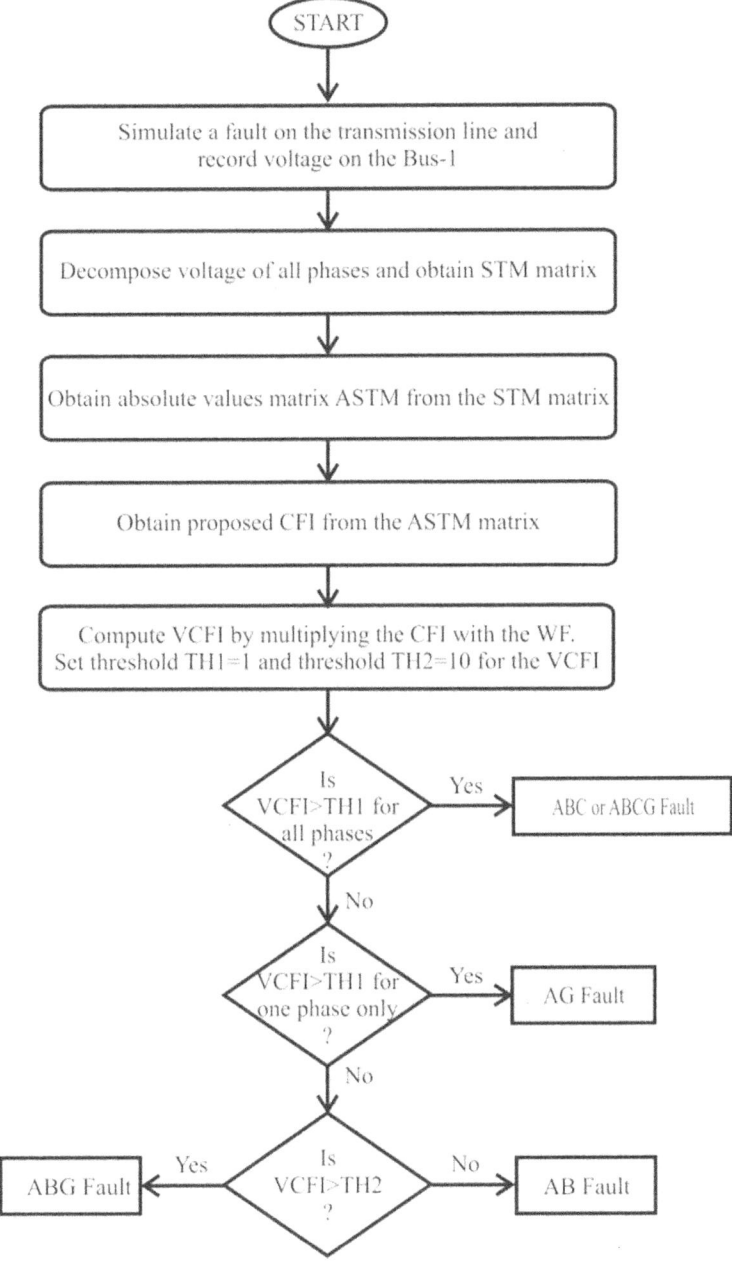

FIGURE 12.13 Classification of faults using voltage-based features.

12.4 CONCLUSION

This research work presents an approach based on the voltage signals for recognition of the faulty events on the transmission line. Voltage signals recorded during the faulty condition are processed using the Stockwell transform to obtain an output matrix. This matrix is used to compute the indices such as median intermediate fault index (MIFI), maximum value intermediate fault index (MVIFI), summation intermediate fault index (SIFI), and co-variance fault index (CFI). A hybrid fault index (HFI) is computed by multiplying the MIFI, MVIFI, SIFI, CFI, and weight factor (WF). A threshold value for the hybrid fault index (HFIT) equal to 25 is selected for the proposed HFI to estimate the faulty condition from the healthy conditions. This HFIT is selected by testing the algorithm for the different conditions such as different values of the fault location, fault impedance, fault incidence angle, reverse power flow, etc. The faults have been effectively classified using the rule-based decision tree. It is concluded that the proposed algorithm is found to be effective in the recognition of faults such as phase to ground, two phases fault, two phases to ground fault, all the three phases fault, and all three phases to ground fault. The voltage-based algorithm is found to be effective in identification of the faults during the conditions such as the variations in the fault incidence angle, variations in the fault impedance, and variations in the fault location. The proposed algorithm has been tested for the two terminal transmission lines. This algorithm may also be tested for protection of the multiple transmission line as well as also in the presence of renewable energy.

REFERENCES

[1] Rathore Bhuvnesh, Om Prakash Mahela, Baseem Khan, Hassan Haes Alhelou, and Pierluigi Siano, "Wavelet-alienation-neural based protection scheme for STATCOM compensated transmission line," *IEEE Transactions on Industrial Informatics*, Vol. 17, No. 4, pp. 2557–2565, April 2021, 10.1109/TII.2020.3001063

[2] Om Prakash Mahela, Jaya Sharma, Bipul Kumar, Baseem Khan, and Hasan Haes Alhelou, "An algorithm for the protection of distribution feeder using Stockwell and Hilbert transforms supported features," *CSEE Journal of Power and Energy Systems*, Vol. 7, No. 6, pp. 1278–1288, November 2021, 10.17775/CSEEJPES.2020.00170

[3] Xiaoyang Tong and Hao Wen, "A novel transmission line fault detection algorithm based on pilot impedance," *Electric Power Systems Research*, Vol. 179, No. 106062, 2020.

[4] Bikash Patel, "A new FDOST entropy based intelligent digital relaying for detection, classification and localization of faults on the hybrid transmission line," *Electric Power Systems Research*, Vol. 157, pp. 39–47, 2018.

[5] Bhuvnesh Rathore and Abdul Gafoor Shaik, "Wavelet-alienation based protection scheme formulti-terminal transmission line," *Electric Power Systems Research*, Vol. 161, pp. 8–16, 2018.

[6] Sayari Das, Shiv P. Singh, and Bijaya K. Panigrahi, "Transmission line fault detection and location using wide area measurements," *Electric Power Systems Research*, Vol. 151, pp. 96–105, 2017.

[7] Hassan Khorashadi Zadeh and Zuyi Li, "Phasor measurement unit based transmission line protection scheme design," *Electric Power Systems Research*, Vol. 81, pp. 421–429, 2011.

[8] Ali Abdali, Kazem Mazlumi, and Reza Noroozian, "High-speed fault detection and location in DC microgrids systems using multi-criterion system and neural network," *Applied Soft Computing Journal*, Vol. 79, pp. 341–353, 2019.

[9] J. Doria-Garcia, C. Orozco-Henao, L. U. Iurinic, and Juan Diego Pulgarin-Rivera, "High impedance fault location: Generalized extension for ground faults," *Electrical Power and Energy Systems*, Vol. 114, No. 105387, 2020.

[10] Érica Mangueira Lima, Núbia Silva Dantas Brito, and Benemar Alencar de Souza, "High impedance fault detection based on Stockwell transform and third harmonic current phase angle," *Electric Power Systems Research*, Vol. 175, No. 105931, 2019.

[11] Wang Xiaowei, Wei Xiangxiang, Yang Dechang, Song Guobing, Gao Jie, Wei Yanfang, Zeng Zhihui, and Peng Wang, "Fault feeder detection method utilized steady state and transient components based on FFT back stepping in distribution networks," *Electrical Power and Energy Systems*, Vol. 114, No. 105391, 2020.

[12] Peng Nan, Zhou Lutian, Liang Rui, and Xu Haoyuan, "Fault location of transmission lines connecting with short branches based on polarity and arrival time of asynchronously recorded traveling waves," *Electric Power Systems Research*, Vol. 169, pp. 184–194, 2019.

[13] B. K. Chaitanya and Anamika Yadav, "Decision tree aided traveling wave based fault section identification and location scheme for multi-terminal transmission lines," *Measurement*, Vol. 135, pp. 312–322, 2019.

[14] Chenhao Sun, Xin Wang, and Yihui Zheng, "Data-driven approach for spatio-temporal distribution prediction of fault events in power transmission systems," *Electrical Power and Energy Systems*, Vol. 113, pp. 726–738, 2019.

[15] Atul Kulshrestha, Om Prakash Mahela, Mukesh Kumar Gupta, Neeraj Gupta, Nilesh Patel, Tomonobu Senjyu, Mir Sayed Shah Danish, and Mahdi Khosravy, "A hybrid protection scheme using Stockwell Transform and Wigner Distribution Function for power system network with solar energy penetration, *Energies*, Vol. 13, No. 14, pp. 3519, 2020, 10.3390/en13143519

[16] Govind Sahay Yogee, Om Prakash Mahela, Kapil Dev Kansal, Baseem Khan, Rajendra Mahla, Hassan Haes Alhelou, and Pierluigi Siano, "An algorithm for recognition of fault conditions in the utility grid with renewable energy penetration," *Energies*, Vol. 13, No. 9, 2383, 10.3390/en13092383

[17] Sheesh Ram Ola, Amit Saraswat, Sunil Kumar Goyal, S. K. Jhajharia, Baseem Khan, Om Prakash Mahela, Hassan Haes Alhelou, and Pierluigi Siano, "A protection scheme for power system with solar energy penetration," *Applied Sciences*, Vol. 10, No. 4, Paper No. 1516, pp. 1–22, Feb. 2020, 10.3390/app10041516

[18] Sheesh Ram Ola, Amit Saraswat, Sunil Kumar Goyal, Virendra Sharma, Baseem Khan, Om Prakash Mahela, Hassan Haes Alhelou, and Pierluigi Siano, "Alienation coefficient and Wigner distribution function based protection scheme for hybrid power system network with renewable energy penetration," *Energies*, Vol. 13, No. 5, Paper No. 1120, March 2020, 10.3390/en13051120

[19] Sheesh Ram Ola, Amit Saraswat, Sunil Kumar Goyal, S. K. Jhajharia, and Om Prakash Mahela. Detection and analysis of power system faults in the presence of wind power generation using Stockwell transform based median, *Springer Lecture Notes in Electrical Engineering Series*, ISSN: 1876-1100, pp. 319–329, 2019. https://link.springer.com/chapter/10.1007/978–981-15–0214-9_36.

[20] Amit Kumar Gangwar, Bhunesh Rathore, and Om Prakash Mahela, "K-means Clustering and Linear Regression Based Protection Scheme for Transmission Line," IEEE 9th Power India International Conference (PIICON 2020) from 28 Feb to 01 March, 2020 – Deenbandhu Chhotu Ram University of Science and Technology, Murthal, India, 10.1109/PIICON49524.2020.9113038

[21] Nikita Tailor, Satyanarayan Joshi, and Om Prakash Mahela, "Transmission Line Protection Schemes Based on Wigner Distribution Function and Discrete Wavelet Transform," IEEE 9th Power India International Conference (PIICON 2020) from 28 Feb to 01 March, 2020 – Deenbandhu Chhotu Ram University of Science and Technology, Murthal, India, 10.1109/PIICON49524.2020.9113011

[22] Jaya Sharma, Bipul Kumar, Om Prakash Mahela, and Akhil Ranjan Garg, "Protection of distribution feeder using Stockwell Transform supported voltage features," IEEE 9th Power India International Conference (PIICON 2020) from 28 Feb to 01 March, 2020, Deenbandhu Chhotu Ram University of Science and Technology, Murthal, India, 10.1109/PIICON49524.2020.9113014

[23] Mohd Zishan Khoker, Om Prakash Mahela, and Gulhasan Ahmad, "A voltage algorithm using discrete wavelet transform and Hilbert Transform for detection and classification of power system faults in the presence of solar energy," 2020 IEEE International Students' Conference on Electrical, Electronics and Computer Science (SCEECS 2020), MANIT Bhopal, India, February 22–23, 2020, 10.1109/SCEECS48394.2020.7

[24] Mohd Zishan Khoker, Om Prakash Mahela, and Gulhasan Ahmad, "A current based hybrid algorithm using discrete wavelet transform and Hilbert transform for detection and classification of power system faults in the presence of solar energy," 2020 IEEE International Students' Conference on Electrical, Electronics and Computer Science (SCEECS 2020), MANIT Bhopal, India, February 22–23, 2020, 10.1109/SCEECS48394.2020.6

[25] Shubhmay Karmakar, Gulhasan Ahmad, Om Prakash Mahela, and Ravi Raj Choudhary, "Algorithm Based on Combined Features of a Stockwell Transform and Hilbert Transform for Detection of Transmission Line Faults with Dynamic Load," First IEEE International Conference on Power, Control and Computing Technologies (ICPC2T), NIT Raipur, India, January 3–5, 2020, 10.1109/ICPC2T48082.2020.9071516

[26] Deepak Gupta, Om Prakash Mahela, and Shoyab Ali, "Voltage Based Transmission Line Protection Algorithm Using Signal Processing Techniques," 2020 IEEE International Students' Conference on Electrical, Electronics and Computer Science (SCEECS 2020), MANIT Bhopal, India, February 22–23, 2020, 10.1109/SCEECS48394.2020.15

[27] Surbhi Thukral, Om Prakash Mahela, and Bipul Kumar, "Detection of Transmission Line Faults in the Presence of Wind Energy Power Generation Source Using Stockwell's Transform," IEEE International Conference on Issues and Challenges in Intelligent Computing Techniques (ICICT 2019), 27–28th September, 2019, KIET Group of Institutions, Delhi-NCR, Ghaziabad, India, 10.1109/ICICT46931.2019.8977695

[28] Rajesh Kumar, Om Prakash Mahela, Mahendra Kumar, Nitin Kumar Suyan, and Neeraj Kumar, "A Current Based Algorithm Using Harmonic Wavelet Transform and Rule Based Decision Tree for Transmission Line Protection," *4th International Conference On Internet of Things: Smart Innovation and Usages (IoT-SIU 2019)*, 18–19 April, 2019, Krishna Engineering College, Ghaziabad, Uttar Pradesh, India, 10.1109/IoT-SIU.2019.8777667

[29] Deepak Gupta, Om Prakash Mahela, and Shoyab Ali, "Current Based Transmission Line Protection Algorithm Using Signal Processing Techniques," 2020 IEEE International Students' Conference on Electrical, Electronics and Computer Science (SCEECS 2020), MANIT Bhopal, India, February 22–23, 2020, 10.1109/SCEECS48394.2020.14

[30] Sheesh Ram Ola, Amit Saraswat, Sunil Kumar Goyal, S. K. Jhajharia, Bhuvnesh Rathore, and Om Prakash Mahela, "Wigner distribution function and alienation coefficient based transmission line protection scheme," *IET Generation, Transmission and Distribution*, Vol. 14, pp. 1842–1853, 10.1049/iet-gtd.2019.1414

[31] O. P. Mahela, B. Khan, H. Haes Alhelou, and S. Tanwar, "Assessment of power quality in the utility grid integrated with wind energy generation," *IET Power Electronics*, Vol. 13, No. 13, pp. 2917–2925, 2020.

[32] Mahela, O. P. et al., "Recognition of power quality issues associated with grid integrated solar photovoltaic plant in experimental framework," *IEEE Systems Journal*, Vol. 15, No. 3, pp. 3740–3748, September 2021, 10.1109/JSYST.2020.3027203

[33] O. P. Mahela, A. G. Shaik, B. Khan, R. Mahla, and H. H. Alhelou, "Recognition of complex power quality disturbances using S-transform based ruled decision tree," *IEEE Access*, Vol. 8, pp. 173530–173547, 2020.

[34] O. P. Mahela, B. Khan, H. H. Alhelou, S. Tanwar, and S. Padmanaban, "Harmonic mitigation and power quality improvement in utility grid with solar energy penetration using distribution static compensator," *IET Power Electronics*. Vol. 14, pp. 912–922, 2021.

[35] Pachauri, R. K. et al., "Impact of partial shading on various PV array configurations and different modeling approaches: A comprehensive review," *IEEE Access*, Vol. 8, pp. 181375–181403, 2020.

13 Algorithm Based on Harmonic Wavelet Transform and Rule-Based Decision Tree for Detection and Classification of Transmission Line Faults

Rajendra Mahla
Department of Electrical Engineering, National Institute of Technology Kurukshetra, India

Neeraj Gupta
Oakland University, Rochester, MI, USA

Akhil Ranjan Garg
Department of Electrical Engineering, Jai Narain Vyas University, Jodhpur, India

CONTENTS

13.1 INTRODUCTION

The power system network is used to transmit and distribute electrical power generated at large generating stations to load centers and consumers. A transmission line is one of the important elements of this electric power system network that is used to supply bulk power from central generating stations to load centers. As a result, power transmission lines have been rapidly developed in number and length. One important factor of an electrical power transmission system is to continuously deliver electrical power to consumers. Sustained fault on these lines may lead to disturbances of power supply in a large area. Hence, isolation of a faulty line helps to maintain uninterrupted power supply in a large area [1–10]. This requires effective protection schemes for transmission lines [11]. Commonly occurred faults on transmission lines include line to ground (LG), double line (LL), double line to ground (LLG), three-phase faults without involvement of the ground (LLL), three-phase fault with involvement of ground (LLLG), and inter-circuit faults [12]. Signal processing and mathematical techniques have been deployed for detection of transmission line faults. A novel approach for detecting, classifying, and locating short-circuit faults in power transmission lines has been reported in [13]. A low cost, fast, and reliable microcontroller-based protection scheme using wavelet transform and artificial neural network has been proposed by authors in [14] and its effectiveness has been evaluated in real time. Hussain *et al.* [15], proposed a simple fault location algorithm for multi-terminal transmission lines using unsynchronized measurements. Developed data synchronization procedure is employed to identify faulted leg before fault location is calculated. A fault location algorithm is independent of fault resistance and source impedance variations. In [16], the authors presented a methodology for fault location based on the theory of state estimation in order to determine location of faults more accurately by considering realistic systematic errors that may be present in measurements of voltage and current. In this methodology, besides calculating most likely fault distance obtained from measurement errors, variance associated with distance found is also determined using the errors theory. Obtained results are relevant to show that a proposed estimation approach works even in adopting realistic variances. He *et al.* [17], proposed a novel technique for fault detection and classification in extremely high-voltage transmission lines using fault transients. A proposed technique, called wavelet singular entropy (WSE), incorporates advantages of wavelet transform, singular value decomposition, and Shannon entropy. WSE is capable of being immune to noise in fault transient and not being affected by transient magnitude so it can be used to extract features automatically from fault transients and express fault features intuitively and quantitatively, even in the case of high-noise and low-magnitude fault transients.

Research work proposed in this chapter is mainly focused on the detection and classification of power system faults to provide voltage-based protection schemes with the help of harmonic wavelet transform and ruled decision tree.

13.2 PROPOSED TEST SYSTEM

An overhead transmission line rated at 765 kV and a line length equal to 200 km is used for the proposed study of detection of transmission line faults, as shown in Figure 13.1. This line is connected to two large area power system networks simulated by two power source blocks of MATLAB®/Simulink. Technical information of a

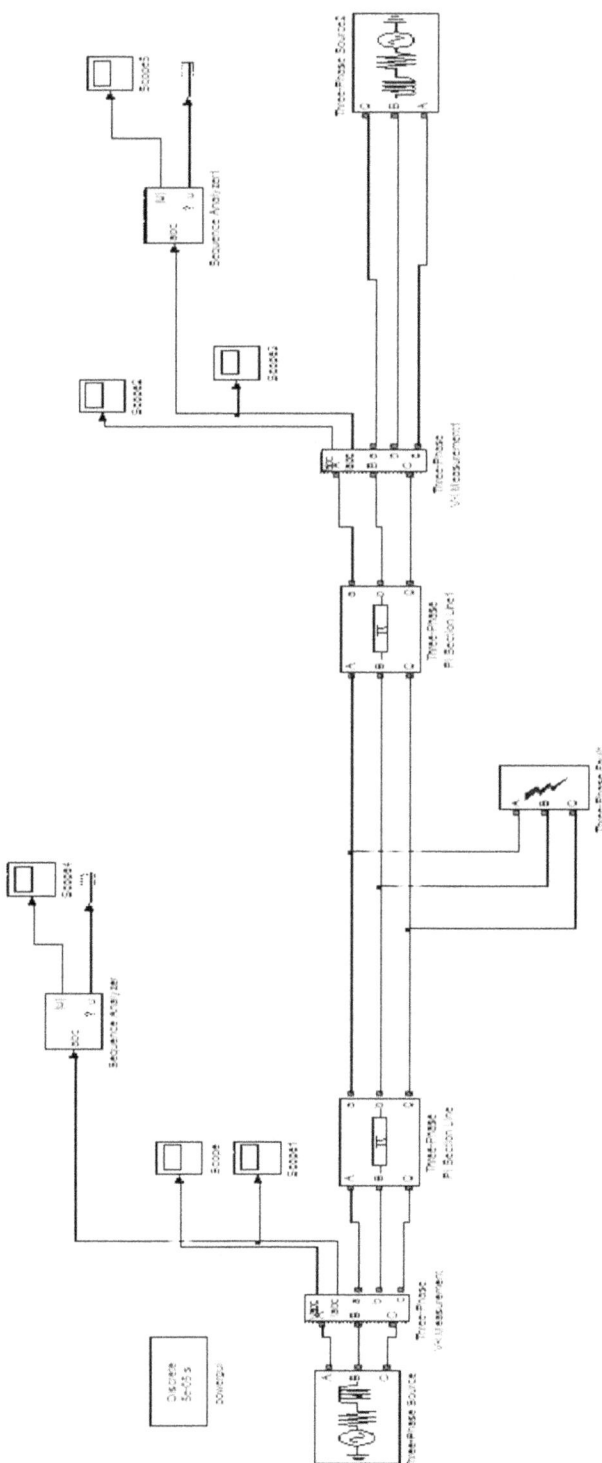

FIGURE 13.1 Proposed test system used for investigation of faults.

TABLE 13.1
Technical Data of the Proposed Test System

Element	Parameter	Value
Sending end source (Generator 1)	Voltage (line)	765 kV
	Phase angle of phase-A	20°
	Source impedance	17.177 + j45.917
Receiving end source (Generator 2)	Voltage (line)	765 kV
	Phase angle of phase-A	0°
	Source impedance	15.31 + j45.917
Transmission Line	Line length	200 km
	Positive sequence impedance	0.01273 + j0.3520 Ω/km
	Zero sequence impedance	0.3864 + 1.5556 Ω/km

proposed test system is provided in Table 13.1. Different types of power system faults have been simulated at the middle of the transmission line. Voltages and currents have been recorded at both ends of a transmission line. Voltage signals recorded at both ends of transmission lines are used to design a voltage-based protection scheme.

Proposed Algorithm

An algorithm using harmonic wavelet transforms and rule-based decision tree for detection and classification of power system faults to provide transmission line protection is illustrated in Figure 13.2. Various types of faults are created at the middle of test transmission line using a three-phase fault MATLAB simulation block. Voltage current signals are recorded at both ends of a transmission line. The voltage signal is decomposed using harmonic wavelet transform. Absolute values of an output matrix of a harmonic wavelet transform has been obtained for voltages recorded at both ends of a transmission line and these values are multiplied to obtain a proposed fault index. Peak values of this fault index are given as input to rule-based decision tree for classification purpose.

13.3 SIMULATION RESULTS AND DISCUSSION

This section presents simulation results related to analysis of power system faults using a proposed fault index based on a harmonic wavelet transform. A voltage signal recorded at both ends of a transmission line are decomposed using a harmonic wavelet

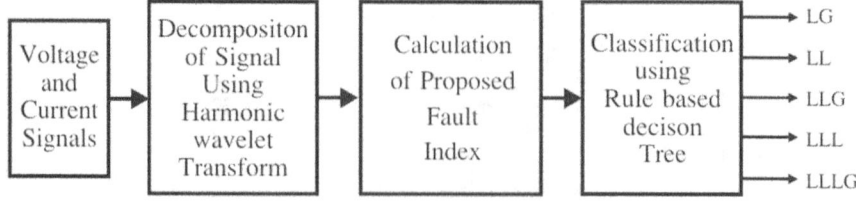

FIGURE 13.2 Proposed algorithm for detection and classification of the transmission line fault.

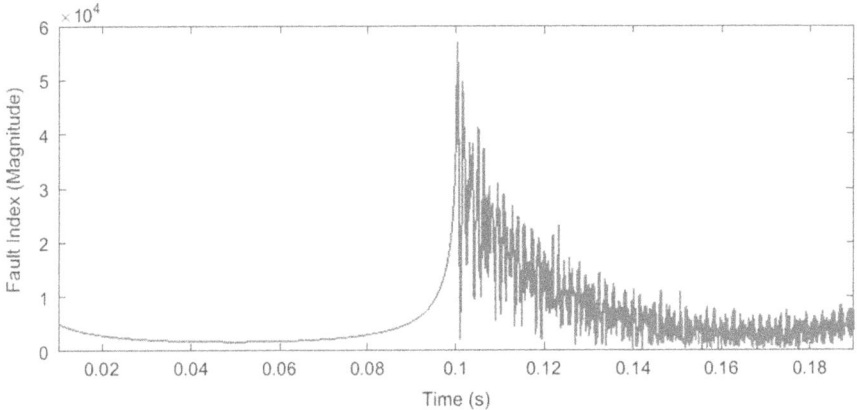

FIGURE 13.3   Voltage-based fault index for phase-A on the first end of the transmission line during the LG fault on phase-A.

transform with a sampling frequency of 1.92 kHz. The absolute value of output of a harmonic wavelet transform is designated as a fault index. Peak values of fault indexes are found to be effective in detection and discrimination of various types of transmission line faults. Detailed simulation results are presented in the following sections.

13.3.1   LINE TO GROUND FAULT

Line to ground (LG) fault is simulated on phase-A of a transmission line at the middle of a transmission line at the sixth cycle from the start of the simulation. Results are taken for 12 cycles, with 6 cycles for pre-fault conditions and 6 cycles for post-fault conditions. Voltage of phase-A is recorded at the first end of the transmission line and decomposed using a harmonic wavelet transform with a sampling frequency of 1.92 kHz. Absolute values of output of harmonic wavelet transform of voltage is calculated and designated as a fault index. A proposed fault index pertaining to voltage of phase-A on the first end of the transmission line during the LG fault is shown in Figure 13.3.

It is observed from Figure 13.3 that values of fault index are very low during a pre-fault condition. These values exceed a very high value of 6×10^4 just after the occurrence of fault indicating the LG fault on phase-A. It is also observed that values of the fault index during the post-fault condition decreases to low value in duration of 0.06 s. Low values in the pre-fault condition and high value just after fault occurrence clearly indicate the presence of the LG fault. Hence, the proposed fault index based on harmonic wavelet transform is found to be effective in detection of transmission line faults.

Line to ground (LG) fault has been simulated on phase-A of transmission line at the middle of the transmission line at the sixth cycle from the start of the simulation. Results are taken for 12 cycles, including 6 cycles of pre-fault conditions and 6 cycles for post-fault conditions. Voltage of phase-A is recorded at a second end of the transmission line and decomposed using a harmonic wavelet transform with a sampling frequency of 1.92 kHz. Absolute values of output of a harmonic wavelet transform of voltage is calculated and designated as a fault index. A proposed fault index pertaining to the voltage of phase-A on the second end of a transmission line during the LG fault is shown in Figure 13.4.

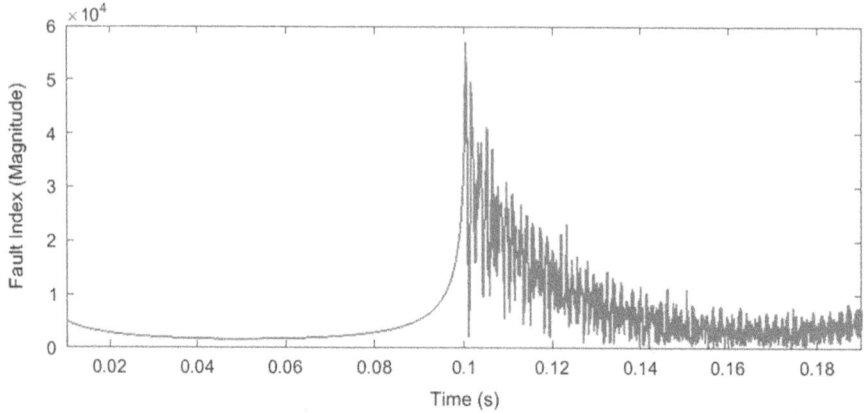

FIGURE 13.4   Voltage-based fault index for phase-A on the second end of a transmission line during the LG fault on phase-A.

It is observed from Figure 13.4 that values of a fault index are very low during the pre-fault condition. These values exceed very high values of 6×10^4 just after the occurrence of fault indicating the LG fault on phase-A. It is also observed that values of a fault index during post-fault conditions decreases to low values in duration of 0.06 s. Low values in a prefault condition and high values just after fault occurrence clearly indicate the presence of the LG fault. Hence, the proposed fault index based on a harmonic wavelet transform is found to be effective in the detection of transmission line faults based on voltages measured at both ends of a transmission line.

The line to ground (LG) fault has been simulated on phase-A of a transmission line at the middle of the transmission line at the sixth cycle from the start of simulation. Results are taken for 12 cycles, including 6 cycles of pre-fault conditions and 6 cycles for post-fault conditions. The voltage of phase-A is recorded at both ends of the transmission line and decomposed using the harmonic wavelet transform with a sampling frequency of 1.92 kHz. Absolute values of output of a harmonic wavelet transform of voltages at both ends are calculated. Multiplication of these values is designated as a fault index. The proposed fault index pertaining to the voltage of phase-A recorded on both ends of a transmission line during the LG fault is shown in Figure 13.5.

It is observed from Figure 13.5 that values of the fault index are zero during the pre-fault condition. These values exceed very high values of 3.4×10^9 just after the occurrence of a fault indicating the LG fault on phase-A. It is also observed that values of a fault index during post-fault conditions decreases to low values in a duration of 0.04 s. Low values in a pre-fault condition and high values just after fault occurrence clearly indicate the presence of a LG fault. Hence, the proposed fault index using a harmonic wavelet transform-based decomposition of voltage on both ends is found to be effective in the detection of transmission line faults.

13.3.2   DOUBLE LINE FAULT

The double line (LL) fault has been simulated by short-circuiting phases-A and B at the middle of the transmission line at the sixth cycle from the start of the simulation. Results are taken for 12 cycles, including 6 cycles of pre-fault conditions and 6

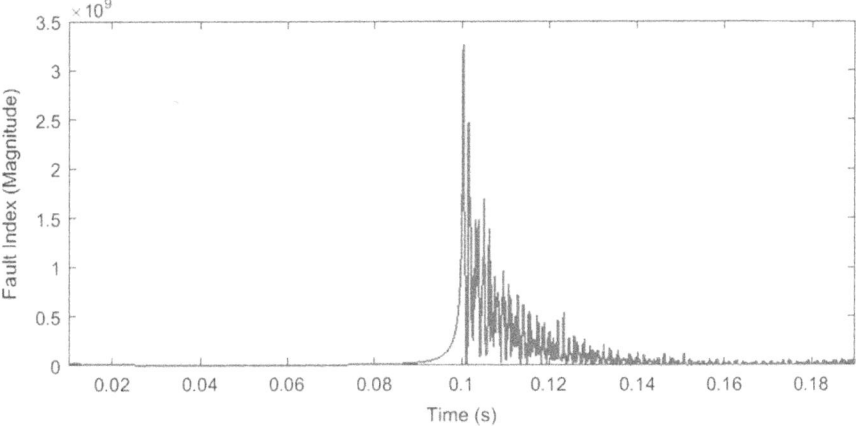

FIGURE 13.5 Fault index for phase-A-based voltages measured at both ends of transmission line during LG fault on phase-A.

cycles for post-fault condition. Voltage of phase-A is recorded at the first end of a transmission line and decomposed using a harmonic wavelet transform with a sampling frequency of 1.92 kHz. Absolute values of the output of a harmonic wavelet transform of voltage is calculated and designated as a fault index. The proposed fault index pertaining to voltage of phase-A on the first end of the transmission line during the LG fault is shown in Figure 13.6.

It is observed from Figure 13.6 that values of a fault index are very low during a pre-fault condition. These values exceed very high values of 2.5×10^5 just after the occurrence of the fault, indicating the LL fault on phases-A and B. It is also observed that values of a fault index during post-fault conditions decrease to low values in a duration of 0.08 s. Low values in a pre-fault condition and high value just after fault occurrence clearly indicate the presence of a LL fault. Hence, the

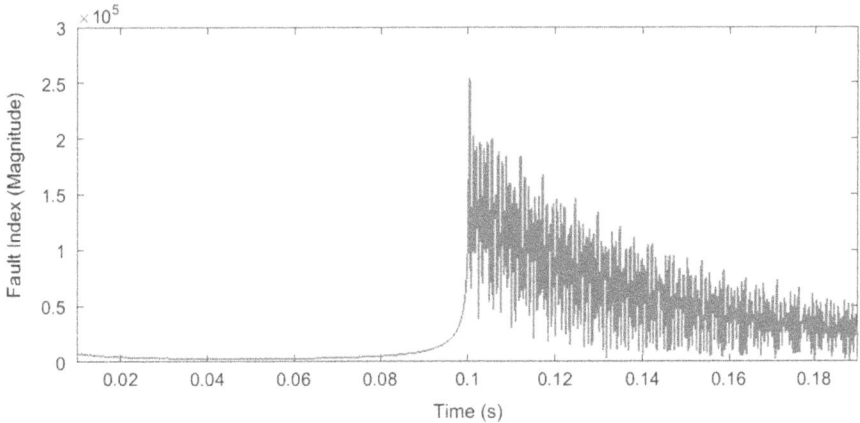

FIGURE 13.6 Voltage-based fault index for phase-A on the first end of transmission line during the LL fault on phases-A and B.

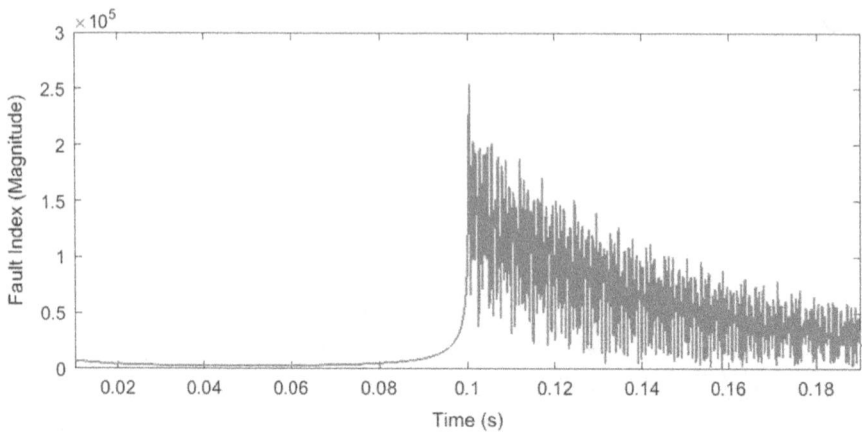

FIGURE 13.7 Voltage-based fault index for phase-A on the second end of the transmission line during the LL fault on phases-A and B.

proposed fault index is based on a harmonic wavelet transform is found to be effective in the detection of a transmission line LL fault.

The double line (LL) fault has been simulated by short-circuiting phases-A and B at the middle of the transmission line at the sixth cycle from the start of the simulation. Results are taken for 12 cycles, including 6 cycles of pre-fault conditions and 6 cycles for post-fault conditions. The voltage of phase-A is recorded at a second end of transmission line and decomposed using a harmonic wavelet transform with a sampling frequency of 1.92 kHz. Absolute values of output of a harmonic wavelet transform of voltage is calculated and designated as a fault index. The proposed fault index pertaining to voltage of phase-A on the second end of the transmission line during the LL fault is shown in Figure 13.7.

It is observed from Figure 13.7 that values of a fault index are very low during the pre-fault condition. These values exceed very high values of 2.5×10^5 just after the occurrence of the fault, indicating the LL fault on phases-A and B. It is also observed that values of a fault index during post-fault conditions decreases to low values in duration of 0.08 s. Low values in a pre-fault condition and high value just after fault occurrence clearly indicate the presence of a LL fault. Hence, the proposed fault index based on a harmonic wavelet transform is found to be effective in the detection of a transmission line LL fault.

The double line (LL) fault has been simulated by short-circuiting phases-A and B of a transmission line at the middle of the transmission line at the sixth cycle from the start of the simulation. Results are taken for 12 cycles, including 6 cycles of pre-fault conditions and 6 cycles for post-fault conditions. The voltage of phase-A is recorded at both ends of a transmission line and decomposed using a harmonic wavelet transform with a sampling frequency of 1.92 kHz. The absolute values of output of a harmonic wavelet transform of voltages at both ends are calculated. Multiplication of these values is designated as the fault index. The proposed fault index pertaining to the voltage of phase-A recorded on both ends of a transmission line during the LL fault is shown in Figure 13.8.

It is observed from Figure 13.8 that values of a fault index are zero during a pre-fault condition. These values exceed a very high value of 6.5×10^{10} just after the

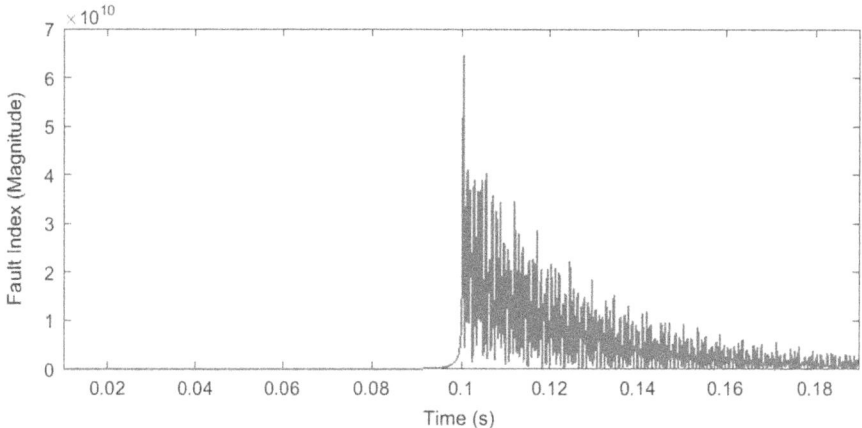

FIGURE 13.8 Fault index for phase-A-based voltages measured at both ends of a transmission line during the LL fault on phases-A and B.

occurrence of a fault indicating the LL fault on phases-A and B. It is also observed that values of a fault index during post-conditions decreases to low values in a duration of 0.06 s. Zero values in a pre-fault condition and high values just after fault occurrence clearly indicate the presence of a LL fault. Hence, the proposed fault index using a harmonic wavelet transform–based decomposition of voltage on both ends is found to be effective in the detection of transmission line LL faults.

13.3.3 DOUBLE LINE TO GROUND FAULT

The double line to ground (LLG) fault has been simulated by simultaneously grounding phases-A and B at the middle of the transmission line at the sixth cycle from the start of the simulation. Results are taken for 12 cycles, including 6 cycles of pre-fault conditions and 6 cycles for post-fault conditions. Voltage of phase-A is recorded at the first end of the transmission line and decomposed using a harmonic wavelet transform with a sampling frequency of 1.92 kHz. Absolute values of output of a harmonic wavelet transform of voltage is calculated and designated as a fault index. The proposed fault index pertaining to voltage of phase-A on the first end of the transmission line during the LLG fault is shown in Figure 13.9.

It is observed from Figure 13.9 that values of a fault index are very low during a pre-fault condition. These values exceed very high values of 2.5×10^5 just after the occurrence of a fault indicating the LL fault on phases-A and B. It is also observed that values of a fault index during a post-fault condition decreases to low values in a duration of 0.08 s. Low values in a pre-fault condition and high value just after a fault occurrence clearly indicate the presence of LLG fault. Hence, the proposed fault index based on a harmonic wavelet transform is found to be effective in the detection of the transmission line LLG fault.

Double line to ground (LLG) fault has been simulated by simultaneously grounding phases-A and B at the middle of the transmission line at the sixth cycle from the start of the simulation. Results are taken for 12 cycles, including 6 cycles of pre-fault conditions and 6 cycles for post-fault conditions. The voltage of phase-A is

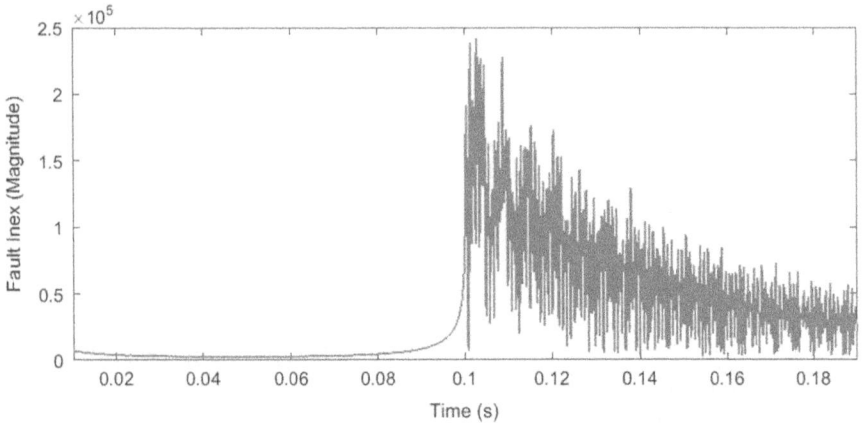

FIGURE 13.9 Voltage-based fault index for phase-A on the first end of the transmission line during the LLG fault on phases-A and B.

recorded at the second end of the transmission line and decomposed using a harmonic wavelet transform with a sampling frequency of 1.92 kHz. Absolute values of output of a harmonic wavelet transform of voltage are calculated and designated as a fault index. The proposed fault index pertaining to the voltage of phase-A on the second end of the transmission line during the LLG fault is shown in Figure 13.10.

It is observed from Figure 13.10 that values of a fault index are very low during the pre-fault condition. These values exceed very high values of 2.5×10^5 just after the occurrence of a fault, indicating the LL fault on phases-A and B. It is also observed that values of a fault index during post-fault conditions decrease to low values in the duration of 0.08 s. Low values in the pre-fault condition and high value just after the fault occurrence clearly indicate the presence of the LLG fault. Hence, the proposed fault index based on a harmonic wavelet transform is found to be effective in the detection of the transmission line LLG fault.

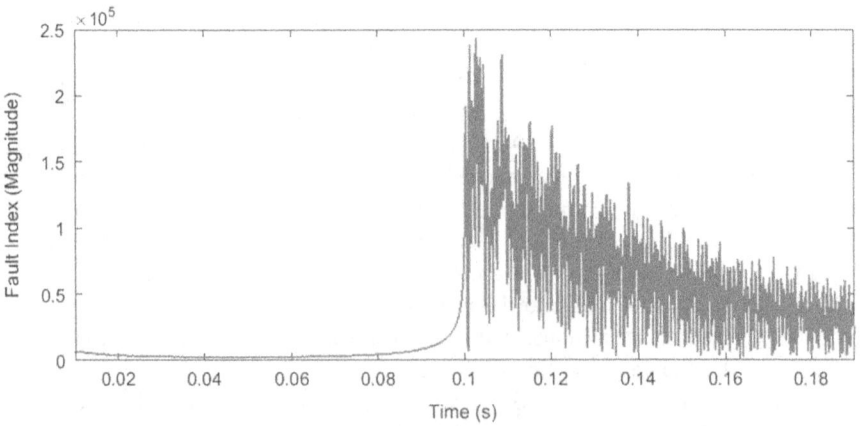

FIGURE 13.10 Voltage-based fault index for phase-A on the second end of the transmission line during the LLG fault on phases-A and B.

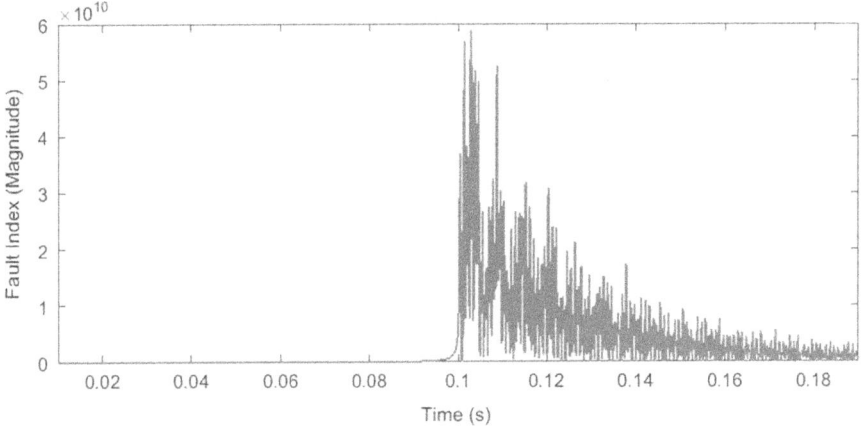

FIGURE 13.11 Fault index for phase-A-based voltages measured at both ends of the transmission line during the LLG fault on phases-A and B.

Double line to ground (LLG) fault has been simulated by short-circuiting phases-A and B of the transmission line at the middle of the transmission line at the sixth cycle from the start of the simulation. Results are taken for 12 cycles, including 6 cycles of pre-fault conditions and 6 cycles for post-fault conditions. The voltage of phase-A is recorded at both ends of the transmission line and decomposed using a harmonic wavelet transform with a sampling frequency of 1.92 kHz. Absolute values of output of a harmonic wavelet transform of voltages at both ends are calculated. Multiplication of these values is designated as a fault index. The proposed fault index pertaining to voltage of phase-A recorded on both ends of the transmission line during the LLG fault is shown in Figure 13.11.

It is observed from Figure 13.11 that values of a fault index are zero during a pre-fault condition. These values exceed a very high value of 6×10^{10} just after the occurrence of a fault, indicating the LLG fault on phases-A and B. It is also observed that values of a fault index during post-conditions decreases to low values in a duration of 0.06 s. Zero values in a pre-fault condition and high values just after a fault occurrence clearly indicate the presence of the LLG fault. Hence, the proposed fault index using a harmonic wavelet transform–based decomposition of voltage on both ends is found to be effective in the detection of the transmission line LLG faults.

13.3.4 THREE-PHASE FAULT WITH THE INVOLVEMENT OF GROUND

The three-phase fault with the involvement of fault (LLLG) is simultaneously grounding all three phases at the middle of the transmission line at the sixth cycle from the start of the simulation. Results are taken for 12 cycles, including 6 cycles of pre-fault conditions and 6 cycles for post-fault conditions. Since three phases are symmetrical in nature, behaviour of all three phases will be similar. However, results related to all three phases have been presented in this section.

The voltage of phase-A is recorded at the first end of the transmission line and decomposed using a harmonic wavelet transform with a sampling frequency of 1.92 kHz.

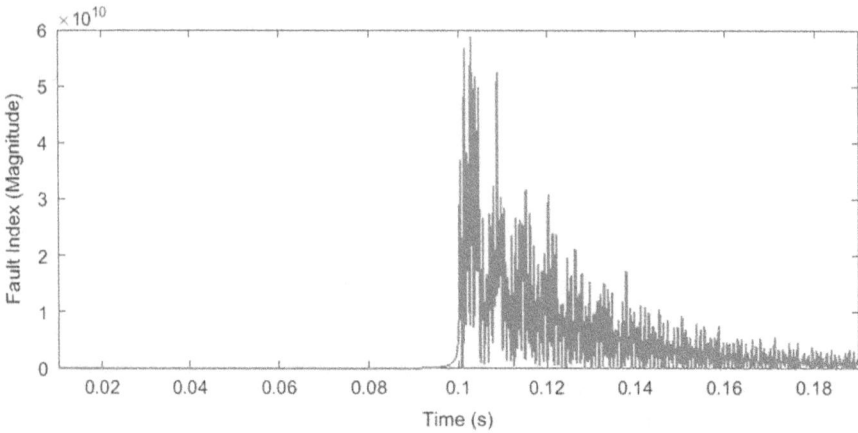

FIGURE 13.12 Voltage-based fault index for phase-A on the first end of the transmission line during the LLLG fault.

The absolute values of output of a harmonic wavelet transform of voltage is calculated and designated as a fault index. The proposed fault index pertaining to voltage of phase-A on the first end of the transmission line during the LLLG fault is shown in Figure 13.12.

It is observed from Figure 13.12 that values of a fault index are very low during a pre-fault condition. These values exceed very high values of 9.5×10^4 just after the occurrence of the fault indicating LLLG fault. It is also observed that values of a fault index during post-fault conditions decreases to low values in duration of 0.08 s. Low values in the pre-fault condition and high value just after a fault occurrence clearly indicate the presence of LLLG fault. Hence, the proposed fault index based on a harmonic wavelet transform is found to be effective in the detection of transmission line LLLG fault.

The voltage of phase-A is recorded at the second end of the transmission line and decomposed using a harmonic wavelet transform with a sampling frequency of 1.92 kHz. Absolute values of output of a harmonic wavelet transform of voltage is calculated and designated as a fault index. The proposed fault index pertaining to voltage of phase-A on the second end of transmission line during the LLLG fault is shown in Figure 13.13.

It is observed from Figure 13.13 that values of a fault index are very low during a pre-fault condition. These values exceed very high values of 9.5×10^4 just after the occurrence of fault indicating LLLG fault. It is also observed that values of a fault index during a post-fault condition decreases to low values in duration of 0.08 s. Low values in a pre-fault condition and high value just after fault occurrence clearly indicate presence of LLLG fault. Hence, the proposed fault index based on a harmonic wavelet transform is found to be effective in the detection of the transmission line LLLG fault.

Three phase fault with the involvement of ground (LLLG) has been simulated by short circuiting all three phases at the middle of the transmission line at the sixth cycle from the start of the simulation. Results are taken for 12 cycles, including 6 cycles of pre-fault conditions and 6 cycles for post-fault conditions. Voltage of phase-A is recorded at both ends of the transmission line and decomposed using a harmonic wavelet transform with a sampling frequency of 1.92 kHz. Absolute

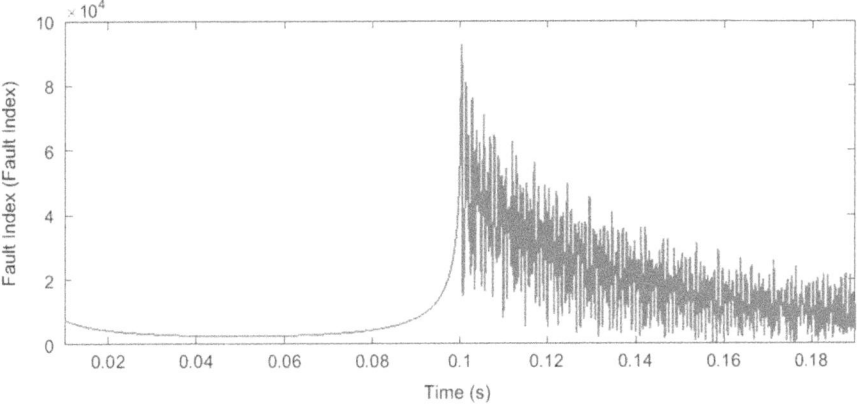

FIGURE 13.13 Voltage-based fault index for phase-A on the second end of the transmission line during the LLLG fault.

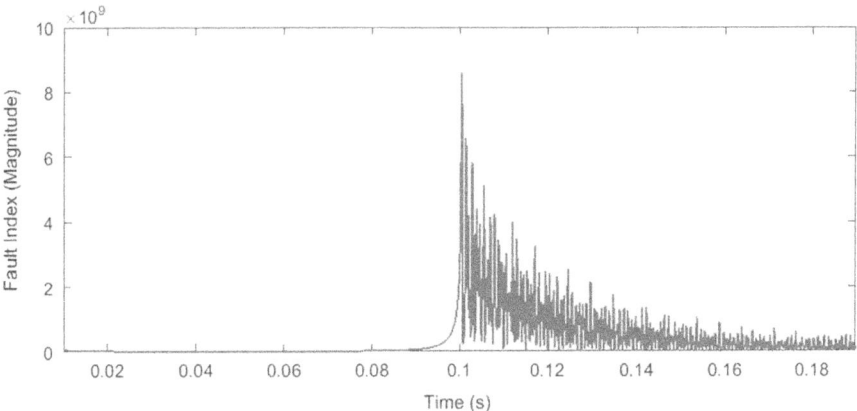

FIGURE 13.14 Fault index for phase-A-based voltages measured at both ends of the transmission line during LLLG fault.

values of output of a harmonic wavelet transform of voltages at both ends are calculated. Multiplication of these values is designated as a fault index. The proposed fault index pertaining to the voltage of phase-A recorded on both ends of the transmission line during LLLG fault is shown in Figure 13.14.

It is observed from Figure 13.14 that values of a fault index are zero during a pre-fault condition. These values exceed very high values of 9.5×10^9 just after the occurrence of fault indicating LLLG fault. It is also observed that values of a fault index during a post-fault condition decreases to low values in a duration of 0.06 s. Zero values in a pre-fault condition and high values just after a fault occurrence clearly indicate the presence of LLLG fault. Hence, the proposed fault index using a harmonic wavelet transform–based decomposition of voltage on both ends is found to be effective in the detection of the transmission line LLLG faults.

TABLE 13.2

Peak Value of the Voltage-Based Fault Indexes

Type of Faults →	LG	LL	LLG	LLL	LLLG
Fault index based on voltage at the first end of transmission line					
Phase-A	6×10^4	2.5×10^5	2.5×10^5	9.5×10^4	9.5×10^4
Phase-B	2.3×10^4	2.5×10^5	3.4×10^5	4.5×10^5	4.5×10^5
Phase-C	2.3×10^4	200	6.5×10^4	3.8×10^5	4.5×10^5
Fault index based on voltage at the second end of transmission line					
Phase-A	6×10^4	2.5×10^5	2.5×10^5	9.5×10^4	9.5×10^4
Phase-B	2.3×10^4	2.5×10^5	3.4×10^5	4.2×10^5	4.2×10^5
Phase-C	2.3×10^4	200	6.5×10^4	3.8×10^5	3.8×10^5
Fault index based on voltages at both ends of the transmission line					
Phase-A	3.4×10^9	6.5×10^{10}	6×10^{10}	9×10^9	9×10^9
Phase-B	9.5×10^8	6.5×10^{10}	10×10^{10}	2×10^{11}	2×10^{11}
Phase-C	5.5×10^8	2×10^4	4×10^9	14×10^{10}	14×10^{10}

13.3.5 CLASSIFICATION OF TRANSMISSION LINE FAULTS

Peak values of the proposed fault index observed during various types of power system faults on the transmission line are provided in Table 13.2. These peak values help to detect and discriminate the various types of faults. These peak values are given as input to the rule-based decision tree for detection and discrimination of different types of faults based on decision rules. Classification using a rule-based decision tree along with different decision rules used for classification purposes is shown in Figure 13.15.

It is observed that the fault index calculated using the voltage at both ends of the transmission line is found to be more effective in detection and classification of transmission line faults. The fault index corresponding to phase-A is found to be more effective in the classification of faults into groups of similar nature. Hovever, the fault index corresponding to phase-C is found to be more effective in the discrimnation of LL and LLG faults. There is no need to classify faults LLL and LLLG from each other because the behavior of these faults are the same and these are considered identical to each other in power system studies. Hence, the protection scheme using a harmonic wavelet transform–based decomposition of voltage recorded at both ends of a transmission line rule-based decision tree is found to be effecctive in detection and discrimination of transmission line faults.

13.4 CONCLUSION

This research work presents a method for detection and classification of transmission line faults. The proposed method is based on a harmonic wavelet transform and rule-based decision tree. The voltage signal recorded at both ends of the transmission line is decomposed using a harmonic wavelet transform and proposed fault index is calculated. Peak values of the proposed fault index are given as input

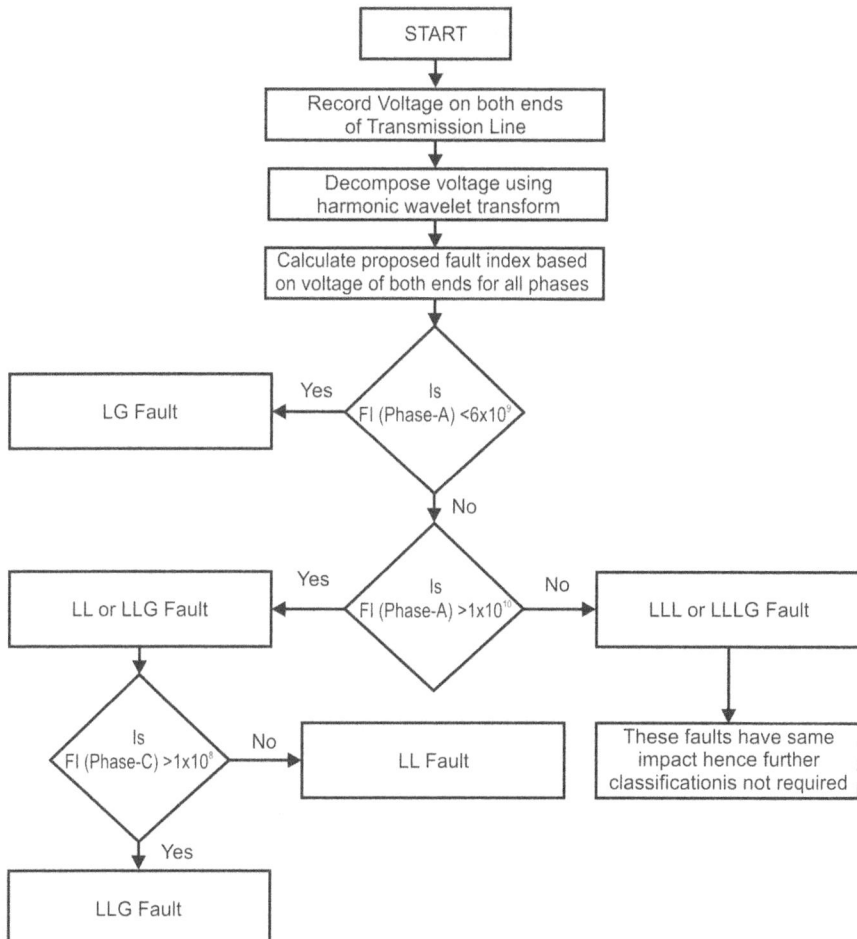

FIGURE 13.15 Rule decision tree–based flow chart for the classification of transmission line faults using voltage-based fault index.

to a rule-based decision tree for classification purposes. It is concluded that the fault index calculated from the voltage recorded at both ends of the transmission line is found to be more effective in detection and classification of transmission line faults.

REFERENCES

[1] B. Rathore, O. P. Mahela, B. Khan, H. Haes Alhelou and P. Siano, "Wavelet-Alienation-Neural Based Protection Scheme for STATCOM Compensated Transmission Line," *IEEE Transactions on Industrial Informatics*, Vol. 17, No. 4, pp. 2557–2565, April 2021, 10.1109/TII.2020.3001063
[2] K. Gangwar, O. P. Mahela, B. Rathore, B. Khan, H. Haes Alhelou and P. Siano, "A Novel K-Means Clustering and Weighted K-NN Regression Based Fast Transmission

Line Protection," *IEEE Transactions on Industrial Informatics*, Vol. 17, No. 9, pp. 6034–6043, September 2021, 10.1109/TII.2020.3037869

[3] B. Rathore, O. P. Mahela, B. Khan and S. Padmanaban, "Protection Scheme using Wavelet-Alienation-Neural Technique for UPFC Compensated Transmission Line," *IEEE Access*, Vol. 9, pp. 13737–13753, 2021, 10.1109/ACCESS.2021.3052315.

[4] R. Kaushik et al., "Recognition of Islanding and Operational Events in Power System With Renewable Energy Penetration Using a Stockwell Transform-Based Method," in *IEEE Systems Journal*, 10.1109/JSYST.2020.3020919

[5] F. M. Gebru, B. Khan, H. H. Alhelou, "Analyzing Low Voltage Ride Through Capability of Doubly Fed Induction Generator Based Wind Turbine," *Computers & Electrical Engineering*, Vol. 86, pp. 106727, 2020.

[6] S. Ram Ola, A. Saraswat, S. K. Goyal, V. Sharma, B. Khan, O. P. Mahela, H. Haes Alhelou, P. Siano, "Alienation Coefficient and Wigner Distribution Function Based Protection Scheme for Hybrid Power System Network with Renewable Energy Penetration," *Energies*, Vol. 13, pp. 1120, 2020.

[7] S. Ram Ola, A. Saraswat, S. K. Goyal, S. K. Jhajharia, B. Khan, O. P. Mahela, H. Haes Alhelou, P. Siano, "A Protection Scheme for a Power System with Solar Energy Penetration," *Applied Sciences*, Vol. 10, pp. 1516, 2020.

[8] G. S. Yogee, O. P. Mahela, K. D. Kansal, B. Khan, R. Mahla, H. Haes Alhelou, and P. Siano, "An Algorithm for Recognition of Fault Conditions in the Utility Grid with Renewable Energy Penetration," *Energies*, Vol. 13, pp. 2383, 2020.

[9] O. P. Mahela, Y. Sharma, S. Ali, B. Khan and S. Padmanaban, "Estimation of Islanding Events in Utility Distribution Grid With Renewable Energy Using Current Variations and Stockwell Transform," *IEEE Access*, Vol. 9, pp. 69798–69813, 2021.

[10] N. K. Swarnkar, O. P. Mahela, B. Khan and M. Lalwani, "Identification of Islanding Events in Utility Grid with Renewable Energy Penetration Using Current Based Passive Method," *IEEE Access*, Vol. 9, pp. 93781–93794, 2021, 10.1109/ACCESS.2021.3092971.

[11] S. Devi, N. K. Swarnkar, S. R. Ola and O. P. Mahela, "Detection of Transmission Line Faults Using Discrete Wavelet Transform," 2016 Conference on Advances in Signal Processing (CASP), Pune, India, 9–11 June 2016.

[12] A. Swetapadma, and A. Yadav, "Directional Relaying Using Support Vector Machine for Double Circuit Transmission Lines Including Cross-country and Inter-circuit Faults," *International Journal of Electrical Power and Energy Systems*, Vol. 81, pp. 254–264, 2016.

[13] H. Fathabadi, "Novel Filter Based ANN Approach for Short-circuit Faults Detection, Classification and Location in Power Transmission Lines," *International Journal of Electrical Power and Energy Systems*, Vol. 74, pp. 374–383, 2016.

[14] E. Koley, R. Kumar, S. Ghosh, "Low Cost Microcontroller Based Fault Detector, Classifier, Zone Identifier and Locator for Transmission Lines Using Wavelet Transform and Artificial Neural Network: A Hardware Co-simulation Approach," *International Journal of Electrical Power and Energy Systems*, Vol. 81, pp. 346–360, 2016.

[15] S. Hussain, A. H. Osman, "Fault Location Scheme for Multi-terminal Transmission Lines Using Unsynchronized Measurements," *International Journal of Electrical Power and Energy Systems*, Vol. 78, pp. 277–284, 2016.

[16] M. C. S. da Cruz, M. A. D. de Almeida, M. F. de Medeiros Júnior, "A State Estimation Approach for Fault Location in Transmission Lines Considering Data Acquisition Errors and Non-synchronized Records," *International Journal of Electrical Power and Energy Systems*, Vol. 78, pp. 663–671, 2016.

[17] Z. He, L. Fu, S. Lin, and Z. Bo, "Fault Detection and Classification in EHV Transmission Line Based on Wavelet Singular Entropy," *IEEE Transactions on Power Delivery*, Vol. 25, No. 4, pp. 2156–2163, 2010.

14 A Voltage-Based Algorithm Using the Gabor Wigner Distribution and Rule-Based Decision Tree for the Detection of Transmission Line Faults

Rajendra Mahla
Department of Electrical Engineering, NIT Kurukshetra, India

Ravi Kishor Ranjan
Production & Quantitative Methods Area-IIM Ahmedabad, India

Baseem Khan
Department of Electrical Engineering, Hawassa University, Ethiopia

Om Prakash Mahela
Power System Study Division, Rajasthan Rajya Vidyut Prasarn Nigam Ltd., Jaipur, India

CONTENTS

DOI: 10.1201/b22884-14

14.1 INTRODUCTION

In the modern industrial era, the faults on transmission lines of a power system network may deteriorate system performances, which may lead to catastrophes. These faults are due to unexpected external disturbances. In order to ensure the safety and reliability of a power system network, fault diagnosis (FDD) problems are continuously receiving research attention [1]. Recently, researchers focused on signal processing techniques such as discrete wavelet transform (DWT), discrete Fourier transform (DFT), short time Fourier transform (STFT), and Hilbert Huang transform for the detection of faults on a transmission line to provide effective protection schemes [2]. Artificial intelligent techniques such as neural network, adaptive neural fuzzy inference system (ANFIS), rule-based decision tree, genetic algorithm, support vector machine, and fuzzy logic have been used for the classification of various types of faults on transmission lines [3]. In [4], the authors proposed an algorithm using discrete wavelet transform (DWT) for the detection of faults on a transmission line. In [5], the authors presented an algorithm using discrete wavelet transformation and wavelet entropy calculations to detect and classify a fault. This is also effective in identifying a fault position in a transmission line with respect to a flexible AC transmission system (FACTS) device placed at the midpoint of a transmission line. In [6], the authors proposed an algorithm using a DWT capable of detecting various types of faults on a power system network in the presence of solar photovoltaic (PV) generation. Different types of faults have been analyzed in detail using key features derived from the current signal. Various techniques are reported for the identification of faults on the power system network. In [7], the authors proposed a wavelet-alienation-neural-based technique for the identification of faults on the STATCOM-compensated transmission line. An approach using the Wigner distribution function (WDF) and alienation coefficient is reported for the protection of a transmission line [8–10]. Fault detection techniques using the hybrid combination of various signal processing methods have been reported to detect faults in the presence of renewable energy [11–15]. Techniques for the detection and classification of faults on transmission lines are reported in [16–29].

14.2 PROPOSED TEST SYSTEM

The test system of transmission lines proposed for this study is described in Figure 14.1. The total line length is 200 km. The line is divided into four sections, each having a length of 40 km (L1), 60 km (L2), 60 km (L3), and 40 km (L4), respectively. The fault is created at 20%, 50%, and 80% of the line length, respectively, at the points F1, F2, and F3. Two generators, G1 and G2, are connected

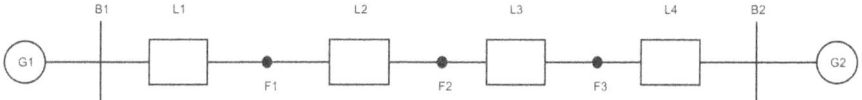

FIGURE 14.1   Proposed test system of the transmission line.

TABLE 14.1
Technical Parameters of Test System

Particulars of Parameters	Values of Parameters
Voltage rating of system	765 kV
Frequency	60 Hz
Source resistance (ohm)	17.77 Ω
Source inductance	0.1218 H
Positive sequence resistance of transmission line	0.01273 Ω/km
Zero sequence resistance of transmission line	0.3864 Ω/km
Positive sequence inductance of transmission line	0.9337e-3 H/km
Zero sequence inductance of transmission line	4.1264e-3 H/km
Positive sequence capacitance of transmission line	12.74e-9 F/km
Zero sequence capacitance of transmission line	7.751e-9 F/km

on the ends of the line with a voltage rating of 765 kV. These generators have a phase difference of 20° to ensure power flow from one end of the transmission line to the other. Voltage signals are captured on buses B1 and B2. Bus B1 is located on the sending end of the transmission line and bus B2 is located on the receiving end of line. The measurement of the voltage is carried out using the three-phase VI measurement block of Simulink.

Technical parameters of the test system, such as system voltage, system frequency, positive and zero sequence resistances of line, positive and zero sequence inductances of line, positive and zero sequence capacitances of line, source resistance, and source inductance, are provided in Table 14.1.

14.3   PROPOSED METHODOLOGY

The following steps are used for the detection and classification of faults on a transmission line using a voltage-based fault index:

- Simulate a type of fault on the proposed test system of a transmission line. Record the voltage on both ends of the line.
- Decompose voltage signals recorded on both ends of the transmission line using the Gabor Wigner distribution.
- Obtain a fault index based on voltage values. First, calculate the maximum values of each column of the Wigner output matrix. Then, calculate the

variance values from the Wigner output matrix. Calculate the voltage-based fault index by multiplying the maximum values and variance obtained from the Wigner output matrix.

- The fault index is given as input to the RBDT for the classification of transmission line faults.
- Repeat the same procedure for all types of faults.

14.4   SIMULATION RESULTS WITH THEIR DISCUSSION

The results related to the detection and classification of transmission line faults using a voltage-based fault index obtained from the Gabor Wigner distribution and RBDT are presented in this section. Transmission line faults like LG, LL, LLG, and LLLG are analyzed. A detailed discussion of simulation results is provided in the following subsections.

14.4.1   LINE TO GROUND FAULT

The LG fault is simulated at the sixth cycle on the middle of the transmission line by grounding phase-A. These results are analyzed for 12 cycles. The voltage signal recorded on phase-A of the sending end of the line is decomposed using the Gabor Wigner distribution and output matrix is obtained. The maximum values (AI) of each column of the matrix are obtained. The variance (VAR) of the output matrix is also calculated. The proposed fault index is obtained by multiplying the above-mentioned maximum values and variance. This fault index based on the voltage for the event of the LG fault is provided in Figure 14.2.

Figure 14.2 indicates that the values of the proposed voltage-based fault index are zero before and after a fault occurrence. It is also observed that during the instance of a fault, the values of the fault index have a high peak value of 7.8097e + 42. The peak values of a voltage-based fault index recorded on phase-A at sending with the LG fault on 20%, 50%, and 80% of the line lengths are provided in Table 14.2. High peak values of a fault index at the instance of fault occurrence and zero during pre-fault and post-fault conditions clearly detect the occurrence of the LG fault. Hence, a proposed voltage-based fault index is effective in the detection of a LG fault.

The voltage signal recorded on the receiving end (second end) of the line is decomposed using the Gabor Wigner distribution and an output matrix is obtained. The maximum values (AI) of each column of the matrix are obtained. The variance (VAR) of the output matrix is also calculated. The proposed fault index is obtained by multiplying the above-mentioned maximum values and variance. This voltage-based fault index is provided in Figure 14.3.

Figure 14.3 indicates that the fault index is zero before and after the fault occurrence. It is also observed that during the instance of the fault, the values of the fault index are high, with a peak value of 9.1882e+42. The peaks of the fault index recorded on phase-A at the receiving end with the LG fault on 20%, 50%, and 80% line lengths are provided in Table 14.3. The peak value of the fault index is high at the instance of the fault occurrence and zero during pre-fault and

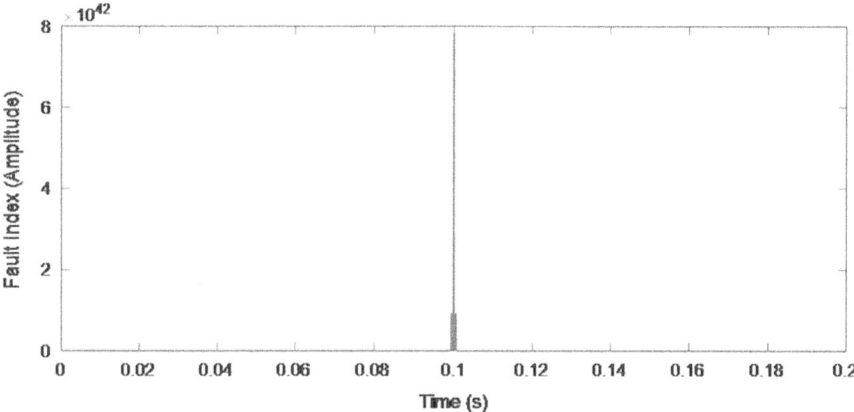

FIGURE 14.2 Voltage-based fault index on the first end of the transmission line during the LG fault at the middle of the line.

TABLE 14.2

Peak Values of Voltage-Based Fault Index on the First End of the Transmission Line

Type of Fault	Peak Value of Proposed Fault Index		
	20% Line Length	*50% Line Length*	*80% Line Length*
LG	8.4056e + 42	7.8097e + 42	7.8550e + 42
LL	7.9432e + 42	7.9254e + 42	7.8089e + 42
LLG	8.8948e + 42	8.1918e + 42	7.8862e + 42
LLLG	9.0983e + 42	8.2170e + 42	7.8351e + 42

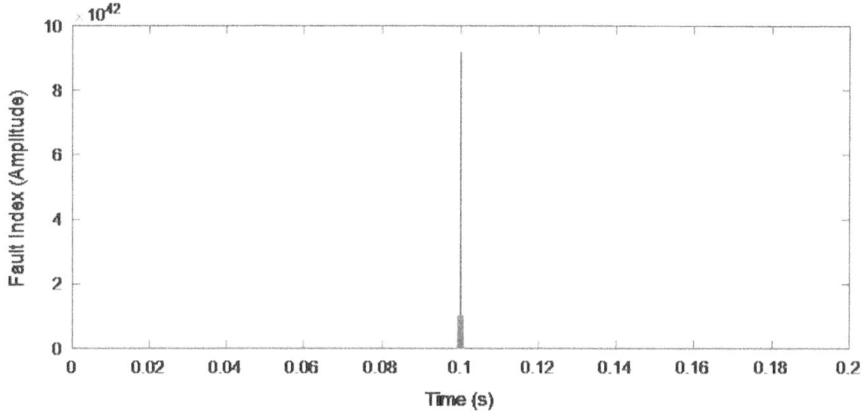

FIGURE 14.3 Voltage-based fault index on the second end of the transmission line during a LG fault at the middle of the line.

TABLE 14.3

Peak Values of Voltage-Based Fault Index on the Second End of the Transmission Line

Type of Fault	Peak Value of Proposed Fault Index		
	20% Line Length	50% Line Length	80% Line Length
LG	9.2491e + 42	9.1882e + 42	9.8692e + 42
LL	9.1369e + 42	9.1200e + 42	9.1308e + 42
LLG	9.1784e + 42	9.4390e + 42	1.0300e + 43
LLLG	9.2971e + 42	9.7504e + 42	1.0751e + 43

post-fault conditions. This clearly detects the occurrence of the LG fault. Hence, the fault index based on voltage is found to be effective in the detection of a LG fault at both ends of the transmission line.

14.4.2 Double Line Fault

The LL is simulated at the sixth cycle on the middle of the transmission line on phases-A and B. These results are plotted for 12 cycles. The voltage signal recorded on phase-A of the sending end of the line is decomposed using the Gabor Wigner distribution and the output matrix is obtained. The maximum values (AI) of each column of the matrix are obtained. The variance (VAR) of the output matrix is also calculated. The proposed fault index is obtained by multiplying the above-mentioned maximum values and variance. This voltage-based fault index during the event of the LL fault is provided in Figure 14.4.

Figure 14.4 indicates that the voltage-based fault index has zero value before and after the fault occurrence. This is also observed at the instance of the LL fault; the fault index is high, with a peak value of 7.9254e + 42. The peak values of a voltage-based fault index recorded on phase-A sent with the LL fault on 20%, 50%, and 80% of line lengths are provided in Table 14.2. A high peak value of the fault index is high at the fault occurrence moment and zero during pre-fault and post-fault conditions and clearly detect the occurrence of a LL fault. Hence, the proposed voltage-based fault index is effective in the detection of a LL fault.

The voltage signal recorded on the receiving end (second end) of the transmission line is decomposed using the Gabor Wigner distribution and the output matrix is obtained. The maximum values (AI) of each column of the matrix are obtained. The variance (VAR) of the output matrix is also calculated. The proposed fault index is obtained by multiplying the above-mentioned maximum values and variance. This voltage-based fault index is provided in Figure 14.5.

Figure 14.5 indicates that the values of the fault index are zero before and after the fault incidence. It is also observed that during the instance of the fault, values of the voltage-based fault index are high, with a peak value of 9.1200e + 42. The

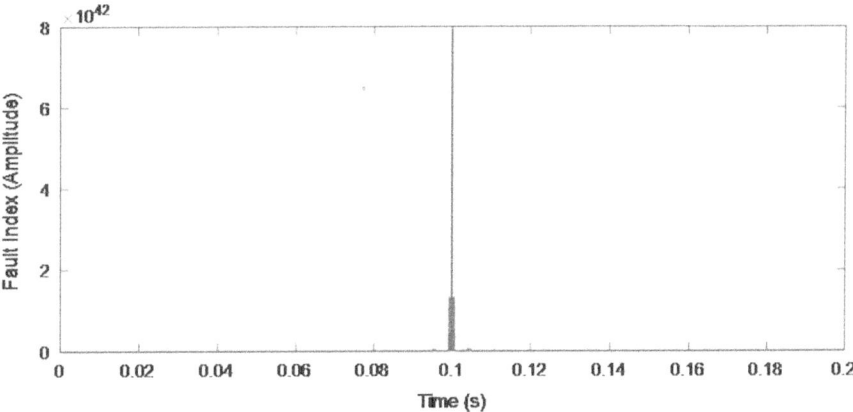

FIGURE 14.4 Voltage-based fault index on the first end of the transmission line during a LL fault at the middle of the line.

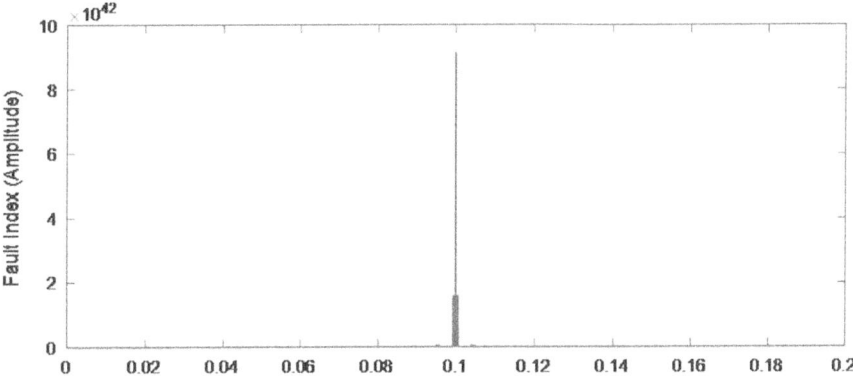

FIGURE 14.5 Voltage-based fault index on the second end of the transmission line during a LL fault at the middle of the line.

maximum values of the fault index recorded on phase-A at the receiving end with the LL fault on 20%, 50%, and 80% of line lengths are provided in Table 14.3. The peak value of the fault index is high at the fault occurrence moment and zero during pre-fault and post-fault conditions, which clearly detects the occurrence of a LL fault. Hence, the proposed fault index is effective in the detection of a LL fault at both ends of the transmission line.

14.4.3 DOUBLE LINE TO GROUND FAULT

The LLG fault is simulated at the sixth cycle on the middle of the test transmission line by grounding phases-A and B simultaneously. These results are plotted for 12 cycles. The voltage signal recorded on phase-A of the sending end of the line is

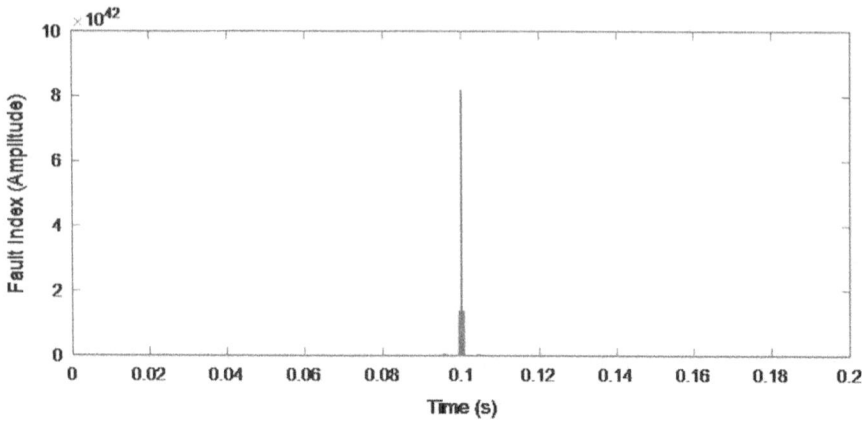

FIGURE 14.6   Voltage-based fault index on the first end of the transmission line during a LLG fault at the middle of the line.

decomposed using the Gabor Wigner distribution and the output matrix is obtained. The maximum values (AI) of each column of the matrix are obtained. The variance (VAR) of the output matrix is also calculated. The proposed fault index is obtained by multiplying the above-mentioned maximum values and variance. This fault index during the event of a LLG fault is provided in Figure 14.6.

Figure 14.6 indicates that the values of the proposed voltage-based fault index are zero before and after a fault occurrence. It is also observed that during the instance of a LLG fault, the values of the fault index are high, with a peak value of 8.1918e + 42. The maximum values of a voltage-based fault index are recorded on phase-A at the sending end with a LLG fault on 20%, 50%, and 80% of line lengths are provided in Table 14.1. The peak value of the fault index is high at the fault occurrence moment and zero during pre-fault and post-fault conditions, which clearly detect the occurrence of a LLG fault. Hence, the proposed voltage-based fault index is effective in the detection of a LLG fault.

The voltage signal recorded on the receiving end (second end) of the line is decomposed using the Gabor Wigner distribution and the output matrix is obtained. The maximum values (AI) of each column of the matrix are obtained. The variance (VAR) of the output matrix is also calculated. The proposed fault index is obtained by multiplying the above-mentioned maximum values and variance. This voltage-based fault index is associated with a LLG fault that is provided in Figure 14.7.

Figure 14.7 indicates that values of a proposed fault index are zero before and after a fault occurrence. It is also observed that during the instance of a fault, the voltage-based fault index is high, with a peak value of 9.4390e + 42. The maximum values of the fault index recoded on phase-A at the receiving end with a LLG fault on 20%, 50%, and 80% of line lengths are provided in Table 14.3. The peak value of the voltage-based fault index is high at the moment of the fault occurrence and zero during pre-fault and post-fault conditions, which clearly detects the occurrence

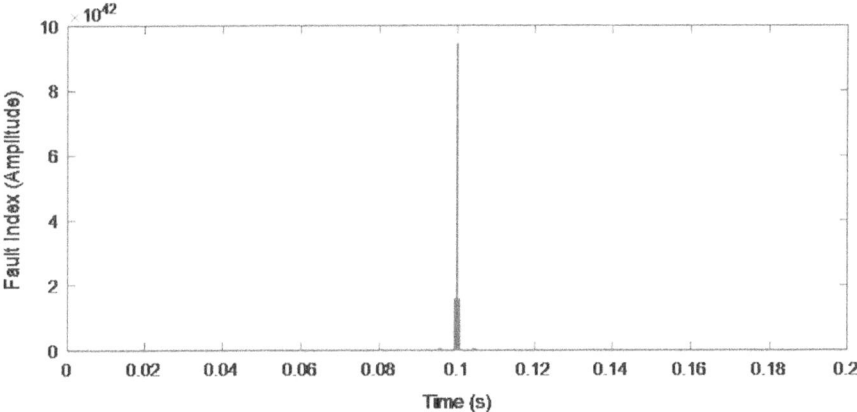

FIGURE 14.7   Voltage-based fault index on the second end of the transmission line during a LLG fault at the middle of the line.

of a LLG fault. Hence, the fault index based on voltage is effective in the detection of a LLG fault at both ends of the transmission line.

14.4.4   THREE-PHASE FAULT INVOLVING GROUND

The LLLG fault is simulated at the sixth cycle on the middle of the transmission line by grounding all phases simultaneously. These results are plotted for 12 cycles. The voltage signal recorded on phase-A of the sending end of the line is decomposed using the Gabor Wigner distribution and the output matrix is obtained. The maximum values (AI) of each column of the matrix are obtained. The variance (VAR) of the output matrix is also calculated. The proposed fault index is obtained by multiplying the above-mentioned maximum values and variance. This fault index during the event of a LLLG fault is provided in Figure 14.8.

Figure 14.8 indicates that the values of the proposed voltage-based fault index are zero before and after the fault occurrence. It is concluded that during the moment of a LLLG fault, the values of the fault index are high, with a peak value of 8.2170e + 42. The maximum values of a fault index recorded on phase-A at the sending end with a LLLG fault on 20%, 50%, and 80% of line lengths are provided in Table 14.2. The maximum value of the fault index is high at the fault occurrence moment and zero during pre-fault and post-fault conditions, which clearly detects the occurrence of a LLLG fault. Hence, the voltage-based fault index is effective in the detection of a LLLG fault.

The voltage signal recorded on the receiving end (second end) of the line is decomposed using the Gabor Wigner distribution and the output matrix is obtained. The maximum values (AI) of each column of the matrix are obtained. The variance (VAR) of the output matrix is also calculated. The proposed fault index is obtained by multiplying the above-mentioned maximum values and variance. This voltage-based fault index associated with a LLLG fault is provided in Figure 14.9.

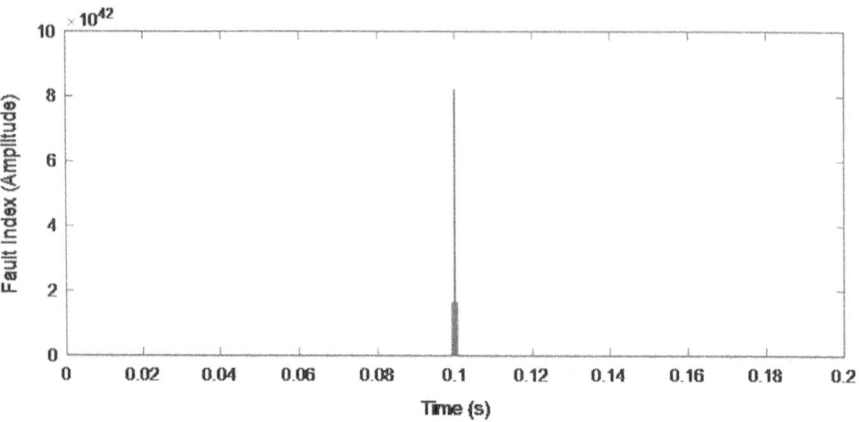

FIGURE 14.8 Voltage-based fault index on the first end of the transmission line during a LLLG fault at the middle of the line.

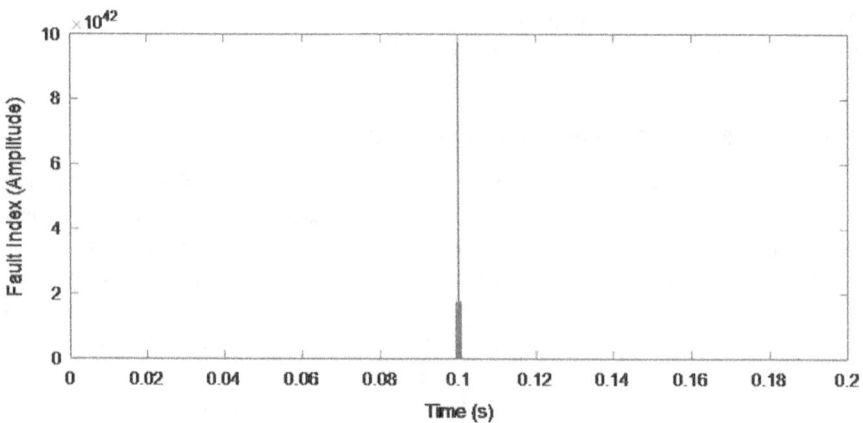

FIGURE 14.9 Voltage-based fault index on the second end of the transmission line during a LLLG fault at the middle of the line.

Figure 14.9 indicates that the values of the proposed fault index are zero before and after the fault occurrence. It is also concluded that during the instance of the fault, values of the voltage-based fault index are high, with a peak value of 9.7504e + 42. The peak values of the fault index for phase-A at the receiving end with a LLLG fault on 20%, 50%, and 80% of line lengths are provided in Table 14.3. The peak value of the fault index is high at the fault occurrence moment and zero during pre-fault and post-fault conditions, which clearly detects the incidence of a LLLG fault. Hence, the fault index based on voltage is effective in the detection of a LLLG fault at both ends of a transmission line.

E. *Classification of the Transmission Line Faults Using a Rule-Based Decision Tree Based on a Voltage-Based Fault Index*

The peak values of the proposed voltage fault index recorded on the sending end (first end) of the line during events of various types of faults at 20%, 50%, and 80% of line lengths are tabulated in Table 6.1. It can be observed from Table 14.2 that the peak value of the fault index is highest during the event of a LLLG fault followed by a LLG fault, a LL fault, and a LG fault. These values are taken as key factors for classification purposes and given as input to the RBDT to design decision rules for classification purposes to provide a voltage-based protection scheme on the sending end of the line.

The maximum values of the voltage-based fault index recorded on the receiving end (second end) of the transmission line during events of different types of faults at 20%, 50%, and 80% of line lengths are tabulated in Table 14.3. It can be observed from Table 14.3 that the peak value of the fault index is highest during the event of a LLLG fault followed by a LL fault, a LLG fault, and a LG fault. This trend is the same as that observed on the first end of the line as expected. These values are taken as key factors for classification purposes and given as input to the RBDT for designing decision rules for classification purposes for a voltage-based protection scheme on the sending end of the transmission line. The classification of faults using a rule-based decision tree based on a voltage-based fault index is shown in Figure 14.10.

14.5   PERFORMANCE COMPARISON

The performance of a proposed algorithm has been evaluated by testing algorithms by creating 100 data sets of each type of fault by varying incidence angles, locations of faults, and fault impedance. The performance is evaluated in terms of correctly classified and misclassified faults. The performance of the proposed algorithm is compared with an algorithm based on the Haar wavelet transform [30] in terms of classification efficiency. The performance of the proposed algorithm and its comparison with an algorithm reported in reference [30] is provided in Table 14.4. It can be observed that the proposed algorithm can detect transmission line faults with a high accuracy compared to that reported in the reference [30]. Hence, efficiency higher than 98% has been achieved with the help of the proposed algorithm.

14.6   CONCLUSION

This chapter proposed an approach based on the voltage for detection and classification of faults on a power transmission line. The study is performed using a test system with two buses and one transmission line with a line length of 200 km and rated at 765 kV and 60 Hz. Two AC sources are connected on two ends of a line with a phase difference of 20° to ensure power flow. The voltages are recorded on two ends of the line. Different types of faults are simulated at 20%, 50%, and 80% of line lengths. The voltage of phase-A is decomposed using the Gabor Wigner distribution and the output matrix is obtained. From this matrix, the maximum values of each column and variance are calculated. These two values are multiplied to obtain a voltage-based fault index. This fault index has zero value in the pre-fault and post-fault conditions. However, during the fault duration, the fault index has

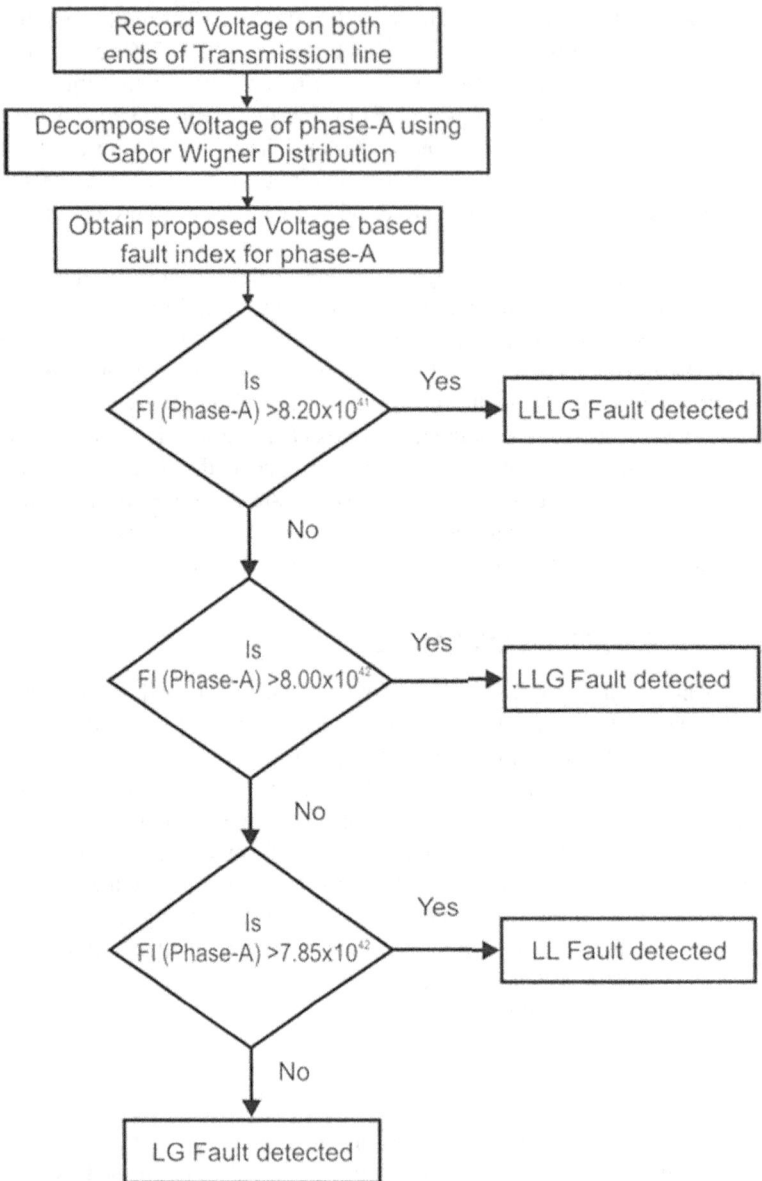

FIGURE 14.10 Rule-based decision tree based on the classification of faults on the transmission line using a voltage-based fault index.

high non-zero values, which help to detect faults. High values of a fault index are used as input for the RBDT to classify various types of faults. The investigated faults include LG, LL, LLG, and LLLG. It is concluded that voltage-based fault indexes have zero values in the absence of the fault and non-zero values during the fault events.

TABLE 14.4

Performance Comparison in Terms of Fault Recognition Efficiency

Type of Fault	Percentage Efficiency (%)	
	Proposed Algorithm	*Reference* [9]
LG	100	94
LL	100	94
LLG	98	89
LLLG	97	84
Overall efficiency	98.75	90.25

REFERENCES

[1] Gulhasan, Ahmad, Om Prakash Mahela, and Sheesh Ram Ola, "A Stockwell Transform Based Approach for Detection of Transmission Line Faults in the Presence of Thyristor Controlled Reactor," 2018 5th IEEE International Conference on Signal Processing and Integrated Networks (SPIN), 22–23 Feb. 2018, Noida, India.

[2] K. R., Krishnanand, P. K. Dash, and M. H. Naeem, "Detection, classification, and location of faults in power transmission lines," *International Journal of Electrical Power and Energy Systems*, Vol. 67, pp. 76–86, 2015.

[3] Moslem, Dehghani, Mohammad Hassan Khooban, and Taher Niknam, "Fast fault detection and classification based on a combination of wavelet singular entropy theory and fuzzy logic in distribution lines in the presence of distributed generations," *International Journal of Electrical Power and Energy Systems*, Vol. 78, pp. 455–462, 2016.

[4] Suman, Devi, Nagendra K. Swarnkar, Sheesh Ram Ola, and Om Prakash Mahela, "Detection of transmission line faults using discrete wavelet transform," IEEE Conference on Advances in Signal Processing (CASP-2016), 9–11 June 2016, Pune, India.

[5] A. M. El-Zonkoly, and H. Desouki, "Wavelet entropy based algorithm for fault detection and classification in FACTS compensated transmission line," *International Journal of Electrical Power and Energy Systems*, Vol. 33, pp. 1368–1374, 2011.

[6] Toshiba, Suman, Om Prakash Mahela, and Sheesh Ram Ola, "Detection of transmission line faults in the presence of solar PV generation using discrete wavelet," 2016 IEEE 7th Power India International Conference (PIICON), 25–27 Nov. 2016, Bikaner, India.

[7] Bhuvnesh, Rathore, Om Prakash Mahela, Baseem Khan, Hassan Haes Alhelou, and Pierluigi Siano, "Wavelet-Alienation-Neural Based Protection Scheme for STATCOM Compensated Transmission Line," *IEEE Transactions on Industrial Informatics*, June 2020, 10.1109/TII.2020.3001063.

[8] Sheesh Ram, Ola, Amit Saraswat, Sunil Kumar Goyal, S. K. Jhajharia, Bhuvnesh Rathore, and Om Prakash Mahela, "Wigner Distribution Function and Alienation Coefficient Based Transmission Line Protection Scheme," *IET Generation, Transmission and Distribution*, Vol. 14, Issue 10, pp. 1842–1853, 22 May 2020, 10.1049/iet-gtd.2019.1414.

[9] Sheesh Ram, Ola, Amit Saraswat, Sunil Kumar Goyal, S. K. Jhajharia, Baseem Khan, Om Prakash Mahela, Hassan Haes Alhelou, and Pierluigi Siano, "A Protection Scheme for Power System with Solar Energy Penetration," *Applied Sciences 2020*, Vol. 10, Issue 4, Paper No. 1516, pp. 1–22, Feb. 2020, 10.3390/app1 0041516.

[10] Sheesh Ram, Ola, Amit Saraswat, Sunil Kumar Goyal, Virendra Sharma, Baseem Khan, Om Prakash Mahela, Hassan Haes Alhelou, Pierluigi Siano, "Alienation Coefficient and Wigner Distribution Function Based Protection Scheme for Hybrid Power System Network with Renewable Energy Penetration," *Energies 2020*, Vol. 13, Issue 5, Paper No. 1120, March 2020, 10.3390/en13051120.

[11] Atul, Kulshrestha, Om Prakash Mahela, Mukesh Kumar Gupta, Neeraj Gupta, Nilesh Patel, Tomonobu Senjyu, Mir Sayed Shah Danish, and Mahdi Khosravy, "A Hybrid Protection Scheme Using Stockwell Transform and Wigner Distribution Function for Power System Network with Solar Energy Penetration," *Energies*, Vol. 13, No. 14, pp. 3519, 2020, 10.3390/en13143519.

[12] Govind Sahay, Yogee, Om Prakash Mahela, Kapil Dev Kansal, Baseem Khan, Rajendra Mahla, Hassan Haes Alhelou, and Pierluigi Siano, "An Algorithm for Recognition of Fault Conditions in the Utility Grid with Renewable Energy Penetration," *Energies*, Vol. 13, No. 9, pp. 2383, 10.3390/en13092383.

[13] Mohd Zishan, Khoker, Om Prakash Mahela, and Gulhasan Ahmad, "A Voltage Algorithm using Discrete Wavelet Transform and Hilbert Transform for Detection and Classification of Power System Faults in the Presence of Solar Energy," 2020 IEEE International Students' Conference on Electrical, Electronics and Computer Science (SCEECS 2020), MANIT Bhopal, India, February 22–23, 2020, 10.1109/ SCEECS48394.2020.7.

[14] Mohd Zishan, Khoker, Om Prakash Mahela, and Gulhasan Ahmad, "A Current Based Hybrid Algorithm using Discrete Wavelet Transform and Hilbert Transform for Detection and Classification of Power System Faults in the Presence of Solar Energy," 2020 IEEE International Students' Conference on Electrical, Electronics and Computer Science (SCEECS 2020), MANIT Bhopal, India, February 22–23, 2020, 10.1109/SCEECS48394.2020.6.

[15] Surbhi, Thukral, Om Prakash Mahela, and Bipul Kumar, "Detection of Transmission Line Faults in the Presence of Wind Energy Power Generation Source Using Stockwell's Transform," IEEE International Conference on Issues and Challenges in Intelligent Computing Techniques (ICICT 2019), 27–28th September, 2019, KIET Group of Institutions, Delhi-NCR, Ghaziabad, India, 10.1109/ICICT46931.201 9.8977695.

[16] Om Prakash, Mahela, Jaya Sharma, Bipul Kumar, Baseem Khan, and Hasan Haes Alhelou, "An Algorithm for the Protection of Distribution Feeder Using Stockwell and Hilbert Transforms Supported Features," *CSEE Journal of Power and Energy Systems*, 10.17775/CSEEJPES.2020.00170.

[17] Sheesh Ram, Ola, Amit Saraswat, Sunil Kumar Goyal, S. K. Jhajharia and Om Prakash Mahela. "Detection and Analysis of Power System Faults in the Presence of Wind Power Generation Using Stockwell Transform Based Median," *Springer Lecture Notes in Electrical Engineering Series*, ISSN: 1876-1100, pp. 319–329, 2019. https://link.springer.com/chapter/10.1007/978–981-15–0214-9_36.

[18] Amit Kumar, Gangwar, Bhunesh Rathore, and Om Prakash Mahela, "K-means Clustering and Linear Regression Based Protection Scheme for Transmission Line," IEEE 9th Power India International Conference (PIICON 2020) from 28 Feb to 01 March, 2020 – Deenbandhu Chhotu Ram University of Science and Technology, Murthal, India, 10.1109/PIICON49524.2020.9113038.

[19] Nikita, Tailor, Satyanarayan Joshi, and Om Prakash Mahela, "Transmission Line Protection Schemes Based on Wigner Distribution Function and Discrete Wavelet Transform," IEEE 9th Power India International Conference (PIICON 2020) from 28 Feb to 01 March, 2020 – Deenbandhu Chhotu Ram University of Science and Technology, Murthal, India, Electronic ISBN: 978-1-7281-6664-3 Print on Demand (PoD) ISBN: 978-1-7281-6665-0, (Available on IEEE Xplore), 10.1109/PIICON4 9524.2020.9113011.

[20] Jaya, Sharma, Bipul Kumar, Om Prakash Mahela, and Akhil Ranjan Garg, "Protection of Distribution Feeder Using Stockwell Transform Supported Voltage Features," IEEE 9th Power India International Conference (PIICON 2020) from 28 Feb to 01 March, 2020 – Deenbandhu Chhotu Ram University of Science and Technology, Murthal, India, Electronic ISBN: 978-1-7281-6664-3 Print on Demand (PoD) ISBN: 978-1-7281-6665-0, (Available on IEEE Xplore), 10.1109/PIICON4 9524.2020.9113014.

[21] Deepak, Gupta, Om Prakash Mahela, and Shoyab Ali, "Voltage Based Transmission Line Protection Algorithm Using Signal Processing Techniques," 2020 IEEE International Students' Conference on Electrical, Electronics and Computer Science (SCEECS 2020), MANIT Bhopal, India, February 22–23, 2020, 10.1109/SCEECS4 8394.2020.15.

[22] Deepak, Gupta, Om Prakash Mahela, and Shoyab Ali, "Current Based Transmission Line Protection Algorithm Using Signal Processing Techniques," 2020 IEEE International Students' Conference on Electrical, Electronics and Computer Science (SCEECS 2020), MANIT Bhopal, India, February 22–23, 2020, Electronic ISBN: 978-1-7281-4862-5 Print on Demand(PoD) ISBN: 978-1-7281-4863-2, (Available on IEEE Xplore), 10.1109/SCEECS48394.2020.14.

[23] Shubhmay, Karmakar, Gulhasan Ahmad, Om Prakash Mahela, and Ravi Raj Choudhary, "Algorithm Based on Combined Features of Stockwell Transform and Hilbert Transform for Detection of Transmission Line Faults with Dynamic Load," First IEEE International Conference on Power, Control and Computing Technologies (ICPC2T), NIT Raipur, India, January 3–5, 2020, Electronic ISBN: 978-1-7281-4997-4 DVD ISBN: 978-1-7281-4996-7 Print on Demand(PoD) ISBN: 978-1-7281-4998-1, (Available on IEEE Xplore), 10.1109/ICPC2T48082.2020.9071516.

[24] Shubhmay, Karmakar, Gulhasan Ahmad, Om Prakash Mahela, and Ravi Raj Choudhary, "Transmission Line Protection Scheme Based on Combined Features of Stockwell Transform and Hilbert Transform," First IEEE International Conference on Power, Control and Computing Technologies (ICPC2T), NIT Raipur, India, January 3–5, 2020, Electronic ISBN: 978-1-7281-4997-4 DVD ISBN: 978-1-7281-4996-7 Print on Demand(PoD) ISBN: 978-1-7281-4998-1, (Available on IEEE Xplore), 10.1109/ICPC2T48082.2020.9071506.

[25] O. P., Mahela, Y. Sharma, S. Ali, B. Khan, and S. Padmanaban, "Estimation of Islanding Events in Utility Distribution Grid With Renewable Energy Using Current Variations and Stockwell Transform," in *IEEE Access*, vol. 9, pp. 69798–69813, 2021.

[26] N. K., Swarnkar, O. P. Mahela, B. Khan, and M. Lalwani, "Identification of Islanding Events in Utility Grid with Renewable Energy Penetration Using Current Based Passive Method," in *IEEE Access*, 10.1109/ACCESS.2021.3092971.

[27] Rajkumar, Kaushik, Om Prakash Mahela, Pramod Kumar Bhatt, Baseem Khan, Akhil Ranjan Garg, Hassan Haes Alhelou, and Pierluigi Siano, "Recognition of Islanding and Operational Events in Power System With Renewable Energy Penetration Using a Stockwell Transform-Based Method," in *IEEE Systems Journal*, 10.1109/JSYST.2020.3020919.

[28] Fsaha Mebrahtu, Gebru, Baseem Khan, and Hassan Haes Alhelou, "Analyzing low voltage ride through capability of doubly fed induction generator based wind turbine," *Computers & Electrical Engineering*, Vol. 86, pp. 106727, 2020.

[29] B., Rathore, O. P. Mahela, B. Khan, and S. Padmanaban, "Protection Scheme using Wavelet-Alienation-Neural Technique for UPFC Compensated Transmission Line," in *IEEE Access*, vol. 9, pp. 13737–13753, 2021, 10.1109/ACCESS.2021.3052315.

[30] Joe-Air, Jiang, Ping-Lin Fan, Ching-Shan Chen, Chi-Shan Yu, and Jin-Yi Sheu, "A fault Detection and Faulted-Phase Selection Approach for Transmission Lines with Haar Wavelet Transform," 2003 IEEE PES-Transmission and Distribution Conference and Exposition, 2003, pp. 285–289.

15 Power Quality Estimation and Event Detection in a Distribution System in the Presence of Renewable Energy

Rajkumar Kaushik and Pramod Kumar Bhatt
Department of Electrical Engineering Amity University,
Jaipur, India

Om Prakash Mahela
Power System Planning Division, Rajasthan Rajya Vidhyut
Prasaran Nigam Ltd., Jaipur, India

CONTENTS

DOI: 10.1201/b22884-15

15.1   INTRODUCTION

Renewable energy sources are emerging as alternative power generation sources that mitigate the problems of CO_2 emission and global warming. RE generators are growing in number and size rapidly. The RE generators have a disadvantage of fluctuating power generation that deteriorates the quality of power [1]. Solar power plants (SPPs) and wind power plants (WPPs) are two commonly used RE generators. A detailed study of WPPs and SPPs is available in [2–4]. Signal processing techniques can effectively used for estimation of PQ issues [5]. Filters and flexible AC transmission systems (FACTS) are used for PQ improvement [6]. A detailed review of the techniques that can be used for detection and mitigation of PQ issues in the presence of RE generation in the utility grids is available in [7,8]. High RE power levels in to the utility grids generates PQ disturbances that include impulses, glitches, notches, wave faults, momentary interruption (MI), sags and swells in voltage, harmonic distortions, and flicker [9–12]. A multi-resolution analysis of voltage signals using DWT is reported in [13]. This technique has the disadvantage that performance is affected in the noisy environment. A method using fuzzy c-means clustering (FCM)–oriented particle swarm optimization (PSO) for recognition of PQ issues is reported in [14]. A technique for recognition of simple nature and complex nature PQ issues using a ST-based technique is reported in [15–17]. A method for estimation of PQ issues associated with grid-integrated SPP in an experimental framework is reported in [18].

A method for identification of PQ issues in the RE-based distribution utility grid using a hybrid combination of the ST and HT is introduced in this chapter. Two factors, PEF and TEF, are introduced in this chapter for estimation and location of PQ issues.

The contents are organized in seven sections of this chapter. Section 15.1 introduces the research work considered in this chapter and review of literature related to the PQ issues. Section 15.2 describes the test system of IEEE-13 nodes, WPP, SPP, and data of the test system used for the study. Section 15.3 describes the algorithm used in this chapter for recognition of PQ issues. Section 15.4 includes the simulation results and their discussion. Section 15.5 describes the PQ results of the classification of the events. Section 15.6 describes the performance comparison of the method with the DWT-based method. Section 15.7 includes the conclusion of the research activity.

15.2   TEST DISTRIBUTION GRID WITH RE GENERATORS

A test distribution grid (TDG) is realized using the IEEE-13 nodes network by integrating a wind power plant (WPP) and a solar power plant (SPP). A test distribution grid with the SPP and WPP is described in Figure 15.1. A WPP is integrated to the TDG on node 680 using a transformer designated as T-WPP. A SPP is also integrated to the same node to form the hybrid grid using the transformer designated as T-WPP. The WPP is rated at 1.5 MW, 575 V, and 60 Hz frequency,

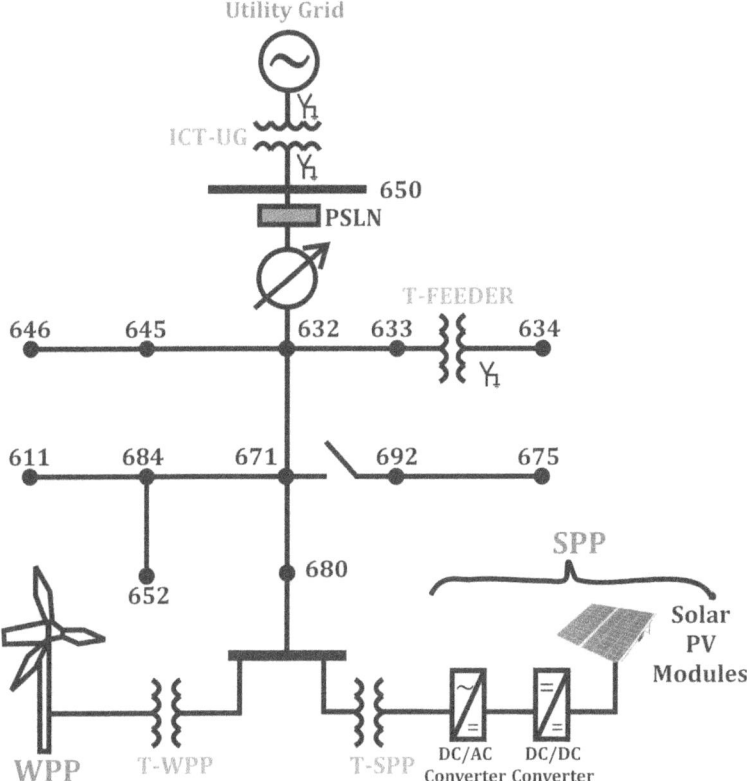

FIGURE 15.1 Utility grid interfaced with WPP.

which uses a doubly fed induction generator (DFIG). Details of wind turbine data, generator data, converter details, and inverter details reported in [19–21] are used for the proposed study. The SPP is rated at 1 MW, 260 V, and 60 Hz frequency. Details of the DC-DC converter, DC-AC converter, and control design of converters, and solar PV plates design reported in [22,23] are used in this study. Details of loads, capacitors, feeders, transformers, and feeder configuration reported in [24–31] are used in this study. The TDG is interfaced to the utility grid (UG) which is rated at 115 kV using an interconnecting transformer (ICT-UG), which is rated at 10 MVA, 115 kV/4.16 kV. The TDG is operated at voltage levels of 4.16 kV and 0.48 kV. A distribution feeder transformer (T-Feeder) is connected between nodes 633 and 634. Details of the transformers are provided in Table 15.1. Node 650 is taken as the PQ estimation node (PQEN) where voltages are recorded.

15.3 PROPOSED PQ ESTIMATION AND EVENT DETECTION ALGORITHM

The PQ estimation and event detection algorithm is described in Figure 15.2. Various steps of the algorithm are detailed below:-

TABLE 15.1

Transformer Details

Transformer	MVA	kV High	kV Low	HV Winding		LV Winding	
				R (Ω)	X (Ω)	R (Ω)	X (Ω)
ICT-UG	10	115	4.16	28.05	210.05	0.1140	0.8305
T-Feeder	1	4.16	0.48	0.3710	275.14	0.0491	0.0041
T-WPP	5	4.16	0.575	0.1728	192.75	0.0006	0.7642
T-SPP	2	4.16	0.260	0.3804	275.42	0.0501	0.0041

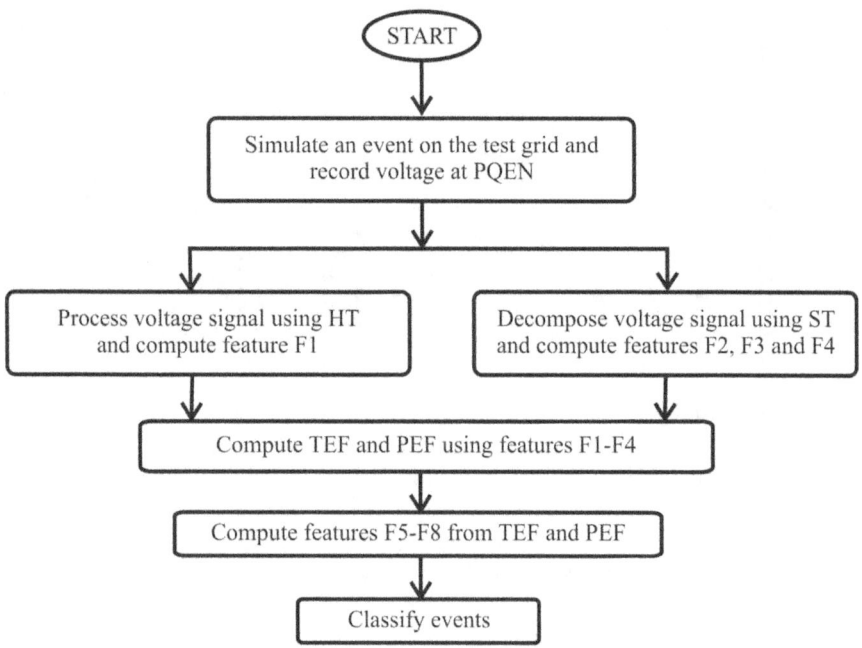

FIGURE 15.2   PQ estimation and event detection algorithm.

Step 1. Simulate an event on a test distribution grid with RE sources and record the voltage signal at node 650, which is considered a PQ estimation node (PQEN).

Step 2. Voltage signal (v) is processed using HT at a sampling frequency of 3.84 kHz and absolute magnitude of output is computed and considered as feature F1, as described below:

$$F1 = abs\,(Hilbert\,(v))$$

Step 3. Voltage signal is processed by ST at a sampling frequency of 3.84 kHz and output absolute magnitude matrix (AMM) is computed.

F2. Compute summation of every columns of AMM as detailed below:

$$F2 = sum(AMM)$$

F3. Maximum amplitude of every column of AMM is computed as detailed below:

$$F3 = max(AMM)$$

F4. Median factor of every column is computed as described below:

$$F4 = median(AMM)$$

Step 4. PQ estimation factor (PEF) is computed using the features F1, F2, and F3, as described below:

$$PEF = F1 \times F2 \times F3$$

Step 5. Time estimation factor (TEF) is computed as detailed below:

$$TEF = F1 \times F2 \times F3 \times F4$$

Step 6. PQ disturbances are classified using rule-based decision tree (RBDT) with the help of features F5 to F8 computed as detailed below:

F5. Variance of the PQEF:

$$F5 = var(PEF)$$

F6. Median of PQEF:

$$F6 = median(PEF)$$

F7. Variance of the time estimation factor:

$$F7 = var(TEF)$$

F8. Median of the time estimation factor:

$$F8 = median(TEF)$$

This method is efficient to estimate PQ disturbances pertaining to various events in a distribution grid with RE penetration having penetration of wind energy.

15.4 DISCUSSION OF SIMULATION RESULTS

The results obtained from the simulation studies for the various case studies considered in this research work are presented in this section.

15.4.1 FEEDER OPENING EVENT

The feeder opening event is simulated at 0.1 s by opening the feeder that has the nodes 692 and 675, where load and capacitors are connected from node 671. A voltage signal is recorded on the PQEN for a period of 12 cycles and processed using the proposed algorithm to compute the PEF and TEF. The voltage signal, PEF, and TEF are illustrated in Figure 15.3. The time elapsed in the computation by the algorithm is provided in Table 15.2.

Figure 15.3(a) shows that there is no visible disturbance associated with the voltage signal. It is inferred from Figure 15.3(b) that the voltage swell is associated with the feeder opening event, which persists for a small time period followed by small magnitude transients that persist for a long duration (0.2 s) after the event incidence. The The TEF plot detailed in Figure 15.3(c) indicates that a high magnitude peak is observed, which helps to localize the feeder opening event.

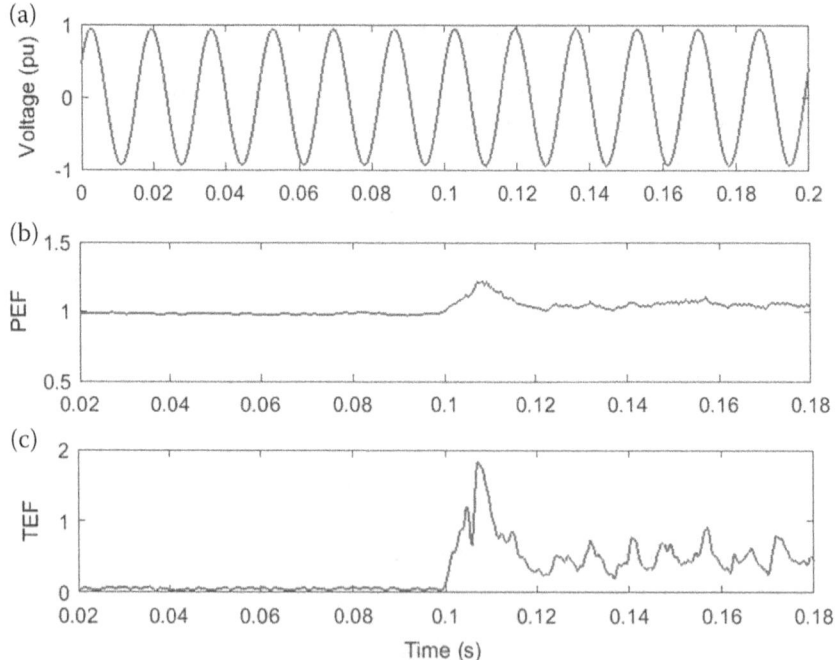

FIGURE 15.3 Feeder opening event: (a) voltage, (b) PQ estimation factor, and (c) time estimation factor.

TABLE 15.2
PQ Disturbances Estimation Time

S. No.	Type of Event Category	Detection Time (s)
1	Feeder Opening Event	5.555130
2	Feeder Closing Event	7.139296
3	Load Switching ON Event	0.563679
4	Load Switching OFF Event	0.404370
5	Capacitor Switching ON Event	0.431764
6	Capacitor Outage Event	0.481350
7	Solar Power Plant Outage Event	3.740406
8	Solar Power Plant Grid Synchronization Event	0.754606
9	Wind Power Plant Outage Event	2.940267
10	Wind Power Plant Grid Synchronization Event	0.836632

Further, it also helps to find out the time duration for which the disturbances are associated in the post-event period by high magnitude. It is also seen from Table 15.2 that the event has been recognized in a time period of 5.55513 s.

15.4.2 Feeder Closing Event

The feeder closing event is simulated at 0.1 s by closing the feeder that has the nodes 692 and 675, where load and capacitors are connected to node 671. This feeder was initially kept open. The voltage signal is recorded at the PQEN for a period of 12 cycles and processed using the proposed algorithm to compute the PEF and TEF. The voltage signal, PEF, and TEF are illustrated in Figure 15.4. The time elapsed in the computation by the algorithm is provided in Table 15.2.

Figure 15.4(a) shows that there are transient disturbances associated with the voltage signal. It is inferred from Figure 15.4(b) that impulsive transient is associated with the feeder closing event, which persists for a small time period followed by small magnitude transients that persist for a small time duration (0.02 s) after the event incidence. Flicker is also seen with the voltage signal in the post-feeder closing period. The The TEF plot detailed in Figure 15.4(c) indicates that a high magnitude peak is observed, which helps to localize the feeder closing event. Further, it also helps to find out the time duration for which the disturbances are associated in the post-event period by a high magnitude. It is also seen from Table 15.2 that the event has been recognized in a time period of 7.139296 s.

15.4.3 Load Switching ON Event

The event of load switching ON is simulated at 0.1 s by connecting the load comprising of 843 kW and 462 kVAr on node 675, which was initially kept open circuited. The voltage signal is recorded at the PQEN for a period of 12 cycles and processed using the proposed algorithm to compute the PEF and TEF. The voltage

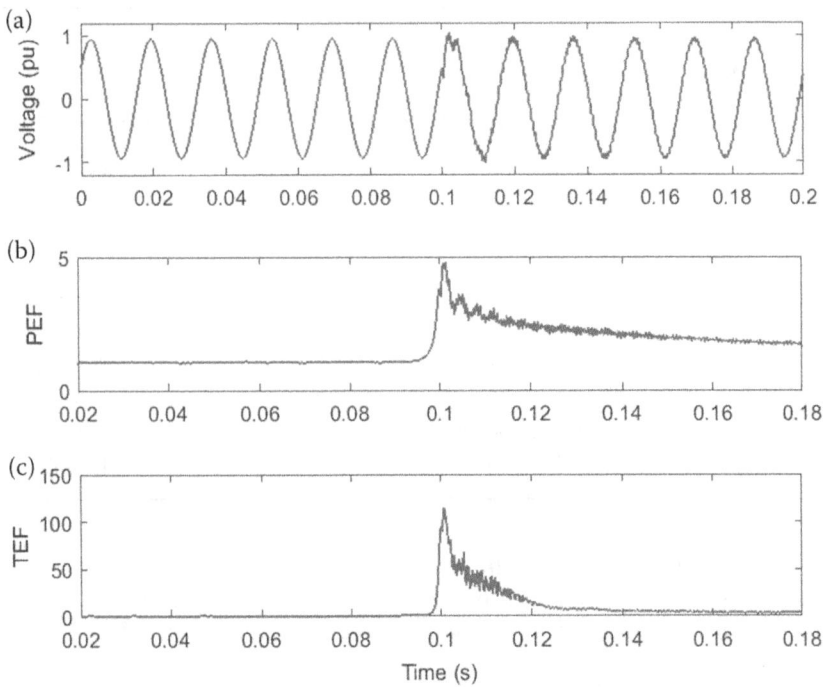

FIGURE 15.4   Event of feeder closing: (a) voltage, (b) PQ estimation factor, and (c) time estimation factor.

signal, PEF, and TEF are illustrated in Figure 15.5. The ime elapsed in the computation by the algorithm is provided in Table 15.2.

Figure 15.5(a) shows that there is no visible disturbance associated with the voltage signal. It is inferred from Figure 15.5(b) that voltage swell is associated with the load switching ON event, which persists for a small time period followed by small magnitude transients at the time of the event. During the post-event period, voltages have reduced by a small magnitude. The The TEF plot detailed in Figure 15.5(c) indicates that a high magnitude peak is observed, which helps to localize the load switching ON event. Further, it also helps to find out the time duration for which the disturbances are associated in the post-event period by high magnitude, which is not observed after the event incidence. It is also seen from Table 15.2 that the event has been recognized in a time period of 0.563679 s.

15.4.4   LOAD SWITCHING OFF EVENT

The event of load switching OFF is simulated at 0.1 s by opening the load comprising of 843 kW and 462 kVAr from node 675, which was initially kept connected. The voltage signal is recorded at the PQEN for a period of 12 cycles and processed using the proposed algorithm to compute the PEF and TEF. The voltage signal, PEF, and TEF are illustrated in Figure 15.6. The time elapsed in computation by the algorithm is provided in Table 15.2.

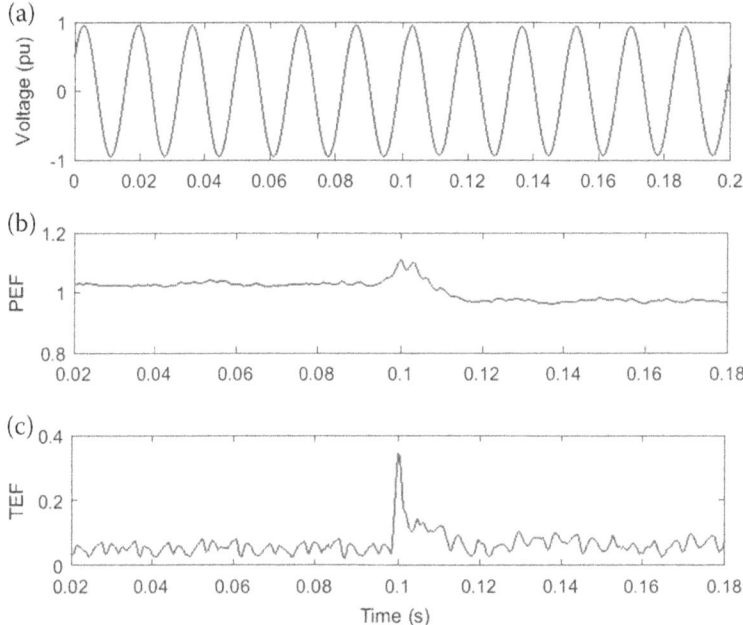

FIGURE 15.5 Load switching ON event: (a) voltage, (b) PQ estimation factor, and (c) time estimation factor.

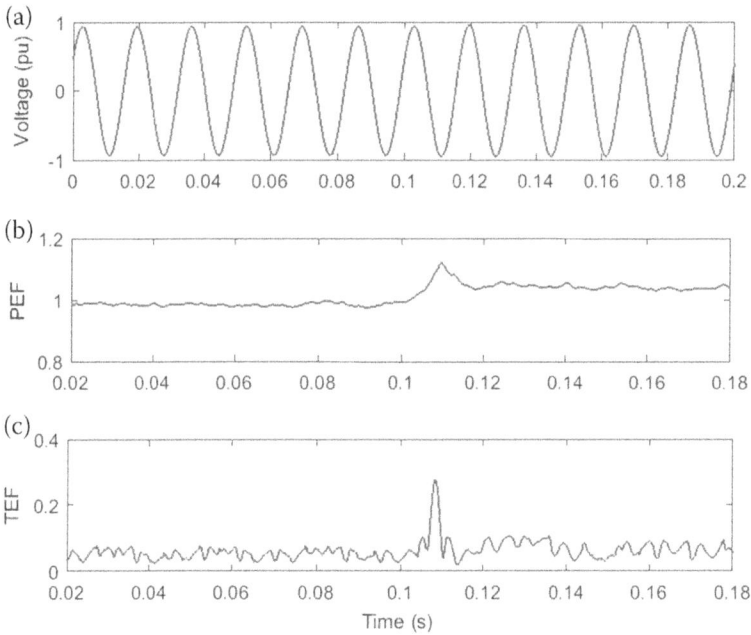

FIGURE 15.6 Load switching OFF event: (a) voltage, (b) PQ estimation factor, and (c) time estimation factor.

Figure 15.6(a) shows that there is no visible disturbance associated with the voltage signal. It is inferred from Figure 15.6(b) that the voltage swell is associated with the load switching OFF event, which persists for a small time period. During the post-event period, voltages have increased by a small magnitude. The The TEF plot detailed in Figure 15.6(c) indicates that a high magnitude peak is observed, which helps to localize the load switching OFF event. Further, it also helps to find out the time duration for which the disturbances are associated in the post-event period by a high magnitude, which is not observed after the event incidence. It is also seen from Table 15.2 that the event has been recognized in a time period of 0. 0.404370 s.

15.4.5 CAPACITOR SWITCHING ON EVENT

The event of capacitor switching ON is simulated at 0.1 s by connecting the capacitor of 600 kVAr to node 675, which was initially kept open circuited. The voltage signal is recorded at the PQEN for a period of 12 cycles and processed using the proposed algorithm to compute the PEF and TEF. The voltage signal, PEF, and TEF are illustrated in Figure 15.7. The time elapsed in the computation by the algorithm is provided in Table 15.2.

Figure 15.7(a) shows that there is no visible disturbance associated with the voltage signal. It is inferred from Figure 15.7(b) that the voltage swell is associated

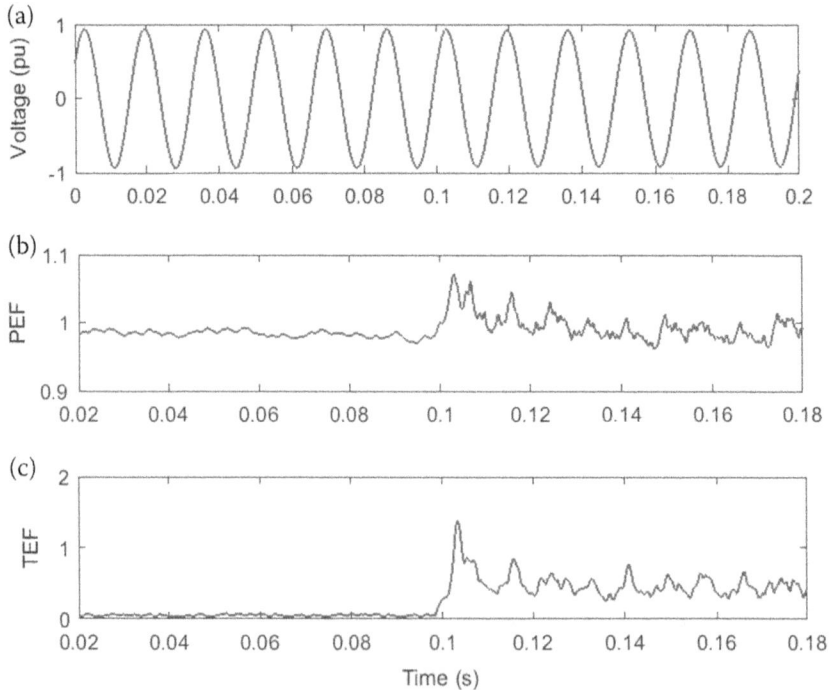

FIGURE 15.7 Capacitor switching ON event: (a) voltage, (b) PQ estimation factor, and (c) time estimation factor.

with the capacitor switching ON event. There are transient disturbances associated with the voltage during the post-capacitor switching ON period, which persists for a period of 0.2 s. During the post-event period, voltages have increased by a small magnitude. The The TEF plot detailed in Figure 15.7(c) indicates that a high magnitude peak is observed just after the event incidence, which helps to localize the capacitor switching ON event. Further, it also helps to find out the time duration for which the disturbances are associated in the post-event period by a high magnitude, which is observed after the event incidence for a period of 0.2 s. It is also seen from Table 15.2 that the event has been recognized in a time period of 0.431764 s.

15.4.6 CAPACITOR OUTAGE EVENT

Event of capacitor switching OFF is simulated at 0.1 s by disconnecting the capacitor of 600 kVAr from the node 675 which was initially kept connected. The voltage signal is recorded at the PQEN for a period of 12 cycles and processed using the proposed algorithm to compute the PEF and TEF. The voltage signal, PEF, and TEF are illustrated in Figure 15.8. The time elapsed in the computation by the algorithm is provided in Table 15.2.

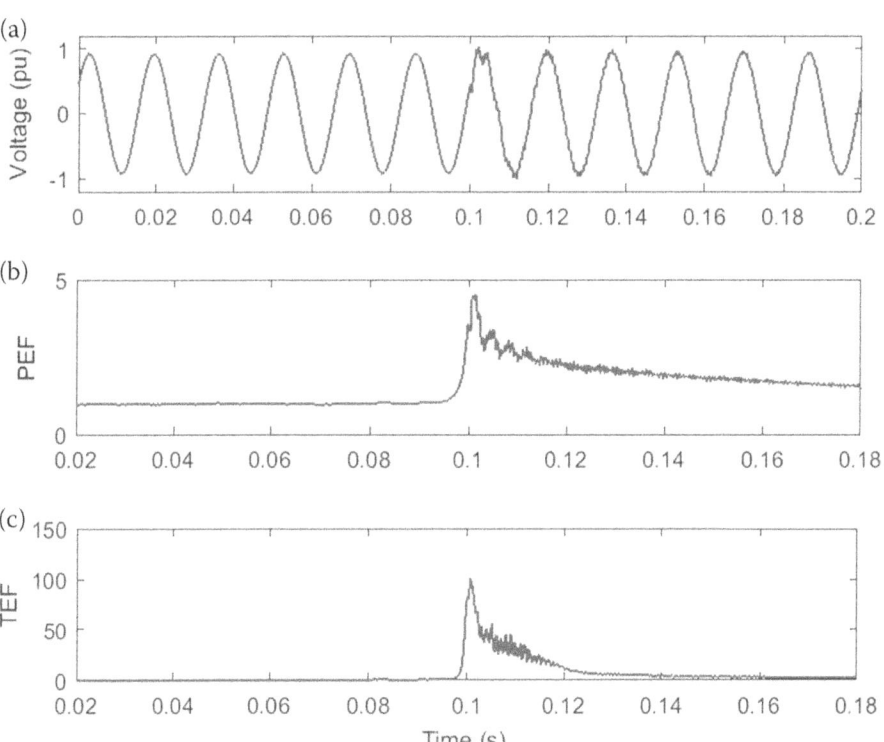

FIGURE 15.8 Capacitor outage event: (a) voltage, (b) PQ estimation factor, and (c) time estimation factor.

Figure 15.8(a) shows that there is a small magnitude transient disturbance associated with the voltage signal. It is inferred from Figure 15.8(b) that an impulsive transient (IT) is associated with the capacitor switching OFF event followed by small magnitude transients. There are transient disturbances associated with the voltage during the post-capacitor switching OFF period, which persists for a small period of 0.04 s. During the post-event period, voltages have decreased by a small magnitude. The The TEF plot detailed in Figure 15.8(c) indicates that a high magnitude peak is observed just after the event incidence, which helps to localize the capacitor switching ON event. Further, it also helps to find out the time duration for which the disturbances are associated in the post-event period by a high magnitude, which is observed after the event incidence for a period of 0.04 s. It is also seen from Table 15.2 that the event has been recognized in a time period of 0.481350 s.

15.4.7 SOLAR POWER PLANT OUTAGE EVENT

The event of a SPP outage is simulated at 0.1 s by opening the circuit breaker (CB) used to integrate the SPP on node 680. The voltage signal is recorded at the PQEN for a period of 12 cycles and processed using the proposed algorithm to compute the PEF and TEF. The voltage signal, PEF, and TEF are illustrated in Figure 15.9. The time elapsed in computation by the algorithm is provided in Table 15.2.

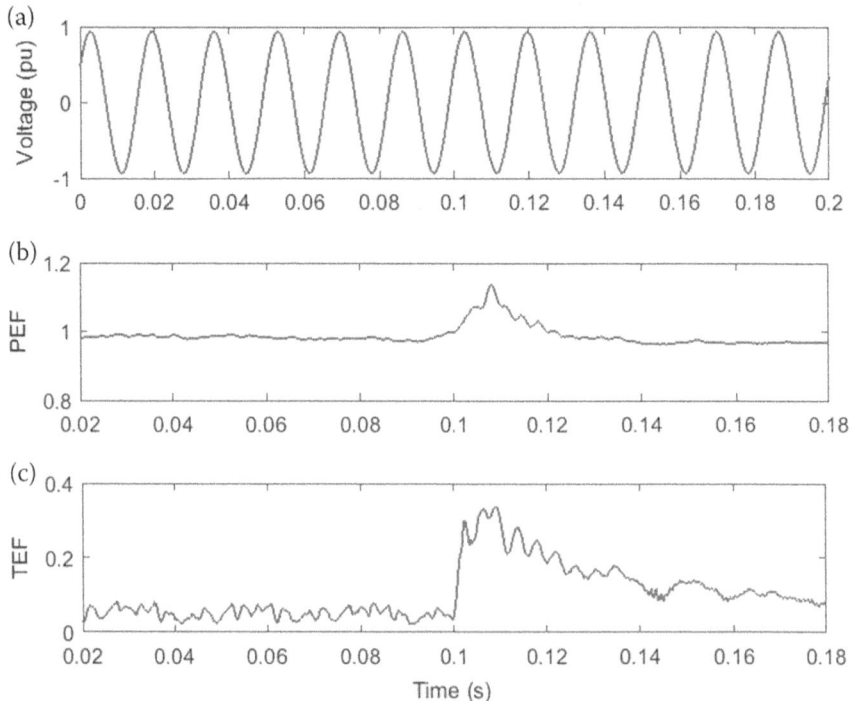

FIGURE 15.9 SPP outage event: (a) voltage, (b) PQ estimation factor, and (c) time estimation factor.

Figure 15.9(a) shows that there are no significant disturbances associated with the voltage signal. It is inferred from Figure 15.9(b) that a voltage swell with a low magnitude impulsive transient (IT) is associated with the event of a SPP outage. Low magnitude transient disturbances persist for a period of 0.08 s after the incidence of the SPP outage event. The TEF plot detailed in Figure 15.9(c) indicates that a high magnitude peak is observed just after the event incidence, which helps to localize the SPP outage event. Further, it also helps to find out the time duration for which the disturbances are associated in the post-event period by a high magnitude, which is observed after the event incidence for a period of 0.08 s. It is also seen from Table 15.2 that the event has been recognized in a time period of 3.740406 s.

15.4.8 SOLAR POWER PLANT GRID SYNCHRONIZATION EVENT

The event of a SPP grid synchronization is simulated at 0.1 s by closing the circuit breaker (CB) used to integrate the SPP on node 680, which was initially kept open circuited. The voltage signal is recorded at the PQEN for a period of 12 cycles and processed using the proposed algorithm to compute the PEF and TEF. The voltage signal, PEF, and TEF are illustrated in Figure 15.10. The time elapsed in computation by the algorithm is provided in Table 15.2.

Figure 15.10(a) shows that there are disturbances associated with the voltage signal. It is inferred from Figure 15.10(b) that a voltage magnitude increases due to the event of the SPP grid synchronization. Further, ripples are observed for period

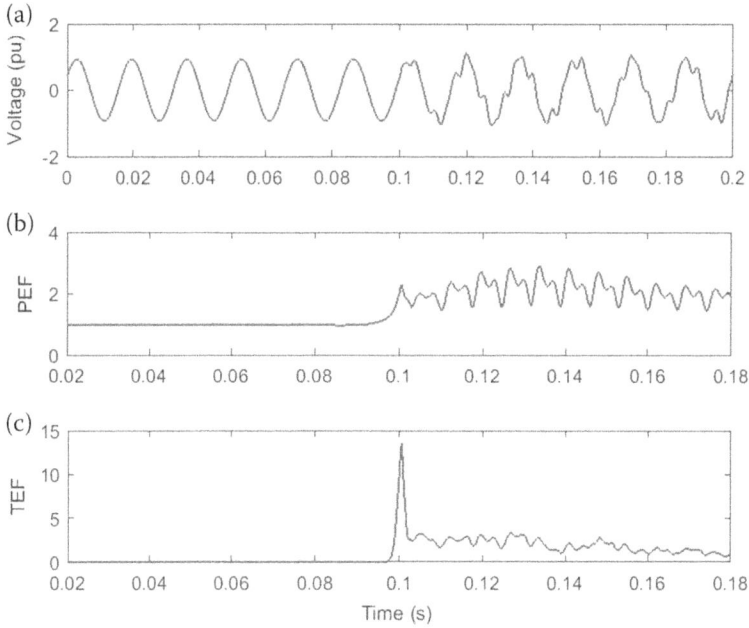

FIGURE 15.10 SPP grid synchronization event: (a) voltage, (b) PQ estimation factor, and (c) time estimation factor.

of 0.2 s after the incidence of the event. Further, low magnitude IT is associated with the event of the SPP grid synchronization. The TEF plot detailed in Figure 15.10(c) indicates that a high magnitude peak is observed just after the event incidence, which helps to localize the SPP grid synchronization event. Further, it also helps to find out the time duration for which the disturbances are associated in the post-event period by a high magnitude, which is observed after the event incidence for a period of 0.2 s. It is also seen from Table 15.2 that the event has been recognized in a time period of 0.754606 s.

15.4.9 WIND POWER PLANT OUTAGE EVENT

The event of a WPP outage is simulated at 0.1 s by opening the circuit breaker (CB) used to integrate the WPP on node 680. The voltage signal is recorded at the PQEN for a period of 12 cycles and processed using the proposed algorithm to compute the PEF and TEF. The voltage signal, PEF, and TEF are illustrated in Figure 15.11. The time elapsed in the computation by the algorithm is provided in Table 15.2.

Figure 15.11(a) shows that there are no significant disturbances associated with the voltage signal. It is inferred from Figure 15.11(b) that a voltage swell is observed due to the event of the WPP grid outage. Further, a low magnitude IT is also associated with the event of the WPP outage. Low magnitude transient disturbances persist for a period of 0.02 s after incidence of the WPP outage event. The TEF plot detailed in Figure 15.11(c) indicates that a high magnitude peak is observed just

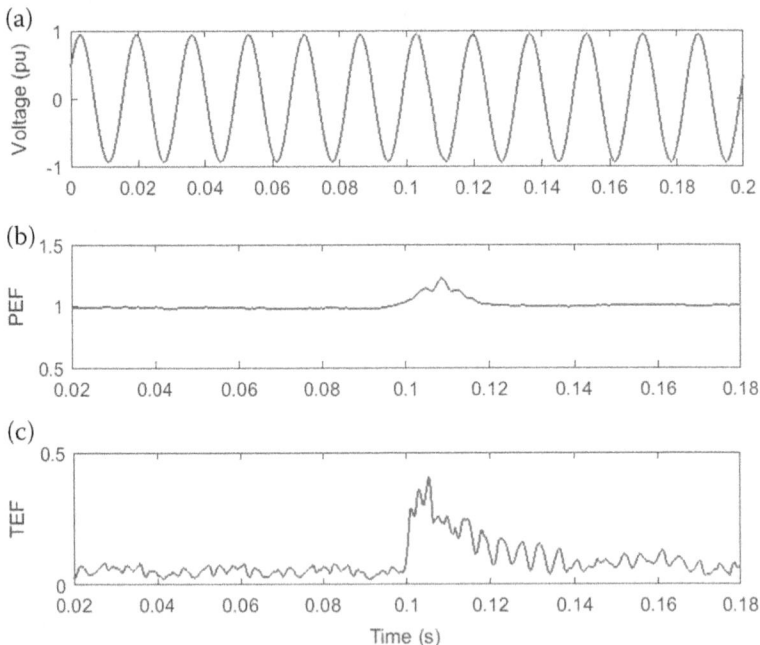

FIGURE 15.11 WPP outage event: (a) voltage, (b) PQ estimation factor, and (c) time estimation factor.

after the event incidence, which helps to localize the WPP outage event. Further, it also helps to find out the time duration for which the disturbances are associated in the post-event period by a high magnitude, which is observed after the event incidence for a period of 0.02 s. It is also seen from Table 15.2 that the event has been recognized in a time period of 2.940267 s.

15.4.10   WIND POWER PLANT GRID SYNCHRONIZATION EVENT

The event of a WPP grid synchronization is simulated at 0.1 s by closing the circuit breaker (CB) used to integrate the WPP on node 680, which was initially kept open circuited. The voltage signal is recorded at the PQEN for a period of 12 cycles and processed using the proposed algorithm to compute the PEF and TEF. The voltage signal, PEF, and TEF are illustrated in Figure 15.12. The time elapsed in the computation by the algorithm is provided in Table 15.2.

Figure 15.12(a) shows that there are transient disturbances associated with the voltage waveform for a period of 0.08 s due to incidence of the WPP grid synchronization event. It is inferred from Figure 15.12(b) that a high magnitude oscillatory transient is associated with the event of the WPP grid synchronization. Further, transient components are observed for a period of 0.08 s after the incidence of the WPP grid synchronization event. The TEF plot detailed in Figure 15.12(c)

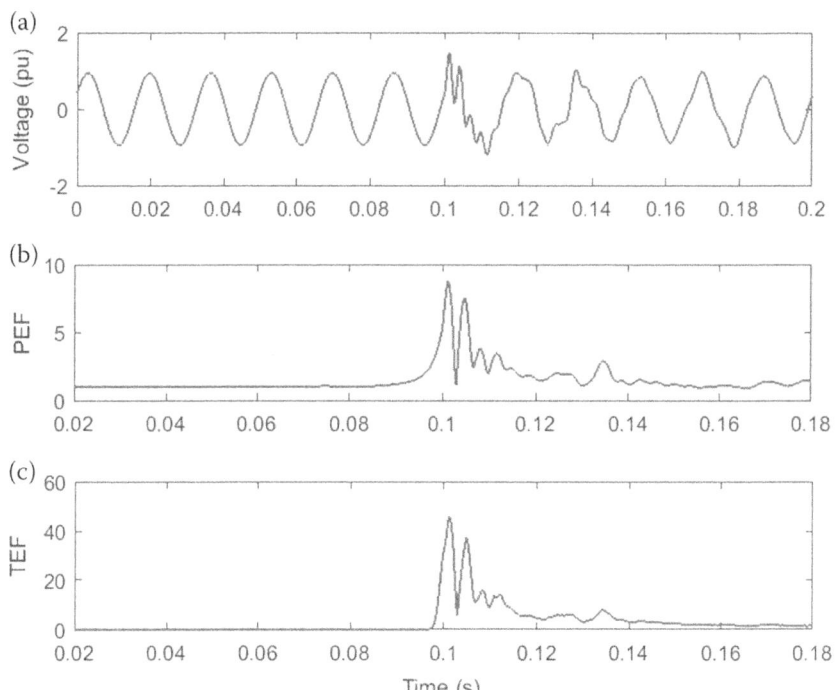

FIGURE 15.12   WPP grid synchronization event: (a) voltage, (b) PQ estimation factor, and (c) time estimation factor.

TABLE 15.3
Features for Classification of Events

S. No.	Symbol of Event Category	Type of Event Category	Features			
			F5	F6	F7	F8
1	EC1	Feeder Opening Event	0.0026	1.0137	0.1667	0.2489
2	EC2	Feeder Closing Event	0.4322	1.5585	206.6783	2.3194
3	EC3	Load Switching ON Event	0.0015	1.0228	0.0448	0.0590
4	EC4	Load Switching OFF Event	0.0012	1.0096	0.0434	0.0570
5	EC5	Capacitor Switching ON Event	0.3762	1.4203	154.3491	1.9369
6	EC6	Capacitor Outage Event	2.3738×10^{-4}	0.9847	0.0606	0.2749
7	EC7	Solar Power Plant Outage Event	0.0015	0.9816	0.0694	0.0812
8	EC8	Solar Power Plant Grid Synchronization Event	0.3679	1.5038	9.4716	0.8725
9	EC9	Wind Power Plant Outage Event	0.0041	0.9957	0.2809	0.0647
10	EC10	Wind Power Plant Grid Synchronization Event	1.0504	1.0457	36.7774	1.2323

indicates that a high magnitude peak is observed just after the event incidence, which helps to localize the WPP grid synchronization event. Further, it also helps to find out the time duration for which the disturbances are associated in the post-event period by a high magnitude, which is observed after the event incidence for a period of 0.08 s. It is also seen from Table 15.2 that the event has been recognized in a time period of 0.836632 s.

15.5 CLASSIFICATION OF EVENTS

Features F5–F8 are computed for all the investigated events and provided in Table 15.3, where symbols EC1 to EC8 are used for the investigated event category (EC). These features are taken as input to the RBDT for classifying the different operational events. The classification of all events using features F5 to F8 with the help of RBDT is illustrated in Figure 15.13. It is observed that events EC2, EC5, EC10, and EC8 are classified one by one using feature F7. The rest of the events are grouped into two groups using the feature F6. Subsequently, the events included in both groups are classified one by one using various features illustrated in Figure 15.13.

15.6 PERFORMANCE COMPARISON

The performance of the technique introduced in this chapter is compared with the DWT-based technique reported in [32,33]. It is established that the performance of

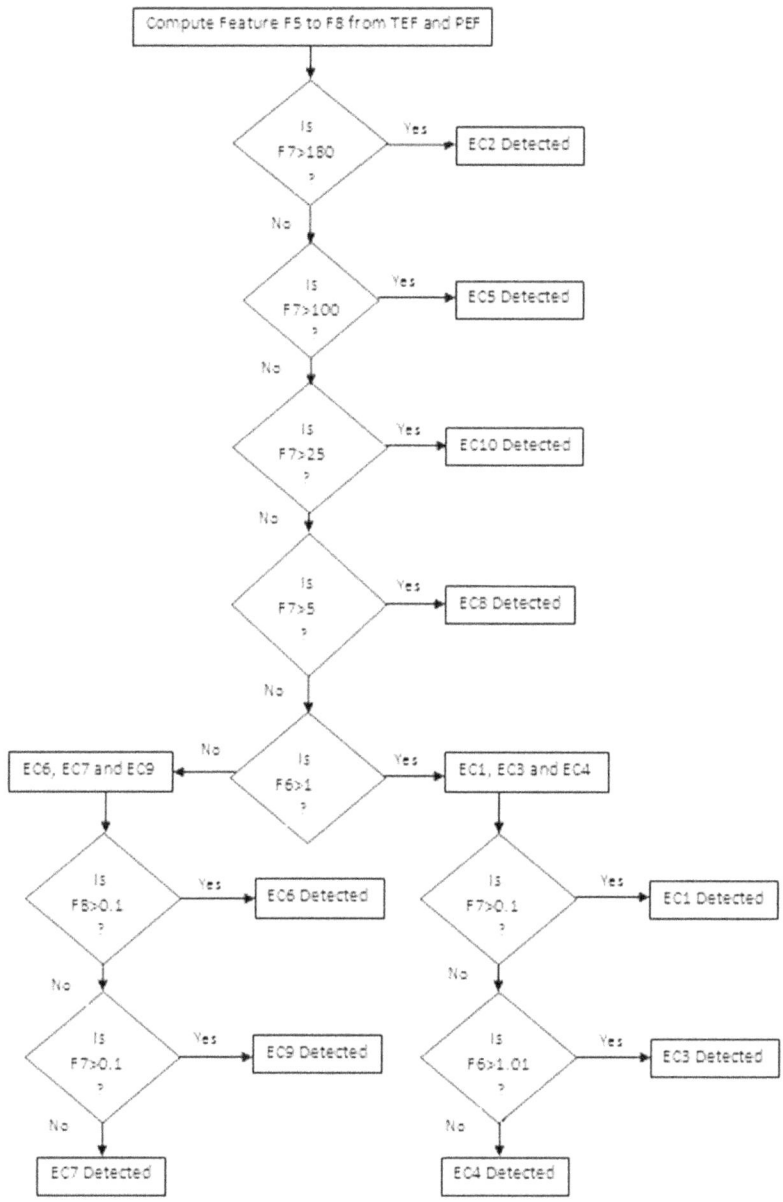

FIGURE 15.13 Event classification.

the proposed method is better in comparison to the DWT-based algorithm reported in [32,33] in terms of the identification of PQ issues, even including the low magnitude transients associated with all the investigated events. However, the DWT is not as effective in the recognition of low magnitude PQ issues.

15.7   CONCLUSION

This chapter introduced a method for the recognition of PQ issues and estimation of events using a hybrid combination of ST and HT in the utility distribution grid with RE penetration. PEF and TEF are introduced, which are effective in the estimation and location of PQ events. It is concluded that PQ issues such as voltage swell, voltage sag, OT, IT, and transients are associated with the operational events of feeder operation, switching of loads, switching of capacitor, and operation of RE generators in the utility distribution grid with penetration of wind and solar energy. Further, the features computed from the PEF and TEF are effective in the identification of the various operational events using RBDT. The performance of the method considered in this chapter is better in comparison to the DWT-based method available in literature. The study is performed on the IEEE-13 nodes test system after integrating a WPP and a SPP in the MATLAB®/Simulink environment.

REFERENCES

[1] H. R., Seo, A. R. Kim, M. Park, and I. K. Yu, "Power quality enhancement of renewable energy source power network using SMES system," *Physica C*, Vol. 471, pp. 1409–1412, 2011.

[2] Om Prakash, Mahela, and Abdul Gafoor Shaik, "Comprehensive Overview of Grid Interfaced Wind Energy Generation Systems," *Renewable and Sustainable Energy Reviews (Elsevier)*, Vol. 57, pp. 260–281, May 2016.

[3] Om Prakash, Mahela, Neeraj Gupta, Mahdi Khosravy, and Nilesh Patel, "Comprehensive Overview of Low Voltage Ride through Methods of Grid Integrated Wind Generator," *IEEE Access*, Vol. 7, Issue 1, pp. 99299–99326, December 2019.

[4] Om Prakash, Mahela, and Abdul Gafoor Shaik. "Comprehensive Overview of Grid Interfaced Solar Photovoltaic Systems," *Renewable and Sustainable Energy Reviews (Elsevier)*, Vol. 68, Part 1, pp. 316–332, February 2017, 10.1016/j.rser.2016.09.096.

[5] Om Prakash, Mahela, Abdul Gafoor Shaik, and Neeraj Gupta. "A critical review of detection and classification of power quality events," *Renewable and Sustainable Energy Reviews (Elsevier)*, Vol. 41, pp. 495–505, January 2015, ISSN No. 1364-0321-0615.

[6] Om Prakash, Mahela, and Abdul Gafoor Shaik. "Topological aspects of power quality improvement techniques: A comprehensive overview," *Renewable and Sustainable Energy Reviews (Elsevier)*, Vol. 58, pp. 1129–1142, May 2016, 10.101 6/j.rser.2015.12.251.

[7] Gajendra Singh, Chawda, Abdul Gafoor Shaik, Mahmood Shaik, P. Sanjeevikumar, Jens Bo Holm-Nielsen, Om Prakash Mahela, and K. Palanisamy, "Comprehensive Review on Detection and Classification of Power Quality Disturbances in Utility Grid with Renewable Energy Penetration," in *IEEE Access*, Vol. 8, pp. 146807–146830, 2020.

[8] Gajendra Singh, Chawda, Abdul Gafoor Shaik, Om Prakash Mahela, Sanjeevikumar Padmanaban, and Jens Bo Holm-Nielsen, "Comprehensive Review of Distributed FACTS Control Algorithms for Power Quality Enhancement in Utility Grid with Renewable Energy Penetration," in *IEEE Access*, vol. 8, pp. 107614–107634, 2020, 10.1109/ACCESS.2020.3000931.

[9] Om Prakash, Mahela, Baseem Khan, Hassan Haes Alhelou, and Sudeep Tanwar, "Assessment of Power Quality in the Utility Grid Integrated with Wind Energy Generation," *IET Power Electronics*, January 2020.

[10] Ashwin, Venkatraman, Kandarpa Sai Paduru, Om Prakash Mahela, and Abdul Gafoor Shaik. "Experimental investigation of power quality disturbances associated with grid integrated wind energy system," In: *4th IEEE International Conference on Advanced Computing and Communication Systems (ICACCS 2017)*, Coimbatore, India, pp. 1–6, January 6–7, 2017.

[11] Kavita, Kumari, Atul Kumar Dadhich, and Om Prakash Mahela. "Detection of Power Quality Disturbances in the Utility Grid with Wind Energy Penetration Using Stockwell Transform," In: *IEEE 7th Power India International Conference on advances in signal processing (PIICON 2016)*, Bikaner, India, November 25–27, 2016.

[12] Om Prakash, Mahela, Pragati Prajapati and Soyab Ali. "Investigation of Power Quality Events in Distribution Network with Wind Energy Penetration," In: *2018 IEEE Students' Conference on Electrical, Electronics and Computer Science (SCEECS 2018)*, MANIT, Bhopal, India, February 24–25, 2018.

[13] Om Prakash, Mahela, Umesh Sharma and Tanuj Manglani." Recognition of Power Quality Disturbances Using Discrete Wavelet Transform and Fuzzy C-means Clustering," In: *2018 IEEE 8th Power India International Conference (PIICON 2018)*, NIT Kurukshetra, India, December 10–12, 2018.

[14] R., Hooshmand, and A. Enshaee, "Detection and classification of single and combined power quality disturbances using fuzzy systems oriented by particle swarm optimization algorithm," *Electric Power Systems Research*, Vol. 80, pp. 1552–1561, 2010.

[15] Mahaveer, Meena, Om Prakash Mahela, Mahendra Kumar and Neeraj Kumar. "Detection and Classification of Complex Power Quality Disturbances Using Stockwell Transform and Rule Based Decision Tree," In: *IEEE PES International Conference on Smart Electric Drives and Power System (ICSEDPS-2018)*, G H Raisoni College of Engineering, Nagpur, India, June 12–13, 2018.

[16] Mahaveer, Meena, Om Prakash Mahela, Mahendra Kumar and Neeraj Kumar, "Detection and Classification of Power Quality Disturbances Using Stockwell Transform and Rule Based Decision Tree," In: *IEEE PES International Conference on Smart Electric Drives and Power System (ICSEDPS-2018)*, G H Raisoni College of Engineering, Nagpur, India, June 12–13, 2018.

[17] Om Prakash, Mahela and Abdul Gafoor Shaik. "Recognition of Power Quality Disturbances Using S-Transform Based Ruled Decision Tree and Fuzzy C-Means Clustering Classifiers," *Applied Soft Computing (Elsevier)*, Vol. 59, pp. 243–257, October 2017.

[18] Om Prakash, Mahela, Abdul Gafoor Shaik, Neeraj Gupta, Mahdi Khosravy, Baseem Khan, Hassan Haes Alhelou, and Sanjeevikumar Padmanaban, "Recognition of the Power Quality Issues Associated with Grid Integrated solar Photovoltaic Plant in Experimental Frame Work," *IEEE Systems Journal*, 2020, 10.1109/JSYST.2020.3 027203.

[19] Om Prakash, Mahela, and Abdul Gafoor Shaik. "Power quality detection in distribution system with wind energy penetration using discrete wavelet transform," In: *2nd IEEE International Conference on Advances in Computing and Communication Engineering (ICACCE-2015)*, Tula's Institute, Dehradun, India, May 1–2, 2015.

[20] Om Prakash, Mahela, Baseem Khan, Hassan Haes Alhelou, and Pierluigi Siano, "Power Quality Assessment and Event Detection in Distribution Network with Wind Energy Penetration Using Stockwell Transform and Fuzzy Clustering," *IEEE Transactions on Industrial Informatics*, Vol. 16, Issue 11, pp. 6922–6932, November 2020.

[21] Sheesh Ram, Ola, Amit Saraswat, Sunil Kumar Goyal, S. K. Jhajharia, Baseem Khan, Om Prakash Mahela, Hassan Haes Alhelou, and Pierluigi Siano, "A Protection Scheme for Power System with Solar Energy Penetration," *Applied Sciences 2020*, Vol. 10, Issue 4, Paper No. 1516, pp. 1–22, Feb. 2020.

[22] Om Prakash, Mahela, and Abdul Gafoor Shaik. "Power Quality Recognition in Distribution System with Solar Energy Penetration Using S-Transform and Fuzzy C-Means Clustering," *Renewable Energy (Elsevier)*, Vol. 106, pp. 37–51.

[23] Om Prakash, Mahela, and Abdul Gafoor Shaik. "Detection of power quality disturbances associated with grid integration of 100 kW solar PV plant," In: *1st IEEE Uttar Pradesh Conference-International Conference on Energy Economics and Environment (ICEEE 2015)*, Noida, India, March 27–28, 2015, 10.1109/EnergyEconomics.2015.7235070.

[24] Govind Sahay, Yogee, Om Prakash Mahela, Kapil Dev Kansal, Baseem Khan, Rajendra Mahla, Hassan Haes Alhelou, and Pierluigi Siano, "An Algorithm for Recognition of Fault Conditions in the Utility Grid with Renewable Energy Penetration," *Energies*, Vol. 13, No. 9, pp. 2383, 10.3390/en13092383.

[25] W. H., Kersting, "Radial distribution test feeders," *IEEE Transation on Power System*, Vol. 6, pp. 975–985, 1991, 10.1109/59.119237.

[26] Abdul Gafoor, Shaik and Om Prakash Mahela, "Power quality assessment and event detection in hybrid power system," *Electric Power Systems Research (Elsevier)*, Vol. 161, pp. 26–44, March 2018.

[27] O. P., Mahela, Y. Sharma, S. Ali, B. Khan and S. Padmanaban, "Estimation of Islanding Events in Utility Distribution Grid With Renewable Energy Using Current Variations and Stockwell Transform," in *IEEE Access*, vol. 9, pp. 69798–69813, 2021.

[28] N. K., Swarnkar, O. P. Mahela, B. Khan and M. Lalwani, "Identification of Islanding Events in Utility Grid with Renewable Energy Penetration Using Current Based Passive Method," in *IEEE Access*, 10.1109/ACCESS.2021.3092971.

[29] Fsaha Mebrahtu, Gebru, Baseem Khan, and Hassan Haes Alhelou, "Analyzing low voltage ride through capability of doubly fed induction generator based wind turbine," *Computers & Electrical Engineering*, Vol. 86, 106727, 2020.

[30] Takele Ferede, Agajie, Baseem Khan, Hassan Haes Alhelou, and Om Prakash Mahela, "Optimal expansion planning of distribution system using grid-based multi-objective harmony search algorithm," *Computers & Electrical Engineering*, Vol. 87, 106823, 2020.

[31] Yishak, Kifle, Baseem Khan, and Pawan Singh, "Assessment and Enhancement of Distribution System Reliability by Renewable Energy Sources and Energy Storage", *Journal of Green Engineering*, Vol. 8, Issue 3 (2), pp. 219–262, July 2018.

[32] Om Prakash, Mahela, Kapil Dev Kansal and Sunil Agarwal. "Detection of Power Quality Disturbances in Utility Grid with Wind Energy Penetration," In: *8th IEEE India International Conference on Power Electronics (IICPE–2018)*, MNIT Jaipur, India, December, 13–14, 2018.

[33] Om Prakash, Mahela, Kapil Dev Kansal and Sunil Agarwal. "Detection of Power Quality Disturbances in Utility Grid with Solar Photovoltaic Energy Penetration," In: *8th IEEE India International Conference on Power Electronics (IICPE–2018)*, MNIT Jaipur, India, December 13–14, 2018, 10.1109/IICPE.2018.8709597.

16 Recognition and Categorization of PQ Disturbances Using a Power Quality Index and Mesh Plots

Om Prakash Mahela
Power System Planning Division, Rajasthan Rajya Vidyut
Prasarn Nigam Ltd., Jaipur, India

Shoyab Ali and Shruti Rathore
Department of Electrical Engineering, Vedant College of
Engineering and Technology, Bundi, India

Baseem Khan
Department of Electrical and Computer Engineering,
Hawassa University, Ethiopia

Krishan Gopal Sharma
Department of Electrical Engineering, Government
Engineering College Ajmer, India

Akhil Ranjan Garg
Department of Electrical Engineering, Faculty of Engineering
and Architecture, J.N.V. University, Jodhpur, India

CONTENTS

DOI: 10.1201/b22884-16

16.1   INTRODUCTION

Power quality (PQ) is a subject of importance/concern in the last 30 years to consumers as well as utilities. PQ deteriorates due to intensive use of power electronic–supported equipment, microprocessor-supported devices, controllers used in the industry, loads of non-linear nature, and extensive use of computers [1]. Power quality issues may lead to damage or improper operation of equipment connected to the grid [2]. Signal processing methods are being employed for the identification and classification of the power quality issues [2]. A method is used for the identification and classification of PQ issues in a utility network with the availability of wind energy using a method supported by the Stockwell transform and fuzzy C-means clustering (FCM) [3]. Recognition of PQ disturbances pertaining to solar energy interfaced with the solar energy using the Stockwell transform is available in [4]. A detailed study of existing methods used for the control algorithms for distributed flexible alternating current–based transmission system (FACTS) devices to improve the quality of power in a utility grid with renewable energy penetration is reported in [5]. A similar study for the comprehensive review of detection and classification methods used for the assessment of power quality disturbances in a utility grid with renewable energy penetration is reported in [6]. Hence, it is established that the reported techniques have merits and demerits for the identification of the PQ issues. Further, there are limitations of the existing methods. To overcome the demerits and limitations of the existing methods for the recognition of the PQ issues, different techniques may be designed by optimizing the use of the features so that the efficiency of the recognition of the PQ issues may be improved. In [7], the authors introduced multi-resolution-based decomposition of signals with PQ disturbances, which is effective for analysis of PQ events having a superimposed transient. This method for decomposition of signals using multi-resolution has the capability for identification and localizes the transient disturbances as well as classification of different PQ issues. In [8], a technique using continuous wavelet transform (CWT) to recognize and analyze sags in a voltage signal and transients associated with the voltage signal is proposed. A recursive approach is implemented for the improvement and computation of time-frequency-based electrical events. The characteristics of analyzed signals are evaluated in the time-frequency plane. In [9], the authors introduced a method for online detection of disturbances in voltage signals using the wavelet transform. In [10], a de-noising technique combined with WT for PQ monitoring scheme is introduced. The developed scheme is effective for detecting and localizing PQ events even in a noisy environment. Hence, the scheme can be used for the storage of proper signals used for encompassing PQ events and subsequent analysis. In [11], performance of the wavelet-based scheme for online detection of a voltage-based PQ disturbance has been evaluated. In [12], an approach for the data compression of PQ transients has been reported. The analyzed data can be utilized for analyzing the classification of PQ events. The original data have been reconstituted from the compressed data of PQ events and subsequently analyzed with the help of an advanced version of WT and ST. In [13], the authors introduced a scheme for the identification as well as classification of PQ events associated with the power system networks using a combination ST and NN. A ST gives resolution based on frequency, which uses simultaneous localization of real and imaginary spectrums. A ST is the same compared to the WT spectrum, but there is a correction of phase in the ST spectrum. In [14], the authors introduced a scheme supported by the representation of power system waveforms in two dimensions (2-D), which can be utilized for automatic analysis as well as detection of transient events associated with the power signals. It is observed that the omission of approximation coefficient of signals of WT and de-noising of detailed signals, inverse 2-D DWT gives fair results of detection and localization of signals even in the conventional techniques. Different methods of recognition of disturbances in the utility grid are available in [15–24]. The output of the PQ recognition of techniques can be used for PQ improvement using the methods reported in [25–38].

After a detailed analysis of the research work discussed in the previous section, it is observed that the feature selection is an important task of the recognition of single-stage PQ disturbances, which needs to be investigated in detail. Hence, this research work has considered a technique using the Stockwell transform and decision rules for classification of the single-stage PQ issues with a minimum number of features. Research work included in this chapter is summarized with the help of the following points:

- A technique supported by the Stockwell transform and decision rules for identification of the single-stage PQ events have been proposed in this work.
- The Stockwell transform and decision rules supported approach is effective in the identification of the single-stage PQ issues.
- A power quality index (PQI) and a PQ time location index (PQTLI) are proposed for the identification of different types of single-stage PQ issues.
- Six statistical features are computed from the PQI and PQTLI, which are considered as the input to the rule-supported decision tree to classify the PQ disturbances.

- Classification accuracy of single-stage PQ issues has been achieved as high as 98.44%.
- The study is performed in the MATLAB®/Simulink environment.

This manuscript is organized as follows: Section 16.2 discusses the methodology utilized in this work. Section 16.3 presents the results with a detailed discussion. Section 16.4 presents the extraction of features from the PQI and PQTLI for classification of PQ events. Section 16.5 presents the classification of PQ disturbances. Section 16.6 discusses the performance validation, followed by the conclusion in Section 16.7.

16.2 RESEARCH METHOD

The proposed algorithm used for recognition and classification of the single-stage PQ events is detailed in this section. Generation of the single-stage PQ events by the use of standard mathematical formulations is also described in this section.

16.2.1 FORMULATION OF SINGLE-STAGE PQ DISTURBANCES

The mathematical relations reported in [39] are utilized for the generation of single-stage PQ issues. These mathematical relations are formulated in MATLAB software. Formulated single-stage PQ issues include pure sine wave, sag associated with voltage signal, swell associated with voltage signal, momentary interruption associated with voltage signal, harmonics associated with voltage signal, oscillatory transient associated with voltage signal, impulsive transient associated with voltage signal, notch associated with voltage signal, and spike associated with voltage signal. These signals with associated single-stage PQ issues are processed using the proposed algorithm.

16.2.2 ALGORITHM ADOPTED TO IDENTIFY AND CATEGORIZE THE PQ DISTURBANCES

The algorithm adopted to identify and categorize the PQ issues is described in Figure 16.1. This algorithm is used for the recognition of the single-stage PQ events. This algorithm can be implemented with the following steps:

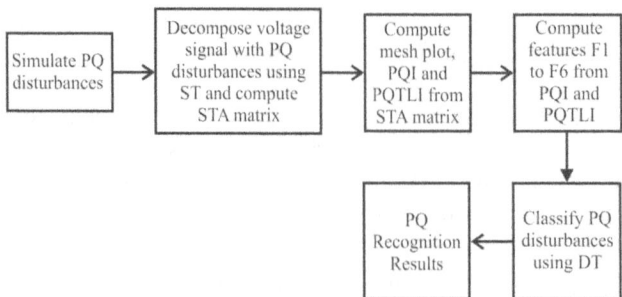

FIGURE 16.1 Proposed PQ recognition algorithm.

- Simulate the PQ events associated with the voltage signal using the mathematical formulation available in the [39].
- Decompose voltage signal with PQ events using the Stockwell transform with 1.6 kHz sampling frequency (32 samples in every cycle) and compute the output matrix STM, which is complex in nature.
- Compute the absolute values of each element of the matrix STM to compute a matrix with absolute values and the same dimensions and designed as STMA. Below-mentioned command is used:

$$STMA = abs(STM)$$

- Compute the maximum magnitude of every column of the STMA matrix using the following command:

$$AP = max(STA)$$

- Compute the summation of each column of the STMA matrix using the following command:

$$SP = sum(STA)$$

- Compute the angle of the magnitude of every column of the STMA matrix using the following command:

$$PP = angle(max(SS))$$

- Compute the median of each column of the STMA matrix using the following command:

$$M = median(STA)$$

- Compute the weight factor using the following command:

$$C = max(SP)100;$$

- Compute the power quality index (PQI) using the following command:

$$PQI2 = (AP. * SP. * PP)/1.75;$$

- Compute the power quality time location index (PQTLI) using the following command:

$$PQI1 = C. * (AP. * SP. * M);$$

- Compute the mesh plot from the STMA matrix.
- Plot mesh plot, PQI, and PQTLI. Compare these plots for signal with PQ disturbance with the respective plots of the pure sine wave to recognize the associated PQ disturbances.
- Extract features from the PQI and PQTLI plots. The values of these features are considered as input to the rule-based decision tree for classification of the PQ disturbances.

16.3 RESULTS AND ANALYSIS

Results of the simulation for the identification of PQ issues using the proposed approach are discussed. Signals with associated PQ events are processed using the Stockwell transform for computing the output matrix. Proposed power quality index (PQI), PQ time location index (PQTLI), and mesh plots are computed from this matrix to identify the PQ disturbances associated with the voltage signals. Features are computed from the PQI and PQTLI and considered as input to the rule-supported decision tree to categories of the different PQ events.

16.3.1 VOLTAGE SIGNAL WITHOUT PQ DISTURBANCE

Voltage signals without a PQ disturbance are processed using the Stockwell transform and the absolute values output matrix STMA is computed. A mesh plot of this matrix is obtained, which represents the time on the x-axis, normalized frequency on the y-axis, and amplitude on the z-axis. It is observed that for the normalized frequency corresponding to the fundamental frequency of 50 Hz, the amplitude is unity for all the time instants between 0 to 0.2s. High magnitudes are not observed for the frequencies other than the fundamental. The voltage signal without a PQ issue is processed using the Stockwell transform and proposed power quality index (PQI) and proposed power quality time location index (PQTLI) are computed. The voltage signal, PQI, and PQTLI are are computed. It is observed that the voltage waveform is pure sinusoidal in nature. The mesh plot, PQI, and PQTLI plots for a pure sine wave are considered as the reference plots to recognize a PQ event.

16.3.2 SAG ASSOCIATED WITH THE VOLTAGE SIGNAL

The voltage signal with sag between the time 0.06 s to 0.14 s is simulated and processed using the Stockwell transform and absolute values output matrix STMA is computed. A mesh plot of this matrix is obtained, which represents the time on the x-axis, normalized frequency on the y-axis, and amplitude on the z-axis, which is shown in Figure 16.2. From this figure, it is observed that for the normalized frequency corresponding to the fundamental frequency of 50 Hz, the amplitude except between the time 0.06 s to 0.14 s is unity. During the time 0.06 s to 0.14 s, the magnitude corresponding to the fundamental frequency decreases, which indicates the presence of the voltage sag. Hence, sag associated with the voltage signal has been detected effectively.

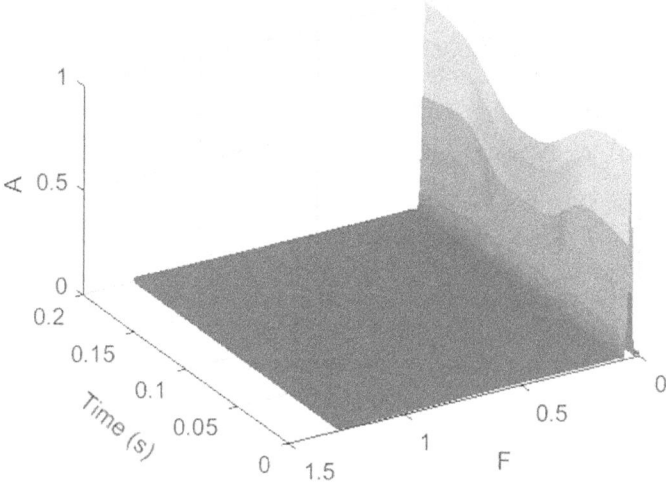

FIGURE 16.2 Mesh plot for the voltage waveform with sag.

The voltage signal with a PQ disturbance of sag is processed using the Stockwell transform the and proposed power quality index (PQI) and proposed power quality time location index (PQTLI) are computed. The voltage signal, PQI, and PQTLI are shown in Figure 16.3(a), (b), and (c), respectively. From Figure 16.3(a), it is observed that the magnitude of the voltage waveform has decreased between the time 0.06 s to 0.14 s. Figure 16.3(b) indicates that the magnitude of the PQI has decreased between the time 0.06 s to 0.14 s, which indicates the presence of the voltage sag. Figure 16.3(c) shows that the PQTLI has sharp magnitude peaks at the

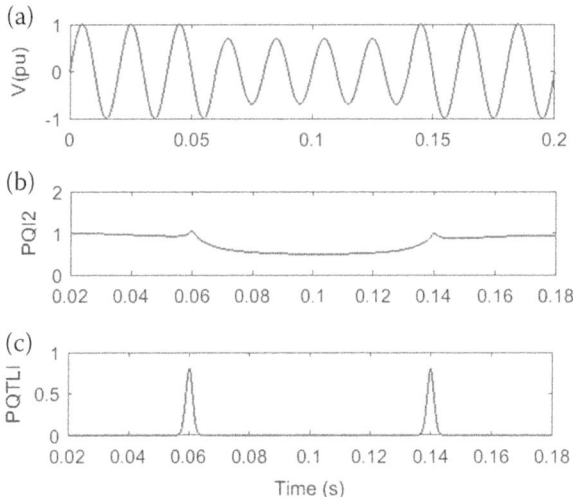

FIGURE 16.3 Voltage waveform with sag: (a) voltage waveform, (b) PQ index, and (c) PQ time location index.

time instants of 0.06 s and 0.14 s, which indicates the start and end of the sag in voltage. Hence, it is established that the proposed approach effectively identified and localized PQ disturbance of sag associated with the voltage signal.

16.3.3 SWELL ASSOCIATED WITH THE VOLTAGE SIGNAL

The voltage signal with a swell between the time 0.06 s to 0.14 s is simulated and processed using the Stockwell transform and the absolute values output matrix STMA is computed. A mesh plot of this matrix is obtained, which represents the time on the x-axis, normalized frequency on the y-axis, and amplitude on the z-axis, which is shown in Figure 16.4. From this figure, it is observed that for the normalized frequency corresponding to fundamental frequency of 50 Hz, the amplitude except between the time 0.06 s to 0.14 s is unity. During the time 0.06 s to 0.14 s, the magnitude corresponding to the fundamental frequency increases, which indicates the presence of the voltage swell. Hence, the swell associated with the voltage signal has been detected effectively. The mesh plot gives the visual inspection of the available voltage swell.

The voltage signal with a PQ disturbance of a swell is processed using the Stockwell transform and the proposed power quality index (PQI) and proposed power quality time location index (PQTLI) are computed. The voltage signal, PQI, and PQTLI are shown in Figure 16.5(a), (b), and (c), respectively. From Figure 16.5(a), it is observed the magnitude of the voltage waveform has increased between the time 0.06 s to 0.14s. Figure 16.5(b) indicates that the magnitude of the PQI has increased between the times 0.06 s to 0.14 s, which indicates the presence of the voltage swell. Figure 16.5(c) shows that the PQTLI has sharp magnitude peaks at the time instants of 0.06 s and 0.14 s, which indicates the start and end of the swell in voltage. Hence, it is established that proposed approach effectively identified and localized the PQ disturbance of the swell associated with the voltage signal.

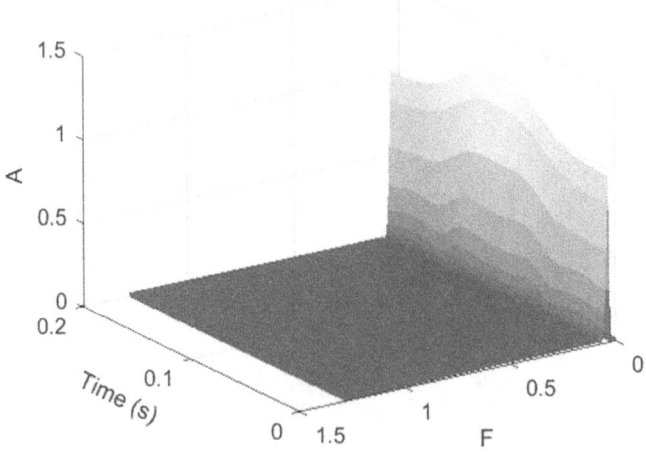

FIGURE 16.4 Mesh plot for the voltage waveform with swell.

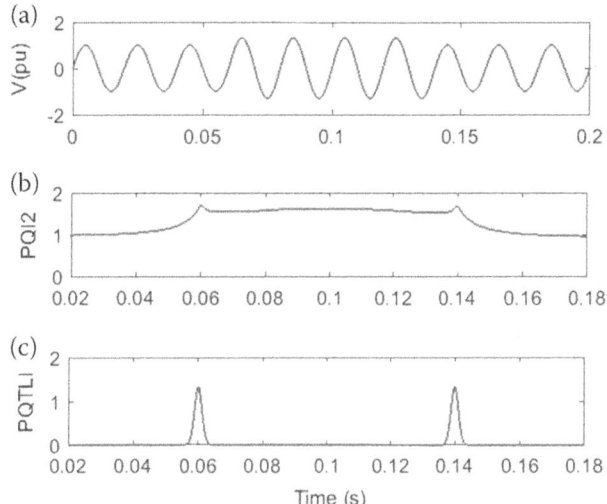

FIGURE 16.5 Voltage waveform with swell: (a) voltage waveform, (b) PQ index, and (c) PQ time location index.

16.3.4 MOMENTARY INTERRUPTION ASSOCIATED WITH THE VOLTAGE SIGNAL

The voltage signal with an interruption between the time 0.06 s to 0.14 s is simulated and processed using the Stockwell transform and the absolute values output matrix STMA is computed. A mesh plot of this matrix is obtained, which represents the time on the x-axis, normalized frequency on the y-axis, and amplitude on the z-axis. It is observed that for the normalized frequency corresponding to a fundamental frequency of 50 Hz, the amplitude except between the time 0.06 s to 0.14 s is unity. During the time 0.06 s to 0.14 s, the magnitude corresponding to the fundamental frequency decreases and goes below 10%, which indicates the presence of momentary interruption (MI). Hence, the MI associated with the voltage signal has been detected effectively. The voltage signal with a PQ disturbance of momentary interruption (MI) is processed using the Stockwell transform and the proposed power quality index (PQI) and proposed power quality time location index (PQTLI) are computed. The voltage signal, PQI, and PQTLI are obtained. It is observed that the magnitude of the voltage waveform has decreased between the time 0.06 s to 0.14 s below 10%. It is also observed that the magnitude of the PQI has decreased between the time 0.06 s to 0.14 s below 10%, which indicates the presence of the MI. Further, it is observed that the PQTLI has sharp magnitude peaks at the time instants of 0.06 s and 0.14 s, which indicates the start and end of the MI in voltage. Hence, it is established that the proposed approach effectively identified and localized the PQ disturbance of MI associated with the voltage signal.

16.3.5 HARMONICS ASSOCIATED WITH THE VOLTAGE SIGNAL

The voltage signal with harmonics of the orders 3rd, 5th, and 7th is simulated and processed using the Stockwell transform to compute the absolute values output

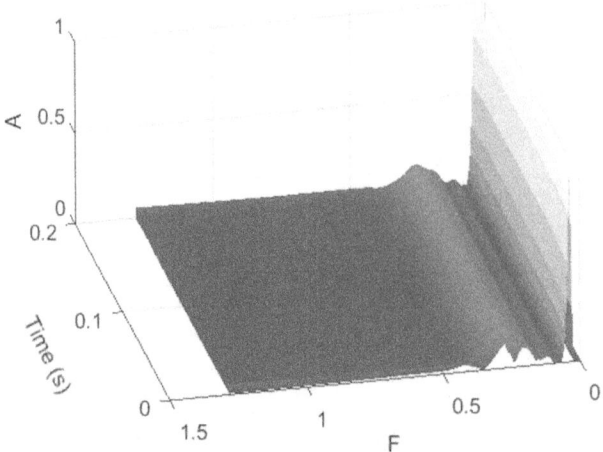

FIGURE 16.6   Mesh plot for the voltage waveform with harmonics.

matrix STMA. A mesh plot of this matrix is obtained, which represents the time on the x-axis, normalized frequency on the y-axis, and amplitude on the z-axis, which is shown in Figure 16.6. From this figure, it is observed that for the normalized frequency corresponding to the fundamental frequency of 50 Hz, the amplitude is constant and equal to unity. Further, the amplitude for the frequency corresponding to the 3rd, 5th, and 7th harmonic components is also high for the entire time range. Hence, the harmonics have been detected effectively using the mesh plot. The mesh plot gives the visual inspection of the available harmonic components.

The voltage signal with a PQ disturbance of harmonics of the orders 3rd, 5th, and 7th is processed using the Stockwell transform and the proposed power quality index (PQI) and proposed power quality time location index (PQTLI) are computed. The voltage signal, PQI, and PQTLI are shown in Figure 16.7(a), (b), and (c), respectively. From Figure 16.7(a), it is observed that the magnitude of the voltage waveform has continuous ripples over the entire time range, which indicates the presence of harmonics. Figure 16.7(b) indicates that magnitude of the PQI has continuous ripples of high magnitude and a fixed pattern over the entire time range, which indicates the presence of the harmonics. Figure 16.7(c) shows that the PQTLI has continuous ripples with sharp tips and a fixed pattern over the entire time range, which indicates the presence of harmonics with the signal. Hence, it is established that the proposed approach effectively identified harmonics associated with the voltage signal.

16.3.6   Oscillatory Transient Associated with the Voltage Signal

The voltage signal with an oscillatory transient between the time duration 0.08 s to 0.10 s is simulated and processed using the Stockwell transform to compute the absolute values output matrix STMA. A mesh plot of this matrix is obtained, which represents the time on the x-axis, normalized frequency on the y-axis, and amplitude on the z-axis, which is shown in Figure 16.8. From this figure, it is observed that for the

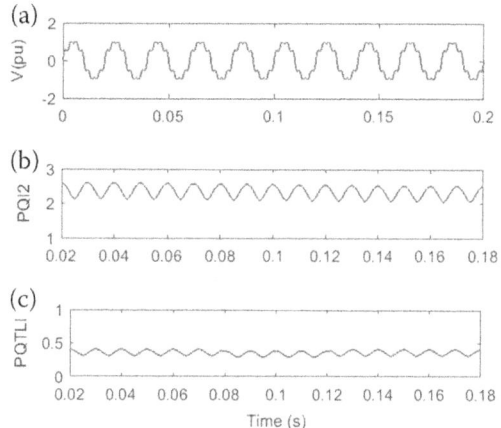

FIGURE 16.7 Voltage waveform with harmonics: (a) voltage waveform, (b) PQ index, and (c) PQ time location index.

normalized frequency corresponding to the fundamental frequency of 50 Hz, the amplitude is constant and equal to unity. Further, a sharp magnitude peak is observed between the time duration 0.08 s to 0.10 s, indicating the presence of the OT. It is also observed that due to the presence of OT, low magnitude finite values are present throughout the time range, which indicates that all the frequencies are associated with the OT. Hence, the OT has been effectively recognized using the mesh plot. The mesh plot gives the visual inspection of the available OT with the signal.

The voltage signal with the oscillatory transient between the time duration 0.08 s to 0.10 s is processed using the Stockwell transform and the proposed power quality index (PQI) and proposed power quality time location index (PQTLI) are computed.

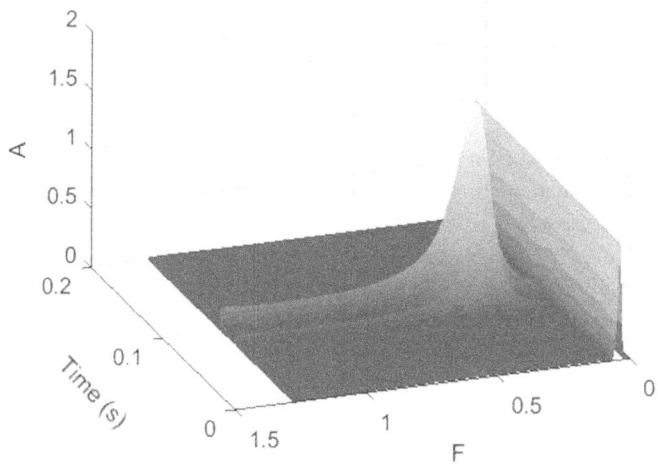

FIGURE 16.8 Mesh plot for the voltage waveform with oscillatory transient.

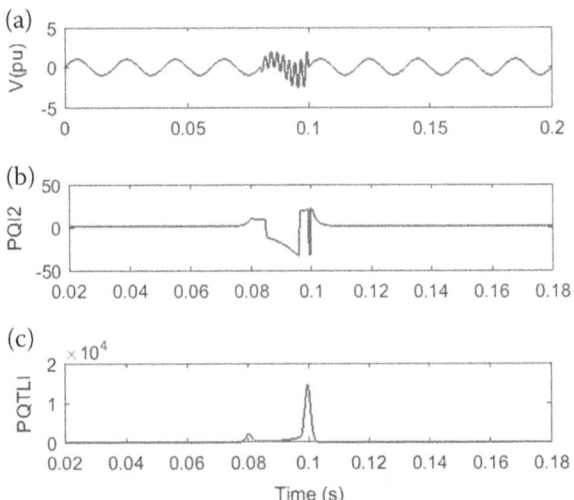

FIGURE 16.9 Voltage waveform with oscillatory transient: (a) voltage waveform, (b) PQ index, and (c) PQ time location index.

The voltage signal, PQI, and PQTLI are shown in Figure 16.9(a), (b), and (c), respectively. From Figure 16.9(a), it is observed that the magnitude of the waveform oscillates for the duration of 0.08 s to 0.10 s, indicating that the OT is associated with the voltage waveform. Figure 16.9(b) indicates that the magnitude of the PQI changes abruptly during the time of 0.08 s to 0.10 s, which indicates the presence of the OT associated with the voltage signal. Figure 16.9(c) shows that the PQTLI has high magnitude peaks at the moments 0.08 s and 0.10 s, which localizes the OT by identifying the starting of the OT and end of the OT.

16.3.7 IMPULSIVE TRANSIENT ASSOCIATED WITH THE VOLTAGE SIGNAL

The voltage signal with the impulsive transient between the time duration 0.085 s to 0.88 s is simulated and processed using the Stockwell transform to compute the absolute values output matrix STMA. A mesh plot of this matrix is obtained, which represents the time on the x-axis, normalized frequency on the y-axis, and amplitude on the z-axis, which is shown in Figure 16.10. From this figure, it is observed that for the normalized frequency corresponding to the fundamental frequency of 50 Hz, the amplitude is constant and equal to unity. Further, finite values are observed over the entire frequency range between the time duration 0.085 s to 0.88 s, indicating the presence of the IT. Hence, the IT has been effectively recognized using the mesh plot. The mesh plot gives the visual inspection of the available IT with the signal.

  The voltage signal with the impulsive transient between the time duration 0.085 s to 0.88 s is processed using the Stockwell transform and the proposed power quality index (PQI) and proposed power quality time location index (PQTLI) are computed. The voltage signal, PQI, and PQTLI are shown in Figure 16.11(a), (b), and (c), respectively. From Figure 16.15(a), it is observed that there is a sharp magnitude peak on the waveform between the time duration of 0.08 s to 0.10 s, indicating that the IT is

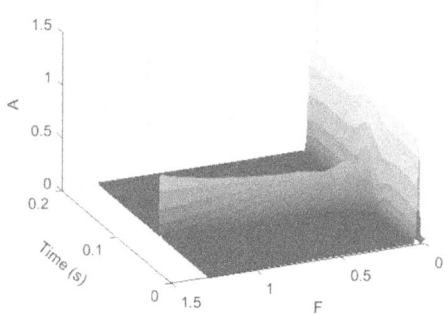

FIGURE 16.10   Mesh plot for the voltage waveform with impulsive transient.

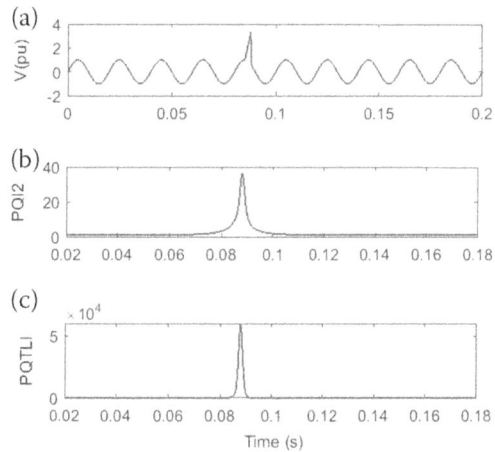

FIGURE 16.11   Voltage waveform with impulsive transient: (a) voltage waveform, (b) PQ index, and (c) PQ time location index.

associated with the voltage waveform. Figure 16.11(b) indicates that the magnitude of the PQI changes abruptly during the time of 0.085 s to 0.88 s and a sharp magnitude peak is obtained, which indicates the presence of the IT associated with the voltage signal. Figure 16.11(c) shows that PQTLI has a high magnitude peak between the time duration of 0.085 s to 0.88 s, which detects and localizes the IT.

16.3.8   Notch Associated with the Voltage Signal

The voltage signal with notches throughout the time range is simulated and processed using the Stockwell transform to compute the absolute values output matrix STMA. A mesh plot of this matrix is obtained, which represents the time on the x-axis, normalized frequency on the y-axis, and amplitude on the z-axis, which is shown in Figure 16.12. From this figure, it is observed that for the normalized frequency corresponding to the fundamental frequency of 50 Hz, the amplitude is constant and equal to unity. Further, a continuous series of ripples is available over

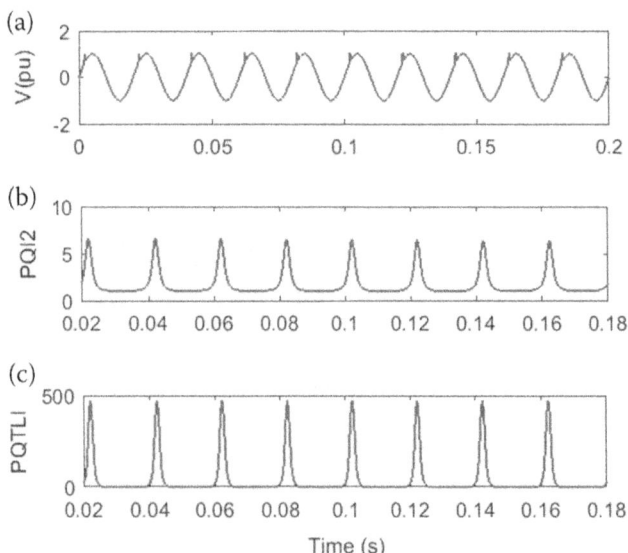

FIGURE 16.12 Voltage waveform with spikes: (a) voltage waveform, (b) PQ index, and (c) PQ time location index.

the entire time range, indicating the presence of notches. Hence, the notches have been effectively recognized using the mesh plot.

The voltage signal with notches throughout the time range is processed using the Stockwell transform and the proposed power quality index (PQI) and proposed power quality time location index (PQTLI) are computed. The voltage signal, PQI, and PQTLI are shown in Figure 16.13(a), (b), and (c), respectively. From Figure 16.13(a), it is observed that there is a series of notches on the waveform throughout the time range. Figure 16.13(b) indicates a continuous train of sharp magnitude peaks over the entire time range, which indicates the presence of the notches. Figure 16.13(c) also indicates the continuous train of the sharp magnitude peaks over the entire time range, which indicates the presence of the notches associated with the voltage signals.

16.3.9 Spike Associated with the Voltage Signal

The voltage signal with spikes throughout the time range is simulated and processed using the Stockwell transform to compute the absolute values output matrix STMA. A mesh plot of this matrix is obtained, which represents the time on the x-axis, normalized frequency on the y-axis, and amplitude on the z-axis, which is shown in Figure 16.14. From this figure, it is observed that for the normalized frequency corresponding to the fundamental frequency of 50 Hz, the amplitude is constant and equal to unity. Further, a continuous series of ripples is available over the entire time range, indicating the presence of spikes. Here, the pattern of this plot is slightly different from the respective plot for the notches. This will be identified using the additional statistical features introduced. Hence, the spikes have been effectively recognized using the mesh plot. The mesh plot gives the visual inspection of the available spikes with the signal.

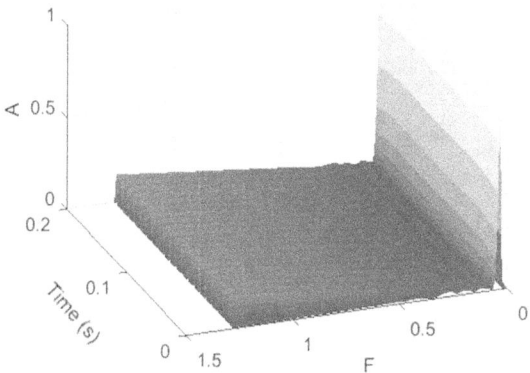

FIGURE 16.13 Mesh plot for the voltage waveform with notches.

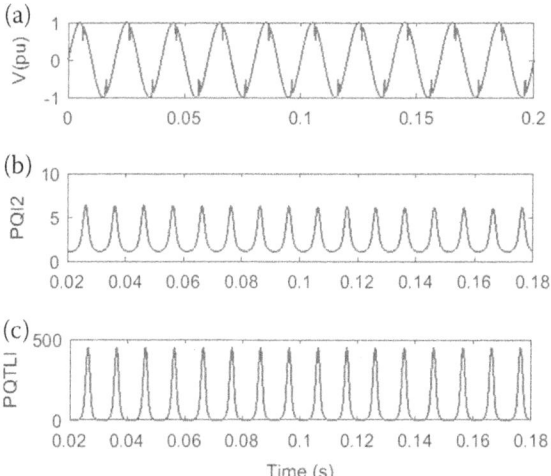

FIGURE 16.14 Voltage waveform with notches: (a) voltage waveform, (b) PQ index, and (c) PQ time location index.

The voltage signal with spikes throughout the time range is processed using the Stockwell transform and the proposed power quality index (PQI) and proposed power quality time location index (PQTLI) are computed. The voltage signal, PQI, and PQTLI are shown in Figure 16.15(a), (b), and (c), respectively. From Figure 16.15(a), it is observed that there is a series of spikes on the waveform throughout the time range with a fixed interval. Figure 16.15(b) indicates that there is continuous train of the sharp magnitude peaks over the entire time range, which indicates the presence of the spikes. Figure 16.15(c) also indicates that there is a continuous train of the sharp magnitude peaks over the entire time range, which indicates the presence of the spikes associated with the voltage signals.

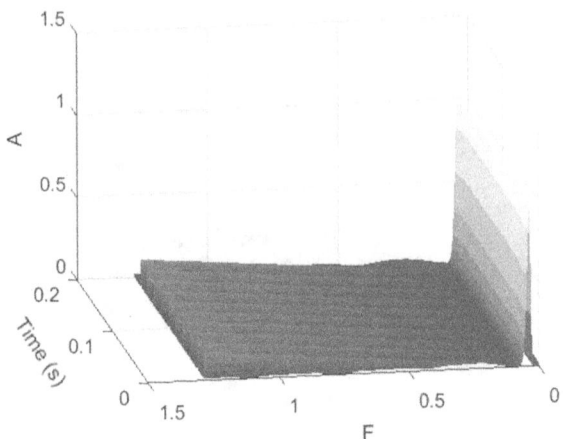

FIGURE 16.15   Mesh plot for the voltage waveform with spikes.

16.4   EXTRACTION OF FEATURES FROM THE PQI AND PQTLI FOR THE CLASSIFICATION OF PQ EVENTS

The features F1 to F6 extracted from the PQI and PQTLI using the statistical techniques are detailed in this section. These features are taken as input to the rule-based decision tree for classifying the PQ disturbances. Definitions of these features are detailed as follows:

F1. Standard deviations of PQI:

$$\sigma = \sqrt{\frac{1}{N-1} \sum_{i=1}^{N} (x_i - \bar{x})^2}$$

where x_i: observed values of sample items, $\bar{x}$: mean value of these observations, N : samples of PQI.
F2. Standard deviations of PQTLI.
F3. Skewness of PQI:

$$S = \frac{E(x - \mu)^3}{\sigma^3}$$

where x: array of data of PQI, μ: mean of x, σ: standard deviation of x, and E: expected value of the quantity.
F4. Skewness of PQTLI.
F5. Kurtosis of PQI:

$$k = \frac{E(x - \mu)^4}{\sigma^4}$$

TABLE 16.1

Features Used for Classification of PQ Disturbances

S. No.	PQ Disturbance	F1	F2	F3	F4	F5	F6
1	Sine wave	0.0954	0.6280	7.0744	8.6011	58.4253	80.7389
2	Voltage sag	0.2320	0.6482	0.3969	8.1844	4.6018	75.1122
3	Voltage swell	0.2792	0.6613	0.1321	7.6760	1.2307	68.3393
4	Momentary interruption	0.4297	0.7384	−0.2979	6.4389	1.9588	50.6393
5	Harmonics	0.2303	8.7890	3.0903	9.0341	23.1372	89.2747
6	Oscillatory transient	6.5559	1.3337×10^3	−2.3205	8.3421	14.2943	77.4863
7	Impulsive transient	3.6749	4.950×10^3	6.6577	9.4919	52.3453	97.8504
8	Notch	1.6355	138.9422	1.2030	1.5594	3.0282	3.9502
9	Spike	1.4136	112.5588	2.1703	2.6926	6.5813	9.0602

where x: array of data of PQI, μ: mean of x, σ: standard deviation of x, and E: expected value of the quantity.

F6. Kurtosis of PQTLI.

The values of features F1 to F6 for the PQ disturbances are tabulated in Table 16.1. From Table 16.2, it is observed that the accuracy of the proposed approach is as high as 98.44% for the classification of the single-stage PQ events.

16.5 CLASSIFICATION OF PQ DISTURBANCES

The single-stage PQ events have been classified using the decision-supported rules. These decision rules are driven by the feature values given in Table 16.1. PQ events

TABLE 16.2

Classification Results of Single-Stage PQ Events

S. No.	PQ Disturbance	Correctly Identified PQ Events (in numbers	Incorrectly Identified PQ Events (in numbers	Accuracy (%)
1	Sine wave	50	0	100
2	Voltage sag	50	0	100
3	Voltage swell	50	0	100
4	Momentary interruption	50	0	100
5	Harmonics	48	2	96
6	Oscillatory transient	50	0	100
7	Impulsive transient	50	0	100
8	Notch	47	3	94
9	Spike	48	2	96
Average Accuracy (%)				98.44%

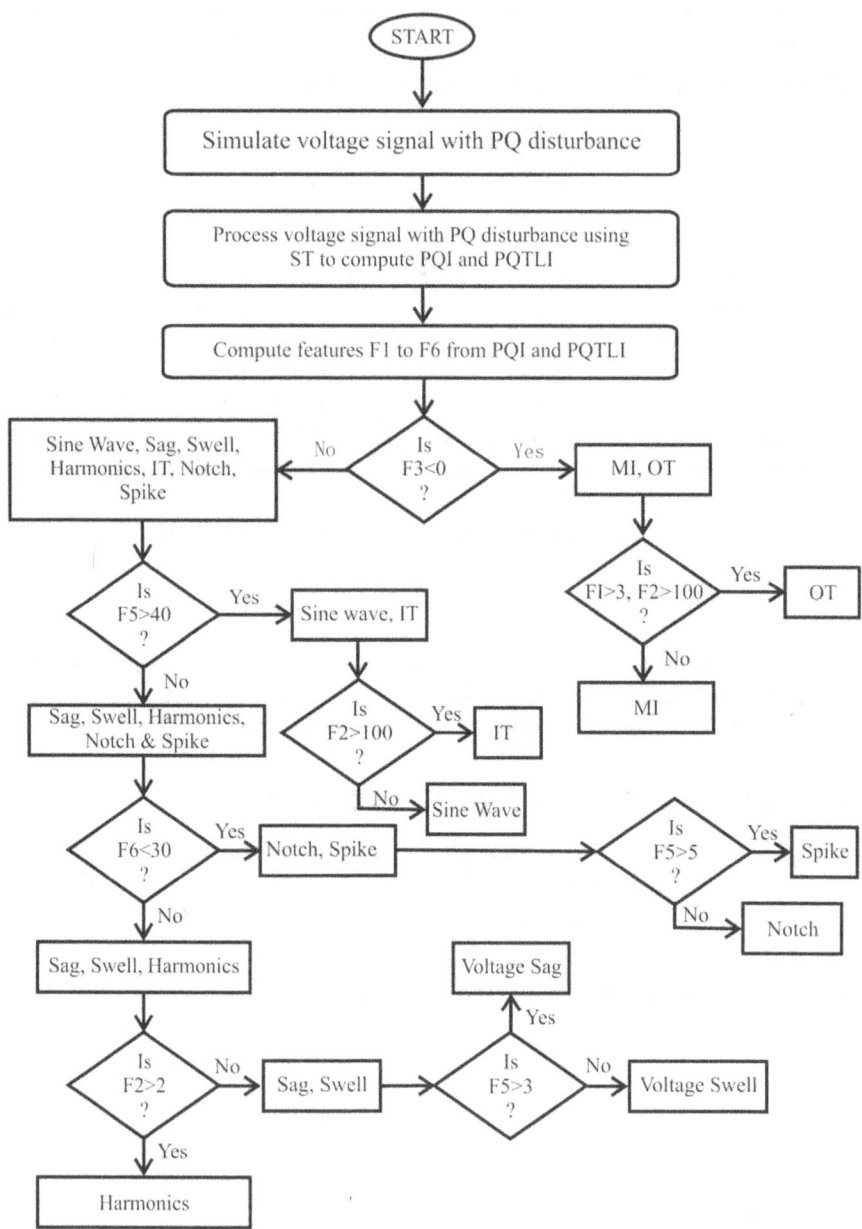

FIGURE 16.16 Classification tree of single-stage PQ events.

are grouped into different clusters using the features and decision rules. Further, the PQ issues are classified one by one from the decision-supported rules from these clusters. The flow chart of the classification of the PQ events is illustrated in Figure 16.16. In this graph, the terminal nodes indicate the final response in terms of the categorized PQ events. It is observed that all the investigated single-stage PQ disturbances are classified effectively using the decision rules driven by the features F1 to F6. Classification accuracy of the approach is evaluated by testing the algorithm on a data set of 50 data computed by changing the various parameters used to define the PQ issues. The correctly identified and incorrectly identified PQ events are described in Table 16.2.

16.6   PERFORMANCE VALIDATION

To establish the performance of the algorithm introduced in this for identification and classification of the PQ events, accuracy of this algorithm is compared with the accuracy of the algorithm reported in the reference [25]. This reference has proposed an algorithm based on the discrete wavelet transform (DWT) and artificial neural network (ANN) and an efficiency of 88% is reported for this algorithm. However, the efficiency of the proposed algorithm based on the ST and DT is 98.44%, which is much higher compared to the algorithm reported in [25].

16.7   CONCLUSION

A technique for the identification of single-stage PQ disturbances is proposed in this work. In the proposed technique, signals with a PQ disturbance are simulated using the mathematical models. These signals are processed using the Stockwell transform to compute an output matrix in frequency domain. A power quality index (PQI) and a PQ time location index (PQTLI) are computed from this matrix and used for the identification of the different types of single-stage PQ issues. Six statistical features are computed from the PQI and PQTLI, which are considered as the input to the rule-based decision tree for classification of the PQ disturbances. It is concluded that the proposed technique based on the Stockwell transform and rule-based decision is found to be effective in the identification of the single-stage PQ issues. Classification accuracy of the single-stage PQ issues has been achieved as high as 98.44%. The performance of the algorithm is compared with the accuracy of the algorithm based on the discrete wavelet transform (DWT) and artificial neural network (ANN), which has an efficiency of 88%. Hence, the proposed algorithm classifies the single stage as well as complex PQ events with accuracy greater than the accuracy of the DWT- and ANN-based approaches reported in literature. The simulation studies of the identification and classification of the PQ issues, which are simple in nature, are presented in this work. However, before application of the algorithm in the PQ monitoring devices, hardware-based validation can be considered as future work.

REFERENCES

[1] Mahela, Om Prakash, and Shaik Abdul Gafoor. "Power Quality Improvement in Distribution Network using DSTATCOM with Battery Energy Storage System," *International Journal of Electrical Power and Energy Systems*, Vol. 83, pp. 229–240, December 2016, 10.1016/j.ijepes.2016.04.011.

[2] Mahela, Om Prakash, and Shaik Abdul Gafoor. "Power Quality Recognition in Distribution System with Solar Energy Penetration Using S-Transform and Fuzzy C-Means Clustering," *Renewable Energy*, Vol. 106, pp. 37–51, June 2017, 10.1016/j.renene.2016.12.098.

[3] Mahela, Om Prakash, Khan Baseem, Alhelou Hassan Haes, and Siano Pierluigi. "Power Quality Assessment and Event Detection in Distribution Network with Wind Energy Penetration Using Stockwell transform and Fuzzy Clustering," *IEEE Transactions on Industrial Informatics*, Vol. 16, Issue 11, pp. 6922–6932, November 2020.

[4] Mahela, Om Prakash, Khan Baseem, Alhelou Hassan Haes, and Tanwar Sudeep."Assessment of Power Quality in the Utility Grid Integrated with Wind Energy Generation," *IET Power Electronics*, Vol. 13, pp. 2917–2925, January 2020, https://doi.org/10.1049/iet-pel.2019.1351.

[5] Chawda, Gajendra Singh, Shaik Abdul Gafoor, Mahela Om Prakash, Padmanaban Sanjeevikumar, and Holm-Nielsen Jens Bo."Comprehensive Review of Distributed FACTS Control Algorithms for Power Quality Enhancement in Utility Grid with Renewable Energy Penetration," in *IEEE Access*, Vol. 8, pp. 107614–107634, 2020.

[6] Gajendra Singh Chawda, Shaik Abdul Gafoor, Shaik Mahmood, Sanjeevikumar P., Holm-Nielsen Jens Bo, Mahela Om Prakash, and Palanisamy K. "Comprehensive Review on Detection and Classification of Power Quality Disturbances in Utility Grid with Renewable Energy Penetration," in *IEEE Access*, Vol. 8, pp. 146807–146830, 2020, 10.1109/ACCESS.2020.3014732.

[7] Gaouda, A. M., M. M. A. Salama, M. K. Sultan, and Chikhani A. Y. "Power Quality Detection and Classification Using Wavelet-Multiresolution Signal Decomposition," *IEEE Transactions on Power Delivery*, Vol. 14, No. 4, pp. 1469–1476, October 1999.

[8] Poisson, Olivier, Rioual Pascal, and Meunier Michel."Detection and Measurement of Power Quality Disturbances Using Wavelet Transform," *IEEE Transactions On Power Delivery*, Vol. 15, No. 3, pp. 1039–1044, July 2000.

[9] Karimi, Masoud, Mokhtari Hossein, and Reza Iravani M. "Wavelet Based On-Line Disturbance Detection for Power Quality Applications," *IEEE Transactions on Power Delivery, VoL.* 15, No. 4, pp. 1212–1220, October 2000.

[10] Yang, Hong-Tzer, and Liao Chiung-Chou. "A De-Noising Scheme for Enhancing Wavelet-Based Power Quality Monitoring System," *IEEE Transactions on Power Delivery*, Vol. 16, No. 3, pp. 353–360, July 2001.

[11] Mokhtari, Hossein, Karimi-Ghartemani Masoud, and Reza Iravani M. "Experimental Performance Evaluation of a Wavelet-Based On-Line Voltage Detection Method for Power Quality Applications," *IEEE Transactions on Power Delivery*, Vol. 17, No. 1, pp. 161–172, January 2002.

[12] Dash, P. K., Panigrahi B. K., Sahoo D. K., and Panda G. "Power Quality Disturbance Data Compression, Detection, and Classification Using Integrated Spline Wavelet and S-Transform," *IEEE Transactions on Power Delivery*, Vol. 18, No. 2, pp. 595–600, April 2003.

[13] Lee, I. W. C., and Dash P. K., "S-Transform-Based Intelligent System for Classification of Power Quality Disturbance Signals," *IEEE Transactions on Industrial Electronics*, Vol. 50, No. 4, pp. 800–805, August 2003.

[14] Ece, Dogan Gökhan, and Gerek Ömer Nezih. "Power Quality Event Detection Using Joint 2-D-Wavelet Subspaces," *IEEE Transactions on Instrumentation and Measurement*, Vol. 53, No. 4, pp. 1040–1046, August 2004.

[15] Masoum, M. A. S., Jamali S., and Ghaffarzadeh N. "Detection and classification of power quality disturbances using discrete wavelet transform and wavelet networks," *IET Science, Measurement and Technology*, Vol. 4, Issue 4, pp. 193–205, 2010.

[16] Yan Cui, Z., and Li Wen hui, "A classification method for complex power quality disturbances using EEMD and rank wavelet SVM," *IEEE Transactions on Smart Grid*, Vol. 6, No. 4, pp. 1678–1685, July 2015.

[17] Mahela, Om Prakash, and Shaik Abdul Gafoor. "Recognition of Power Quality Disturbances Using S-transform and Rule-Based Decision Tree," In: 2016 IEEE First International Conference on Power Electronics, Intelligent Control and Energy Systems (ICPEICES 2016), New Delhi, India, July 4–6, 2016.

[18] Mahela, Om Prakash, and Shaik Abdul Gafoor. "Recognition of Power Quality Disturbances Using S-Transform and Fuzzy C-Means Clustering," In: IEEE International Conference and Utility Exhibition on Co-generation, small power plants and district energy (ICUE 2016), BITEC, Bang Na, Bangkok, Thailand, September 14–16, 2016.

[19] Shareef, Hussain, Mohamed Azah, and Ibrahim Ahmad Asrul. "An image processing based method for power quality event identification," *Electric Power Systems Research*, Vol. 46, pp. 184–197, 2013.

[20] Hong, Ying-Yi, and Wang Cheng-Wei."Switching Detection/ Classification Using Discrete Wavelet Transform and Self-Organizing Mapping Network," *IEEE Transactions On Power Delivery*, Vol. 20, No. 2, pp. 1662–1668, April 2005.

[21] Gupta, Neeraj, Khosravy Mahdi, Patel Nilesh, Dey Nilanjan, and Mahela Om Prakash. "Mendelian Evolutionary Theory Optimization Algorithm,"*Springer Soft Computing*, Vol. 24, pp. 14345–14390, 2020.

[22] Gupta, N., Khosravy M., Patel N., Mahela O. P., and Varshney G. "Plant Genetics-Inspired Evolutionary Optimization: A Descriptive Tutorial," In: Khosravy M., Gupta N., Patel N., Senjyu T. (eds), *Frontier Applications of Nature Inspired Computation*, pp. 53–77, 2020.

[23] Gupta, N., Khosravy M., Mahela O. P., and Patel N. "Plant Biology-Inspired Genetic Algorithm: Superior Efficiency to Firefly Optimizer," In: Dey N. (eds), *Applications of Firefly Algorithm and its Variants: Case Studies and New*, Springer, 2020.

[24] Hussein, Ahmed Salam, and Hawas Majli Nema. " Power quality analysis based on simulation and MATLAB/Simulink," *Indonesian Journal of Electrical Engineering and Computer Science*, Vol. 16, No. 3, pp. 1144–1153, December 2019.

[25] Saleh, Kamel, and Hantouli Naeil, "A photovoltaic integrated unified power quality conditioner with a 27-level inverter," *TELKOMNIKA Telecommunication, Computing, Electronics and Control*, Vol. 17, No. 6, pp. 3232–3248, December 2019.

[26] Alhamrouni, Ibrahim, Hanafi F. N., Salem Mohamed, Rahman Nadia H. A., Jusoh Awang, and Sutikno T. "Design of shunt hybrid active power filter for compensating harmonic currents and reactive power," *TELKOMNIKA Telecommunication, Computing, Electronics and Control*, Vol. 18, No. 4, August 2020, pp. 2148–2157.

[27] Tamil Selvi, M., and Gunapriya D. "A Power Quality Improvement for Microgrid Inverter Operated In Grid Connected and Grid Disconnected Modes," *Bulletin of Electrical Engineering and Informatics*, Vol. 3, No. 2, June 2014, pp. 113–118.

[28] Ajitha, P., and Jananisri D. "Voltage Sag Mitigation and Load Reactive PowerCompensation by UPQC," *Bulletin of Electrical Engineering and Informatics*, Vol. 3, No. 2, June 2014, pp. 109–112.

[29] Rathore, B., Mahela O. P., Khan B., Haes Alhelou H., and Siano P. "Wavelet-Alienation-Neural Based Protection Scheme for STATCOM Compensated Transmission Line," in *IEEE Transactions on Industrial Informatics*, Vol. 17, No. 4, pp. 2557–2565, April 2021, 10.1109/TII.2020.3001063

[30] Gangwar, K., Mahela O. P., Rathore B., Khan B., Haes Alhelou H., and Siano P. "A Novel K-Means Clustering and Weighted K-NN Regression Based Fast Transmission Line Protection," in *IEEE Transactions on Industrial Informatics*, Vol. 17, No. 9, pp. 6034–6043, September 2021, 10.1109/TII.2020.3037869

[31] Rathore, B., Mahela O. P., Khan B., and Padmanaban S. "Protection Scheme using Wavelet-Alienation-Neural Technique for UPFC Compensated Transmission Line," in *IEEE Access*, Vol. 9, pp. 13737–13753, 2021, 10.1109/ACCESS.2021.3052315.

[32] Agajie, Takele Ferede, Khan Baseem, Alhelou Hassan Haes, and Mahela Om Prakash. "Optimal expansion planning of distribution system using grid-based multi-objective harmony search algorithm," *Computers & Electrical Engineering*, Vol. 87, p.106823, 2020.

[33] Gebru, Fsaha Mebrahtu, Khan Baseem, and Alhelou Hassan Haes, "Analyzing low voltage ride through capability of doubly fed induction generator based wind turbine," *Computers & Electrical Engineering*, Vol. 86, p. 106727, 2020.

[34] Ram Ola, S., Saraswat, A., Goyal, S. K., Sharma, V., Khan, B., Mahela, O. P., Haes Alhelou, H., and Siano, P. "Alienation Coefficient and Wigner Distribution Function Based Protection Scheme for Hybrid Power System Network with Renewable Energy Penetration," *Energies*, Vol. 13, p. 1120, 2020.

[35] Kiros, S., Khan, B., Padmanaban, S., Haes Alhelou, H., Leonowicz, Z., Mahela, O. P., and Holm-Nielsen, J. B. "Development of Stand-Alone Green Hybrid System for Rural Areas," *Sustainability*, Vol. 12, p. 3808, 2020.

[36] Yogee, G. S., Mahela, O. P., Kansal, K. D., Khan, B., Mahla, R., Haes Alhelou, H., and Siano, P. "An Algorithm for Recognition of Fault Conditions in the Utility Grid with Renewable Energy Penetration," *Energies*, Vol. 13, p.2383, 2020.

[37] Mahela, O. P., Sharma Y., Ali S., Khan B., and Padmanaban S. "Estimation of Islanding Events in Utility Distribution Grid With Renewable Energy Using Current Variations and Stockwell transform," in *IEEE Access*, Vol. 9, pp. 69798–69813, 2021.

[38] Swarnkar, N. K., Mahela O. P., Khan, and Lalwani M. "Identification of Islanding Events in Utility Grid with Renewable Energy Penetration Using Current Based Passive Method," in *IEEE Access*, Vol. 9, pp. 93781–93794, 2021, 10.1109/ACCESS.2021.3092971.

[39] Mahela, Om Prakash and Shaik Abdul Gafoor. "Recognition of Power Quality Disturbances Using S-Transform Based Ruled Decision Tree and Fuzzy C-Means Clustering Classifiers," *Applied Soft Computing*, Vol. 59, pp. 243–257, October 2017, 10.1016/j.asoc.2017.05.061.

Index

Note: Locators in *italics* represent figures in the text.

For Product Safety Concerns and Information please contact our EU
representative GPSR@taylorandfrancis.com
Taylor & Francis Verlag GmbH, Kaufingerstraße 24, 80331 München, Germany